THE CICHLIDS OF SURINAM

THE CICHLIDS OF SURINAM

Teleostei: Labroidei

BY

SVEN O. KULLANDER AND HAN NIJSSEN

Swedish Museum of Natural History, Stockholm
Institute of Taxonomic Zoology, Amsterdam

E.J. BRILL
LEIDEN • NEW YORK • KØBENHAVN • KÖLN
1989

Cover design: Roel van Dijk

Library of Congress Cataloging-in-Publication Data

Kullander, Sven O.
The cichlids of Surinam: Teleostei, Labroidei / by Sven O. Kullander and Han Nijssen.
p. cm.
Bibliography: p.
ISBN 90-04-09077-0
1. Cichlidae—Surinam. I. Nijssen, Han. II. Title.
QL638.C55K83 1989
597'.58—dc20 89-33883
CIP

ISBN 90 04 09077 0

PRINTED IN THE NETHERLANDS BY E.J. BRILL

TABLE OF CONTENTS

SYNOPSIS

The Surinamese cichlid fauna includes twelve genera. Eleven were initially covered in this paper. *Crenicichla* Heckel was added—while this paper was submitted—after publication of recent studies by Ploeg (1986a, 1986b, 1987).

The genera and their Surinamese species are, with recorded maximum size and distribution in Surinamese river drainages:

Cichla Schneider:

Cichla ocellaris Schneider; 414 mm SL; Corantijn, Nickerie, Saramacca, Suriname, probably Marowijne drainage, extralimital distribution in the Guianas.

Chaetobranchus Heckel:

Chaetobranchus flavescens Heckel; 162 mm SL; Corantijn, Saramacca, Suriname drainages, wide extralimital distribution.

Geophagus Heckel:

Geophagus surinamensis (Bloch); 148 mm SL; Saramacca, Suriname, Marowijne drainages, to which restricted.

Geophagus brokopondo new species; 123 mm SL; Suriname drainage, known from two localities only.

Geophagus brachybranchus new species; 138 mm SL; Corantijn and Nickerie drainages, possibly extralimital in Guyana.

Geophagus harreri Gosse; 174 mm SL; Marowijne drainage, to which restricted.

Satanoperca Günther:

Satanoperca leucosticta (Müller & Troschel); 98 mm SL; Nickerie drainage, extralimital in Guyana.

Apistogramma Regan:

Apistogramma steindachneri (Regan); 65 mm SL; Corantijn, Nickerie, Coppename, Saramacca, Suriname drainages, extralimital in Guyana.

Apistogramma ortmanni (Eigenmann); 41 mm SL; Corantijn drainage, extralimital in Guyana and Venezuela.

Guianacara new genus, type species *G. owroewefi*, with subgenera *Guianacara* new subgenus, *Oelemaria* new subgenus, type species *G. oelemariensis*;

Guianacara (G.) owroewefi new species; 107 mm SL; Coppename, Saramacca, Suriname, Marowijne drainages, to which restricted.

Guianacara (G.) sphenozona new species; 85 mm SL; Corantijn drainage, possibly extralimital in Guyana.

Guianacara (Oelemaria) oelemariensis new species; 81 mm SL; known only from the Oelemari Rivier, Marowijne drainage.

Cichlasoma Swainson:

Cichlasoma bimaculatum; 123 mm SL; Corantijn, Nickerie, Suriname, Cottica, Marowijne drainages, extralimital in Venezuela, Guyana, French Guyana and Rio Branco.

Aequidens Eigenmann & Bray:

Aequidens tetramerus (Heckel); 125 mm SL; Corantijn, Nickerie, Coppename, Suriname, Marowijne drainages, wide extralimital distribution.

Aequidens paloemeuensis new species; 95 mm SL; known only from the Paloemeu Rivier, Marowijne drainage.

Krobia new genus, type species *Acara guianensis* Regan:

Krobia guianensis (Regan); 128 mm SL; Corantijn, Nickerie, Coppename, Saramacca, Suriname, Cottica drainages, extralimital in Guyana.

Krobia itanyi (Puyo); 125 mm SL; Marowijne drainage, to which restricted.

Cleithracara new genus, type species *Acara maronii* Steindachner:

Cleithracara maronii (Steindachner); 71 mm SL; Corantijn, Nickerie, Suriname, Commewijne, Cottica, Marowijne drainages, extralimital along Guianan coast.

Nannacara Regan:

Nannacara anomala Regan; 56 mm SL; Corantijn, Nickerie, Saramacca, Suriname, Marowijne drainages, extralimital in Guyana.

Crenicichla Heckel:

Crenicichla saxatilis (Linnaeus); 232 mm SL; Nickerie, Suriname, Marowijne drainages, extralimital distribution in the Guianas.

Crenicichla lugubris Heckel; 231 mm SL; Corantijn drainage, extralimital distribution in Guyana and Brazil.

Crenicichla multispinosa Pellegrin; 225 mm SL; Suriname, Marowijne drainages.

Crenicichla albopunctata Pellegrin; 164 mm SL; Marowijne drainage, extralimital in French Guiana.

Crenicichla coppenamensis Ploeg; 179 mm SL; Coppename, Saramacca drainages.

Crenicichla nickeriensis Ploeg; 191 mm SL; Corantijn, Nickerie drainages.

Crenicichla sipaliwini Ploeg; 173 mm SL; Corantijn drainage, to which restricted.

Oreochromis mossambicus (Peters) is an introduced African cichlid.

Guianacara is a group of geophagine like cichlids including also *Acara geayi* Pellegrin, and probably most closely related to *Acarichthys* Eigenmann which shares derived characters such as anteroventral projection of entopterygoid, dorsal crest on parasphenoid and basisphenoid, and reduced fenestra between coracoid and cleithrum. *Guianacara* is distinguished by wide dorsal shelf on urohyal, prominent nasad process of lateral ethmoid, wide lateral commissure, first pelvic fin ray with inner branch longer than outer, and mediad curved anterior edge of distal postcleithrum. Nominotypical *Guianacara* have two supraneurals, a bar across side and produced anterior dorsal fin lappets; subgenus *Oelemaria* has one supraneural, a midlateral spot and dorsal fin without produced lappets. The genus is widely distributed in northeastern South America, occurring throughout the Guianas and also in the Trombetas, Caroni and Branco drainages.

Krobia may be closely related to *Bujurquina* Kullander and *Tahuantinsuyoa* Kullander, having a similar ceratohyal shape. Included species have uniserial predorsal scale pattern, narrowly scaled dorsal and anal fin, blotchy band from head toward end of dorsal fin base, dorsal accessory caudal fin lateral line between rays D3 and D4. The genus includes also undescribed species in French Guiana and Brazil.

Cleithracara is monotypic; similar to *Laetacara* Kullander, though uniquely distinguished by very wide anterior arm of epibranchial 1 and process anteroventrally on quadrate.

Nannacara is rediagnosed with reference to its 14 principal caudal fin rays, fenestra in lateral ledge of quadrate, loss of second from anteriormost frontal lateralis foramen, two infraorbital ossicles, long pharyngobranchiad 2 arm and short pharyngobranchiad 1 arm of epibranchial 1, and membrane between body and last ray of dorsal and anal fins.

INTRODUCTION

The Cichlidae

The family Cichlidae is one of the major groups of teleost fishes, comprising about 1,400 species. Nearly all of them are freshwater, only a few species frequenting brackish water. South America has probably more than 300 species of cichlids. That would be between 6 and 10% of the South American fresh water fish fauna (estimates vary between 3,000-5,000 species). Cichlids are a proportionally more important faunal component numerically in Africa and Central America, but are nevertheless the major non-otophysan group in South America and one of the four largest families, the Characidae, Pimelodidae and Loricariidae having more or about the same number of species.

The Cichlidae are one of few tropical fish families whose monophyly is virtually uncontested—see Stiassny (1981), Zihler (1982), Kaufmann & Liem (1982), Gaemers (1984), Stiassny & Jensen (1987), Stiassny (1987) on the phylogeny of the family.

Externally, cichlids can be identified by a suite of characters including single nostril, usually divided flank lateral line, 16 (exceptionally 14) principal caudal fin rays, and the spines in the dorsal, anal and pelvic fins. Sizes range from ca 30 mm to at least 600 mm. Habitat and food specializations are still largely uninvestigated, although it seems that the spectrum is considerable. There are examples of rheophilic to extreme lentic forms; and herbivorous, planktivorous, molluscivorous and piscivorous specialists have been distinguished, although the majority of the South American cichlids may be opportunistic omnivores. Cichlids have a varied behaviour and are established study objects of ethologists. Reproductive behaviour includes mouth brooding, of which several strategies are recorded, although the majority of the American cichlids are substrate spawners. All cichlids practise parental care.

Economically, only the about eleven *Cichla* species are important in fishery statistics. A much larger number of smaller species are consumed regularly, however. The flesh of South American cichlids is firm and generally appreciated for the lack of intermuscular bones.

The last revision of the South American cichlids was published over 80 years ago (Regan 1905a-c, 1906a-b). Since then, no attempt has been made at a comprehensive revision, although the number of nominal species and genera has increased considerably. Kullander (1986) gave an estimate of 275 species in 40 genus group taxa; and the first figure is

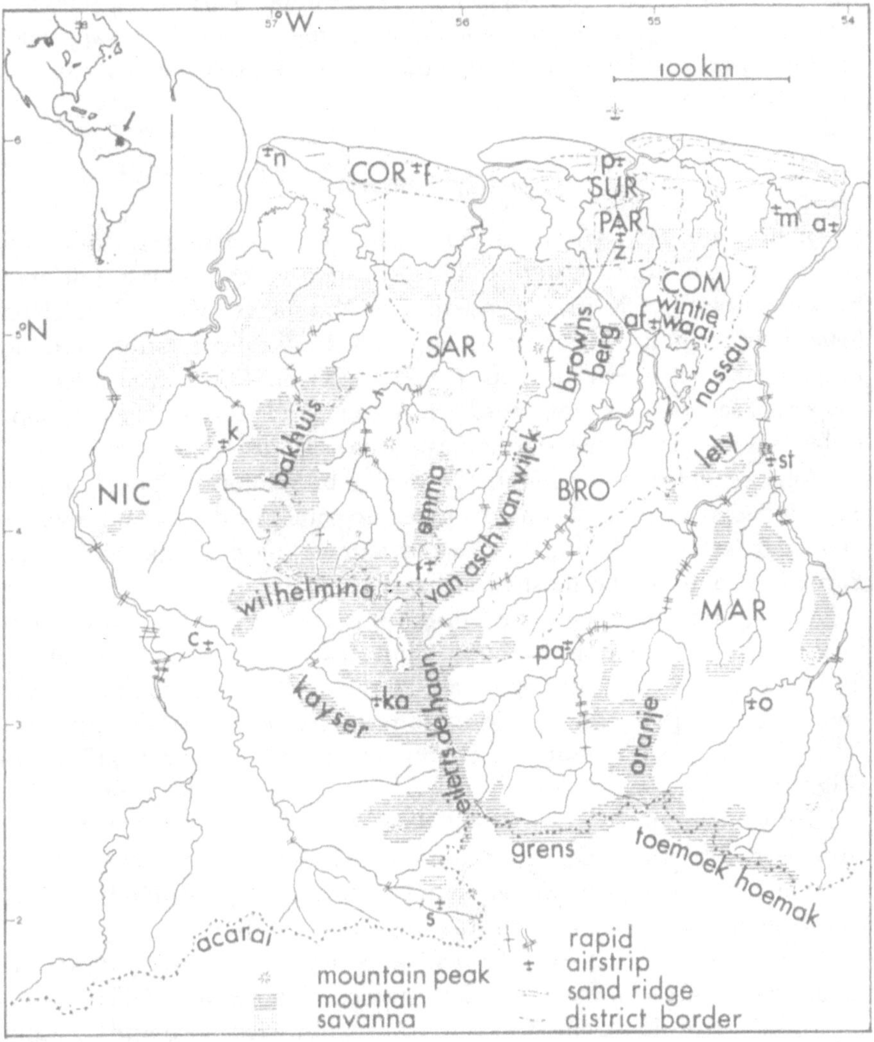

Fig. 1. Map of Surinam showing districts, mountains, and other geographical features. For rivers, see Fig. 2. Modified from a map provided by RMNH. *a* Albina airstrip, *af* Afobaka airstrip, *BRO* Brokopondo district, *c* Coeroeni airstrip, *COM* Commewijne district, *COR* Coronie district, *f* Friendship airstrip, *k* Kabalebo airstrip, *ka* Kayser airstrip, *m* Moengo airstrip, *MAR* Marowijne district, *p* Paramaribo airstrip, *pa* Paloemeu airstrip, *PAR* Para district, *s* Sipaliwini airstrip, *SAR* Saramacca district, *st* Stoelmanseiland airstrip, *SUR* Suriname district, *t* Tafelberg airstrip, *z* Zanderij airfield.

certainly low. 233 nominal species and 33 nominal genera are currently recognized as valid. In this paper we describe another six new species and three new genera, and do this in a faunal survey. Although this approach hardly enlightens the general taxonomic picture, both the geographical and taxonomic coverage is considerable.

Surinam (Fig. 1)

Surinam, up to 1975 known as Dutch Guiana, is a small country on the Guiana coast of South America. The country's 163,820 km^2 are drained by several more or less independent river drainages. Most of the rivers have their tributaries in the strongly eroded crystalline Tumac Humac range and its northern outcrops (to ca 1,200 masl), which also has the sources of several important northern Amazon tributaries. The coastland is flat and marginally marshy, built up from silt carried by currents from the Amazon. Most of the country is covered by rainforest, though a west-east extended band of savannahs takes up much of the coastal lowland and there are also smaller interior savannahs of which the Sipaliwini-Paroe Savannah is the most extensive.

The major drainages are the Corantijn in the west and the Marowijne in the east. Together these rivers drain nearly half of the Surinamese highland area. Smaller drainages are largely lowland and to some extent anastomose in the coastal region: from west to east they comprise the Nickerie, Coppename, Saramacca, Suriname and Commewijne systems (Fig. 2).

Surinam has only one lake, the Brokopondomeer, or van Blommestein stuwmeer, a barrage lake in the Suriname River, but seasonal flooding of savannahs provides temporary lentic conditions of considerable extension. Otherwise, rivers are notable for numerous rapids.

Waters are generally slightly acid and transparent, and creeks and savannah waters commonly humus stained (blackwaters). The waters may become turbid during the rainy season, however. Two rainy seasons are distinguished; one lesser from December to February, one principal from April to July.

History of Surinamese cichlids

The oldest known preserved specimens of South American fishes, collected in the first half of the 18th century, obviously came from Surinam, although locality information is not preserved in all instances. These include some of the first cichlids to be described and illustrated by scientists.

The earliest contributions to Surinamese ichthyology, which also dealt with cichlids, are integrated in more general treatises of specific collections. Gronovius (1754, 1756) described and illustrated *Crenicichla saxatilis* as *Sparus rostro plagioplateo*..., and *Cichlasoma bimaculatum* as *Labrus bruneus*, both said to be from Surinam. The same species were described and figured by Linnaeus (1754) and eventually received their present species epithets by Linnaeus (1758).

Bloch (1792) also had material of *Crenicichla saxatilis* and *Cichlasoma bimaculatum* and published coloured drawings of these species (as *Perca saxatilis* and *Labrus punctatus*). Bloch added two cichlid species to the Surinamese fauna, *Sparus surinamensis* (Bloch 1791) and *Labrus melagaster* (Bloch 1792), the first a valid *Geophagus* species, the latter an African species with obviously erroneous locality data (p. XXXI). Bloch's collection contained yet another Surinamese cichlid, eventually described as *Cichla ocellaris* (Schneider 1801), but with erroneous locality information (East Indies).

Already in 1822 appeared the first checklist of the Surinamese fauna (Collin 1822). It lists, however, only the Linnaean cichlid species.

During the rest of the 19th century several small fish collections were made, chiefly near Paramaribo, but little was published. Steindachner (1882) described *Acara maronii* and Sauvage (1882) *Mesonauta surinamensis*, the latter, however, not a cichlid (p. XXXI).

Kappler (1887) described the Surinamese fish fauna from the geographer-naturalist's viewpoint, and Klinckowström (1892) listed four cichlid species obtained during an expedition to the lower Suriname R. system (*Acara tetramerus* = *Krobia guianensis*; *Acara* sp.? = *Guianacara owroewefi*; *Crenicichla saxatilis*; and *Crenicichla reticulata* = *C. multispinosa*).

Two early 20th century revisers of the family Cichlidae, Regan (1905-1906) and Pellegrin (1904), included some Surinamese cichlid material in their work.

Eigenmann (1912) monographing the Guyana fish fauna, listed seven Surinamese cichlid species: *Cichlasoma bimaculatum*, *Geophagus surinamensis*, *Cichla ocellaris*, *Crenicichla saxatilis*, *Crenicichla multispinosa*, *Mesonauta festivus*, and *Aequidens vittatus*. No *Mesonauta* have, however, been collected in Surinam either before or after Eigenmann's list. Steindachner (1915b) reported *Crenicichla saxatilis* from the Marowijne River.

Although during the 19th and early 20th century the knowledge of the Surinamese cichlid fauna remained at the unreliable checklist level, that of the neighbouring countries French Guiana (Pellegrin 1908, 1909) and especially Guyana (Schomburgk 1849, Eigenmann 1912) received more attention. However, fisheries activities and rather incidental collecting close to the Surinamese coast were reported on in short papers by Lijding

(1959a) and Boeseman (1952, 1954, 1956), and Puyo (1943, 1949) described French Guianan cichlids some of which are reported from rivers in dispute between Surinam and French Guiana.

Boeseman (1952) added 7 species to Eigenmann's (1912) checklist, viz. *Chaetobranchus flavescens*, *Nannacara anomala*, *Aequidens tetramerus*, *Aequidens geayi*, *Geophagus jurupari*, *Apistogramma steindachneri*, and *Crenicichla alta*. Boeseman (1954) identified *Aequidens potaroensis* as an addition to the Surinamese fauna; his material is here redetermined as *Krobia guianensis*. Hoedeman (1951) described *Apistogramma cacatuoides* from near Paramaribo, but the locality is apparently erroneous, the species having only since been recorded from western Amazonia (Kullander 1980a, 1986).

Most recently Gosse (1976) revised the genus *Geophagus* and besides confirming the occurrence of *G. surinamensis* and *G. jurupari* in Surinam, added a distinctive species, *G. harreri*, from the Marowijne drainage. Richter & Nijssen (1980) listed as a faunal addition the cichlid *Crenicichla lugubris*. Ploeg (1986a, 1987) revised *Crenicichla* using Surinamese material.

Fourteen cichlid species have thus been more or less reliably reported from Surinam. We describe 26 species below. The genus *Crenicichla*, which we have added after recent studies by Ploeg (1986a, 1986b, 1987) is represented in Surinam by *C. multispinosa*, *C. saxatilis*, *C. lugubris*, *C. albopunctata*, *C. coppenamensis*, *C. nickeriensis* and *C. sipaliwini*.

Vast materials assembled by early Dutch expeditions (historical accounts in Holthuis 1959, Hoogmoed 1973), the Brokopondo survey (Nijssen 1976; Hoogmoed 1973) and the West-Suriname Expeditie 1971 (Geijskes 1973) resulted in large collections in the museums of Leiden and Amsterdam. With the recent border river collections by R. P. Vari (Corantijn drainage 1979-1980), and J.-P. Gosse (Marowijne drainage 1966, 1969), abundant material is now available for an attempt at a general revision of the Surinamese cichlid fauna.

Despite the numerous collecting expeditions in Surinam during the 20th century, still only a few and limited groups of Surinamese fishes have been reported on. Principal works on non-cichlids include those on Callichthyidae (*Corydoras*) by Nijssen (1970), Loricariidae by Boeseman (1968, 1971), Characidae (Serrasalminae) by Géry (1972), and Pimelodidae and Auchenipteridae by Mees (1974).

Fig. 2. Map of Surinamese river drainages.

ACKNOWLEDGMENTS

We wish to express our sincere thanks to curators, collection assistants, photographers, and reviewers that contributed to this study. Marinus Boeseman and Martien J. P. van Oijen (RMNH), Jean-Pierre Gosse and Louis Walschaerts (ISNB), Richard P. Vari and Susan L. Jewett (USNM), Hans-Joachim Paepke (ZMB), Harald Ahnelt and the late Rainer Hacker (NMW), Marie-Louise Bauchot, Guy G. Teugels and Martine Desoutter (MNHN), Gordon Howes and Bernice Brewster (BMNH) made available their holdings of Surinamese cichlids. Pierre-Yves Le Bail donated French Guianan material that has been included in this paper.

Photographs are by SOK (Fig. 22), Anita Hogeborn-Kullander (Figs 15, 17-19, 23, 25-26, 29, 32-34, 43-44, 66-68, 70, 73, 77, 80-81, 92, 94, 122-123), Göran Frisk and Stefan Rosengren (Fig. 41), and L. A. van der Laan (Figs 6, 10-13, 38, 71, 126-135). Maps were prepared by HN. All line drawings are by SOK.

The manuscript was reviewed by Tyson Roberts, Humphry Greenwood, Bo Fernholm and Ingo Koslowski (*Apistogramma* section).

Financial support from the Swedish Natural Science Research Council enabled SOK to work on the Amsterdam collection and make short visits to ISNB and RMNH in 1982 (NFR grant no. R-RA 4568-102); the manuscript was revised and completed with support from NFR grant no. B-BU 4568-105-107, and printed with support of the Netherlands Organization for Scientific Research (NWO grant no. P 87-213).

This paper is part of a long term, broad revision of the South American Cichlidae by the senior author, and links with other papers: Kullander (1980a, 1983a, 1986, 1987a, 1987b). The paper has been cited in others as 'in prep.' or 'ms', with variable authorship and title (Kullander 1982a, 1986); we are delighted to see it finally completed.

MATERIAL AND METHODS

Descriptive data

Measurements (Fig. 3) were obtained with a dial caliper reading to 0.05 mm, rounded to the nearest 0.1 mm. Particular measurements have been explained in Kullander (1979a, 1980a, 1980b, 1983a, 1986), but the following particulars should be observed for the present paper.

Standard length (SL), snout length and head length are measured, as usual, from the tip of the upper jaw (= premaxillary symphysis). In *Cichla*, head and snout lengths were measured directly from the tip of the upper jaw to the posteriormost point of the osseous operculum (head length) or anteriormost point on the orbital rim (snout length). In all other groups these measurements are linear, from the same posterior points as in *Cichla*, but to a measuring block against which the upper jaw is pressed. In *Cichla*, the length of the fifth dorsal fin spine was measured, but in all other groups the dorsal fin spine measured is the last.

Counts were obtained under a binocular dissection microscope. The methods have been explained in Kullander (1979a, 1980a, 1980b, 1983a). Fin rays were generally counted in transmitted light. The count of vertical scale series (*squ. long.*, Fig. 3) is taken along the series above that containing the lower lateral line, and excludes any scale positioned to any extent over the cleithrum. Vertebral counts include the preural halfcentrum.

Colour pattern. Vertical dark flank markings are termed *bars*, and are usually numbered as specified for each taxonomic group; when such bars are modified into particular markings they are recognized as such. A dark stripe or series of continuous blotches along the middle of the side is referred to as *lateral band.* Dark markings on the head include stripes or spots named with reference to their position relative to the eye: *suborbital*, *preorbital*, *supraorbital*, *postorbital*; in many groups a light rather than dark preorbital stripe is recognized. Dark stripes across the snout and interorbital space are called *facial stripes*. The *cheek spot*, characterizing *Cichlasoma* and some *Aequidens* species is a relatively large dark spot posterodorsally on the cheek; *buccal dots* and *buccal stripes* are smaller spots or lines on the cheek, preorbital, and sometimes gill cover, which are blue in life. Any kind of dark spot at the base of the caudal fin is called *caudal spot*. A spot on the middle of the side is called *midlateral spot* in cichlids other than *Apistogramma* in which the term is *lateral spot*. The

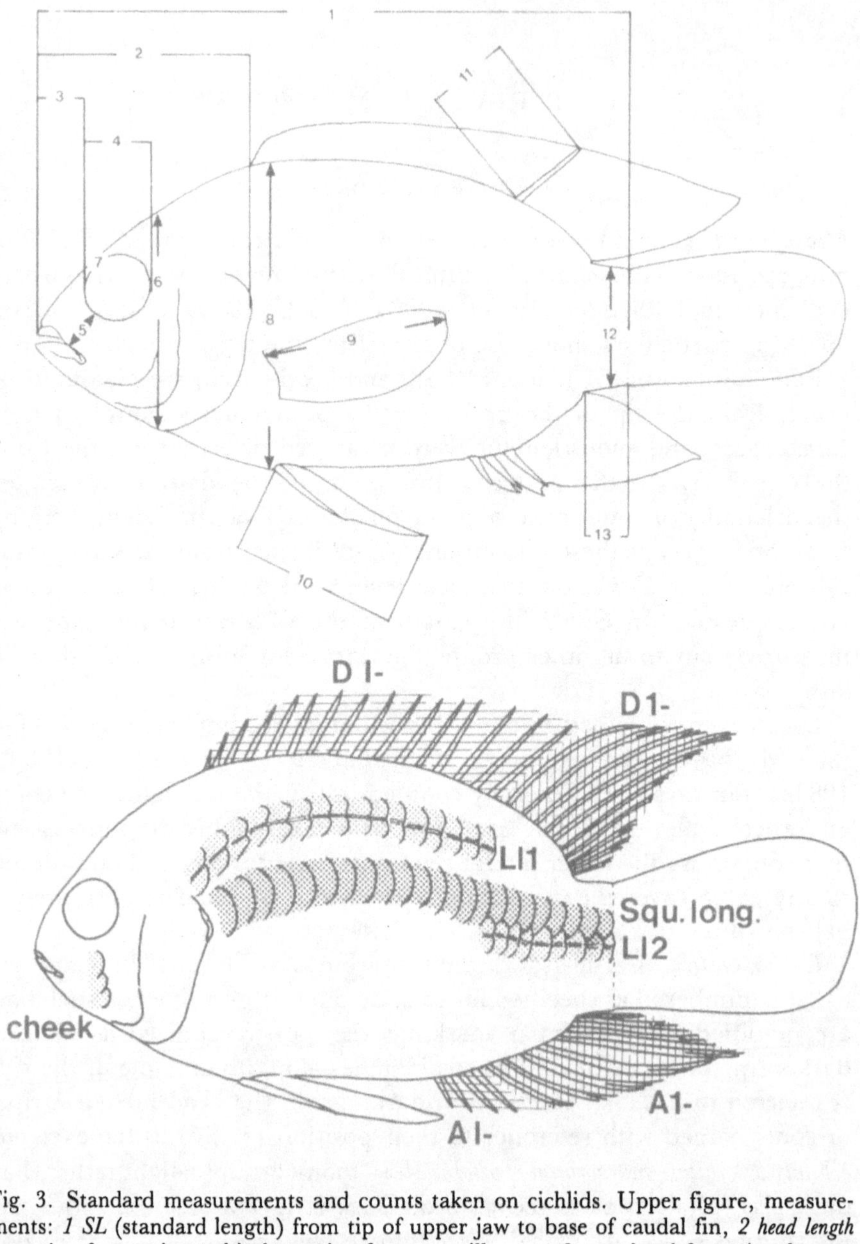

Fig. 3. Standard measurements and counts taken on cichlids. Upper figure, measurements: *1 SL* (standard length) from tip of upper jaw to base of caudal fin, *2 head length* from tip of upper jaw to hind margin of osseous gill cover, *3 snout length* from tip of upper jaw to anterior margin of orbit, *4 orbital diameter*, *5 preorbital depth*, *6 head depth* immediately posterior to orbit, *7 interorbital width* across interorbital space, *8 body depth* at level of insertion of pelvic fin spine, *9 pectoral fin length* from base of first to tip of longest ray, *10 pelvic fin length* from base to tip of longest ray, *11 length of last dorsal fin spine*, *12 caudal peduncle depth* across middle, *13 caudal peduncle length* along longest edge. Lower figure, counts: *Squ. long.* scales in horizontal series next above that including lower lateral line, *Ll1* upper lateral line scales, *Ll2* lower lateral line scales, *cheek* horizontal scale series below orbit, *AI-* (horizontal hatching) anal fin spines (roman numerals), *A1-* (vertical hatching) anal fin (soft) rays (arabic numerals), *DI-* (horizontal hatching) dorsal fin spines (roman numerals), *D1-* (vertical hatching) dorsal fin (soft) rays.

terminology of markings in *Apistogramma* is complex; refer to Kullander (1980a) and Koslowski (1985a) for description.

Osteological terminology follows Kullander (1983a), and generally does not differ from that of most authors.

Material list

The material examined is listed at the end of the paper (p. 224), in the same order as in the descriptive section.

Museum acronyms

ANSP	Academy of Natural Sciences of Philadelphia
BMNH	British Museum (Natural History), London
ISNB	Institut royal des Sciences naturelles de Belgique, Brussels
MBUCV-V	Museo de Biologia, Universidad Central de Venezuela, Caracas
MHNLS	Museo de Historia Natural, Sociedad de Ciencias Naturales La Salle, Caracas
MNHN	Muséum national d'Histoire naturelle, Paris
MZUSP	Museu de Zoologia da Universidade de Saõ Paulo
NMW	Naturhistorisches Museum, Wien
NRM	Swedish Museum of Natural History, Stockholm
RMNH	Rijksmuseum van Natuurlijke Historie, Leiden
USNM	United States National Museum, Smithsonian Institution, Washington, D.C.
ZMA	Zoölogisch Museum, Universiteit van Amsterdam
ZMB	Museum für Naturkunde, Berlin, DDR

Geographical scope and nomenclature

This paper covers all cichlids known to occur in the Corantijn and Marowijne drainages, and all other drainages between these which run to the Atlantic (Fig. 2). We thus include material from western French Guiana localities and would have included Guyanan Corantijn material if we had had any. The borders of Suriname with French Guiana and Guyana are not definitely settled; inclusion here of the New River (Boven Corantijn) and the Rivière Marouini (Marowini Rivier) drainages may differ from their representation on non-Dutch/Surinamese maps. In material lists, country of origin is given as stated on labels. Collecting localities are shown in Fig. 4. The abbreviation *R.* signifies river.

An imprecisely defined region encompassing Atlantic versant drainages between the mouths of the Amazon and Orinoco rivers is here termed Guiana. The states of Guyana (*ex* British Guiana), Surinam (= Suriname; *ex* Dutch Guiana) and French Guiana (Guyane française) are referred to as the Guianas. Use of the terms Amazon, Amazonas, Amazonia follows Kullander (1986); the Rio Tocantins drainage is regarded as separate from the Amazon drainage.

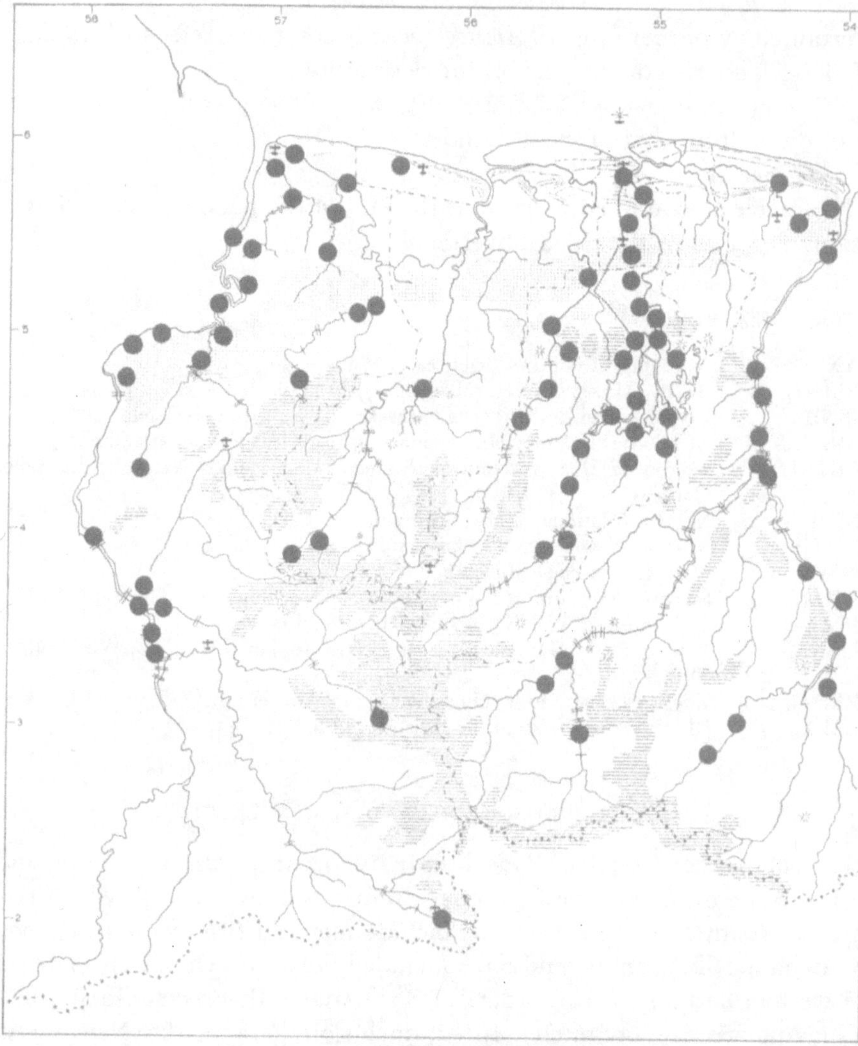

Fig. 4. Map of Surinam, with collecting sites of cichlids reported in this paper. One symbol may represent several adjacent localities.

We use the Surinamese spelling of all locality names as long as they are within Surinamese territory; in the text, however, we employ the English variant spelling of the name of the country, i.e. Surinam instead of Suriname. We note that the adjective of Surinam can be both Surinam and Surinamese and prefer the latter for the sake of clarity.

Several of the rivers dealt with here are border rivers and their names are spelt differently in each country. Our principle is to employ the spelling used in the country from which the particular collection referred to came, but in discussing river systems as such we use the Surinamese variant (e.g. Marowijne instead of Maroni) and the French Oyapock is preferred before the Brazilian Oiapoque in referring to the border river between French Guiana and Brazil.

More of an obstacle may be the repeated use of the same or similar place names in different parts of the country. For instance, there is a Gran Kreek both in the Suriname drainage and in the Marowijne drainage. These might easily be memorized, but to add to the confusion the Suriname drainage Gran Kreek is also called Marowijne Kreek. One of the Marowijne tributaries is called Marowini (Marouini in French Guiana). The Saramacca Kreek, also called Para Kreek is a tributary of the Suriname Rivier, and has nothing to do with the Saramacca Rivier, which in its turn has little to do with the more westerly Saramacca Distrikt. Different maps and authors give different spellings of some names ending in -i, -y or -ie, such as Tapanahony/Tapanahoni, Stondansi/Stondansie.

The following table lists the most important hydrographical terms used in Surinam, with their French and English equivalents:

Dutch/Surinamese	French	English
Kreek, kreekje	Crique	Creek
Rivier, Rio	Rivière, Fleuve	River
Val (pl. vallen)	Chute	Cataract
Hoeka, soela	Saut	Rapids
Meer	Lac	Lake

Taxonomic objective

The objective of this paper is primarily the recognition and description of all Surinamese cichlid species, with mapping of their geographical distribution.

This paper describes 26 naturally occurring Surinamese cichlid species, 6 of which are new to science. Another cichlid species has been collected in our territory, viz. the introduced African cichlid *Oreochromis mossambicus*.

Of the 12 genera that we deal with, 8 have been treated in other recent papers (Kullander 1983a, 1986; Ploeg 1987) and are given only a brief introduction here. Three genera are new and are given fuller descriptions. We also take this opportunity to present a more adequate diagnosis of *Nannacara*.

Treatment of widespread species here may suffer from disregard of variation in extralimital populations. We also made very simple cases of typological species by not subjecting our material to more refined tests of geographical variation. Even though the chosen species level goal of presentation seems adequately conceived, the rich potential of geographical variation provided by individual species whose distribution ranges over two or more separate river drainages, remains untapped.

All the genera dealt with here should be revised in a broader context than a country fauna. We think that a, somewhat expanded, species catalogue is an adequate vehicle for presenting the results of our taxonomic inventory of the Surinamese cichlids.

GEOGRAPHICAL DISTRIBUTION

Considering the artifactual geographical scope, only some notes are given on biogeography. More than half (14 of 26) of our species are known or presumed to occur outside Surinamese rivers.

The Guiana region. Our inventory lends ambiguous support to the recognition of a Guianan ichthyofaunal region. Such a region has been recognized by, e.g., Géry (1969) though never described in any detail. Other authors (Lowe-McConnell 1969) have stressed the 'Amazonian' species in Guianan rivers, especially in the Essequibo.

Endemism is the mark of biogeographical provinces, and 60% of the approximately 58 cichlid species occurring in Guiana are endemic to the area, including Atlantic versant rivers from Amapá to the Barima, and ignoring probable dispersal to the upper Rio Branco. If 'Amazonian' likely dispersalists in the Essequibo and Oyapock are disregarded, endemism is 90%. Unfortunately, the taxonomy and distribution of extra-Surinamese Guianan cichlids is still so poorly known that the above figures can only be a rough indication and an inspiration to deeper investigation of the Guianan ichthyofauna.

We also think that the historical explanation of the seemingly highly endemic Guianan ichthyofauna may be quite complex. Biogeographical provinces must also consider taxonomic relationships to gain value as indicators of topographic history.

The non-monotypic endemic genera and species groups in Guiana do indicate some degree of common history for the Guianan river drainages: *Guianacara*, *Nannacara*, *Aequidens paloemeuensis* + *A. potaroensis*. Each group, however, shows its own pattern of distribution.

It is hardly possible to consider pooled endemism a justification for recognizing Guianan rivers as a faunal region, although (a) genus level endemism, (b) vicariance of closely related species in adjacent drainages, and (c) ranges of regional endemic species through several parallel running drainages, does suggest an unspecifiable level of conformity which may be higher than expected for parallel running rivers. The nonconformity in Surinamese cichlid distribution patterns is, however, so striking that severe doubts must be expressed about the validity of the concept of a Guianan region

A Guianan-NE Venezuelan-Trinidadian track has been discussed by Mago (1970) and Kullander (1983a). It is best characterized by the nandid *Polycentrus schomburgkii*/*punctatus*, though remains a working hypothesis in the absence of sufficient collections especially in the Venezuelan part of the track. The distribution of *Nannacara* may conform to such a pattern.

Other cichlid species reported on in this paper are related to species in Central Amazonia (*Apistogramma ortmanni*, *A. regani*), the Branco-Negro basin (*Apistogramma steindachneri* group), or the Brazilian shield (*Krobia*). A faunal relationship between Guiana and Brazilian highland rivers is evident also in the distribution of *Retroculus*, with one species in the Oyapock, the Xingu, and the Tocantins drainages, respectively.

Distribution patterns

The tentatively recognized patterns of Surinamese cichlid species distribution include:

(1) Wide ranging dispersalists. Three species occur throughout the Amazon basin and the rest of the Guianas: *Aequidens tetramerus* is found even in upper parts of Surinamese drainages, and its variation over an enormous area of distribution should be studied in detail. *Chaetobranchus flavescens* is limited to coastlands. *Crenicichla lugubris* is the third wide ranging species; it is still uncertain whether *Crenicichla saxatilis* belongs to this group.

(2) Western 'Guianan' endemics. Some species occur or are presumed to occur in the Essequibo drainage and have their eastern limit of distribution in Surinam:

- *Cichlasoma bimaculatum* occurs also in the Rio Branco; its eastern Guianan limit is obscure. It reaches the French Guianan coast but does not extend to the Oyapock or Amapá where replaced by the Amazonian *C. amazonarum*.
- *Nannacara anomala* ranges east to the Marowijne; eastwards it is replaced by other species of the same genus. Upper Marowijne material possibly represents a different species.
- *Satanoperca leucosticta* is known only from the lower Nickerie and the Essequibo drainage.
- *Cichla ocellaris* apparently ranges from the Essequibo to the Marowijne, and is predominantly a lowland species. The collecting record seems biased, however.
- *Apistogramma ortmanni* occurs in the Cuyuni, Essequibo, and Corantijn drainages, and is collected only in inland forest creeks.
- *Apistogramma steindachneri* ranges from the Essequibo to the Suriname and perhaps lower Marowijne. It seems more euryoic than *A. ortmanni*.
- *Krobia guianensis* is mainly a coastal form, except for the Sipaliwini locality, ranging from the Essequibo to the lower Marowijne; replaced by

K. itanyi in the Marowijne, and by other species in more eastward rivers.
- *Crenicichla nickeriensis* occurs in the Corantijn and Nickerie drainages.
- *Crenicichla sipaliwini* is restricted to the Corantijn drainage.
- *Crenicichla coppenamensis* occurs in the Coppename and Saramacca drainages.
- *Geophagus brachybranchus* needs more study, but seems to range from the Essequibo to the Nickerie.
- *Guianacara sphenozona* probably belongs in this category of species.

(3) 'Guianan' endemics. *Cleithracara maronii* is the single cichlid species that is widely distributed throughout Guiana, although it is a lowland species in French Guiana and Guyana.

(4) Marowijne endemics. Only three species are endemic to Surinam, viz. *Aequidens paloemeuensis*, and *Guianacara oelemariensis*, which are found only in upper Marowijne drainage tributaries; and *Geophagus brokopondo* which is collected only in the Suriname R. Other species are limited in distribution to the Marowijne drainage in French Guiana and Surinam: *Geophagus harreri* and *Krobia itanyi*.

Endemic Marowijne species are:
- *Krobia itanyi* representing a Guianan-Brazilian highland genus.
- *Guianacara oelemariensis*, of unrecognized closer relationships, representing the sister group of a Guianan genus.
- *Geophagus harreri*, a very specialized member of a genus widespread in particularly the Guianas and Brazilian highlands.
- *Aequidens paloemeuensis*, most closely related to a Guyanan endemic with restricted distribution.

Non-endemic species in the Marowijne are:
- *Aequidens tetramerus*, a wide ranging species.
- *Cleithracara maronii*, a Guianan endemic genus and species.
- *Nannacara anomala* (reserved identification), representing a Guianan genus extending westwards into the Golfo de Paria area and southwards to the Belém area.
- *Cichla ocellaris*, a western Guianan species (reservations about identification of upper Marowijne material).
- *Guianacara owroewefi*, taken also in the neighbouring Suriname, Saramacca and Coppename drainages, member of a Guianan group.
- *Geophagus surinamensis*, taken also in the Suriname and Saramacca drainages; its relationships obscure.
- *Crenicichla multispinosa*, taken also in the Suriname drainage.
- *Crenicichla albopunctata* occurring also in neighbouring streams in French Guiana.

Guianacara owroewefi and *Geophagus surinamensis* distributions tend to duplicate the drainage area bordered in the west by the ranges of the Eilerts de Haan Gebergte, Wilhelmina Gebergte and Bakhuis Gebergte, and less well defined extensions of the Tumac Humac range dividing southern French Guiana. This 'Greater Marowijne' basin (Fig. 5) would include the Coppename, Saramacca, Suriname and Marowijne drainages, and perhaps the Mana drainage. We postulate that these rivers form a faunistic unit defined by endemism which is most marked in the Marowijne.

This area includes also non-endemics, although it is notable that some western and eastern species do not enter. Non-endemics, except the ubiquitous *Cleithracara maronii* and *Aequidens tetramerus*, have their additional ranges more to the west, and do not reach eastwards beyond the Marowijne. None of the French Guianan endemic cichlids reach into the Marowijne drainage.

The second major drainage in Surinam, the Corantijn, stands in bright contrast to the Marowijne, having no endemic species. The only other Surinamese river with an endemic cichlid species is the Suriname R. with *Geophagus brokopondo*.

Biogeographers usually distinguish one or more Pleistocene forest refugia in the Guianan highlands, indicating not only climate stability in those areas throughout the Pleistocene, but also speciation centers (papers in Prance 1982).

With the reservation that Surinamese cichlid distributions are not readily characterized as correlated with any particular terrestrial vegetation type, cichlid endemism in the Greater Marowijne might be taken as support for a Pleistocene forest refugium in the area; the possibility of a continuous Pleistocene rain forest cover is not excluded, however.

Maintenance of a largely endemic Marowijne fish fauna reasonably presupposes long term stable environmental conditions. Reduced gene flow across presumably more arid and wider coastlands may have produced speciation by isolation of upland populations in groups such as the nominotypical *Guianacara*. The absence of lowland endemism, and high incidence of dispersalist species in the coastal area of Surinam would support the view that lowland conditions have been unfavourable to cichlids until relatively recently, perhaps in drier glacial times.

Nevertheless, the distribution pattern of nominotypical *Guianacara* and *Krobia* would be equally well explained by drainage vicariance events. Distribution patterns of genera are also so different that it seems worthwhile to search for a number of independent contributing factors to explain the historical biogeography of Surinam. The diverse taxonomic composition and the absence of otherwise so widespread genera such as

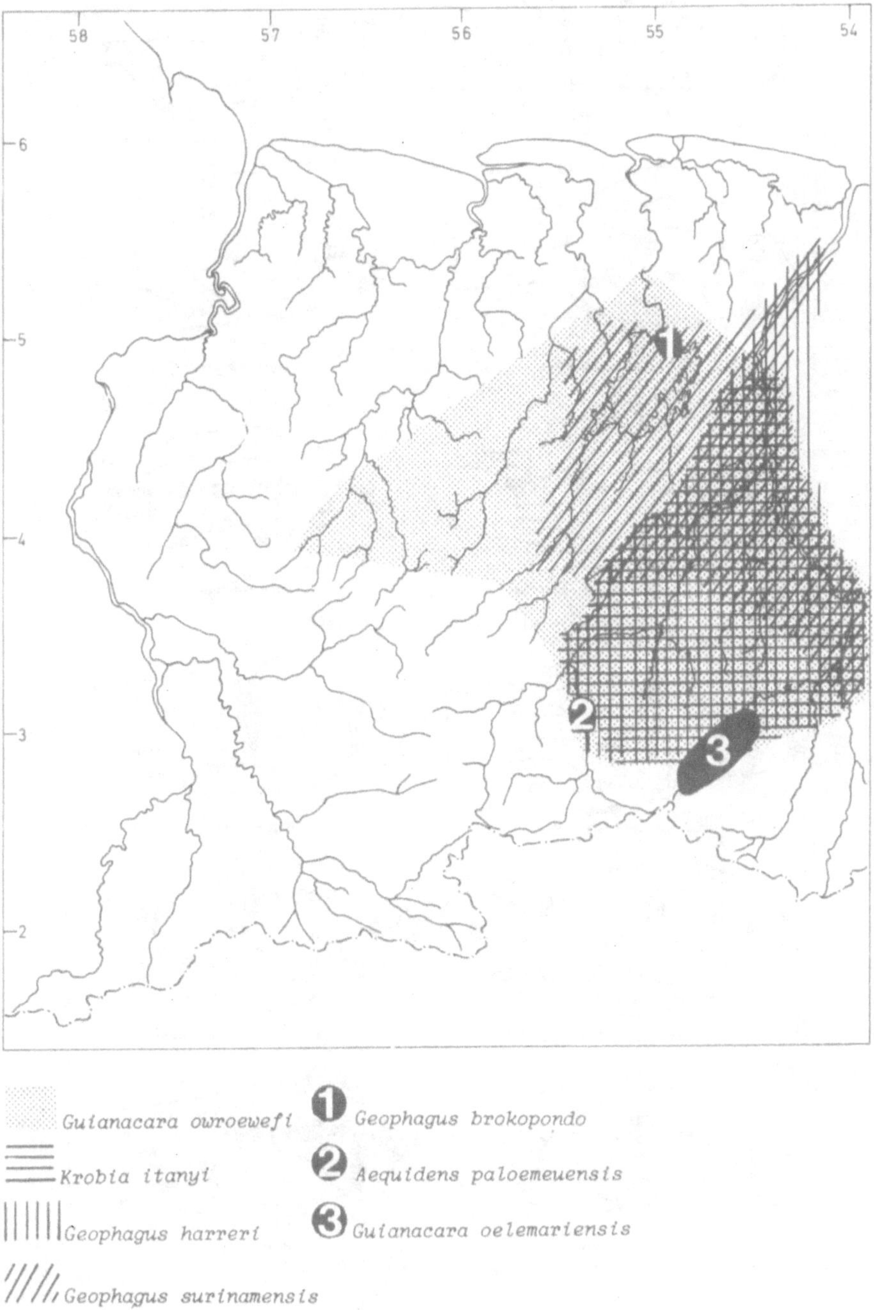

Fig. 5. Surinamese ranges of cichlid species endemic in the 'Greater Marowijne basin'.

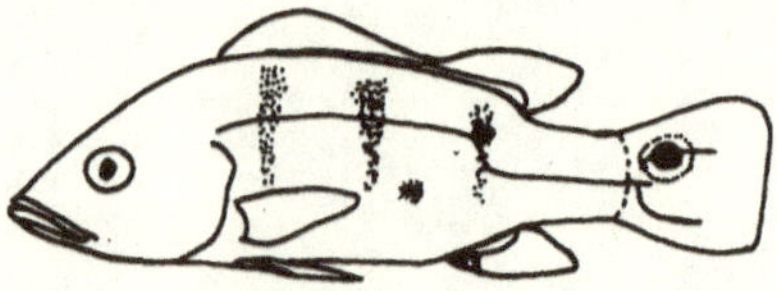

Cichla ocellaris

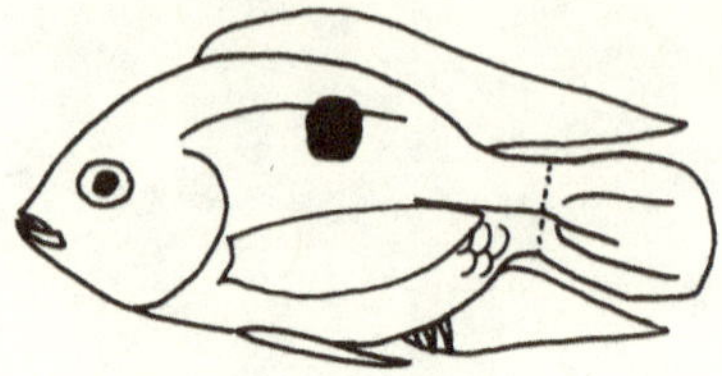

Chaetobranchus flavescens

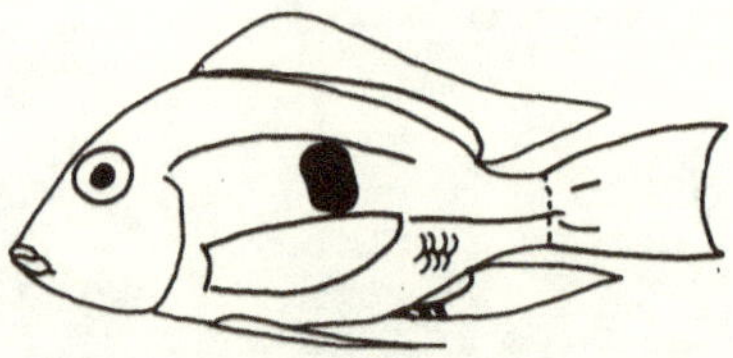

Geophagus surinamensis

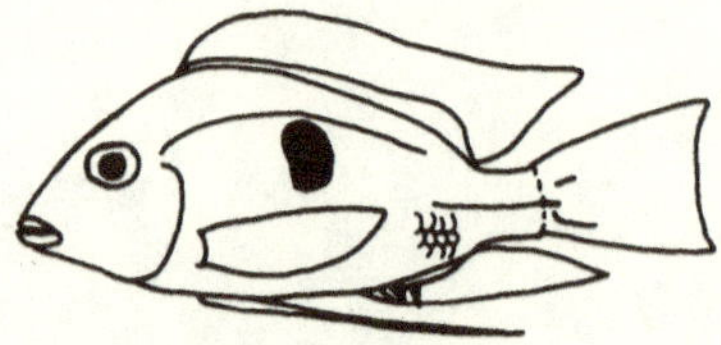

Geophagus brokopondo

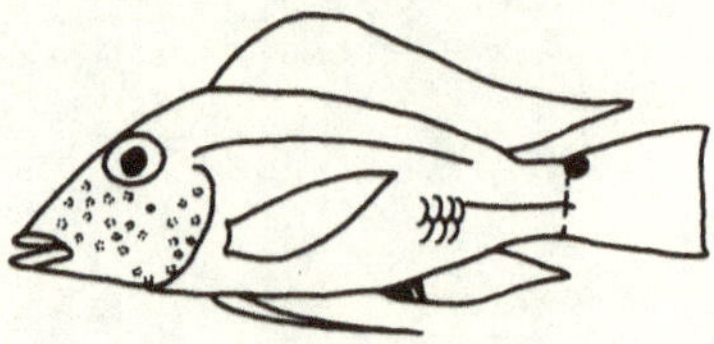

Satanoperca leucosticta

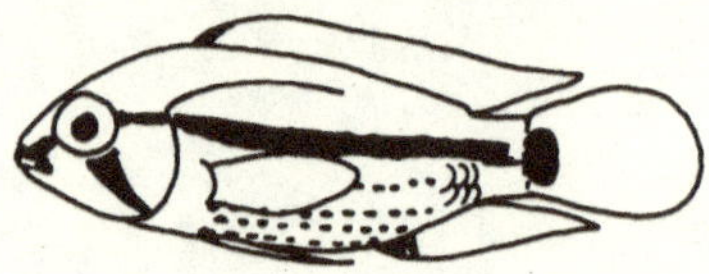

Apistogramma ortmanni

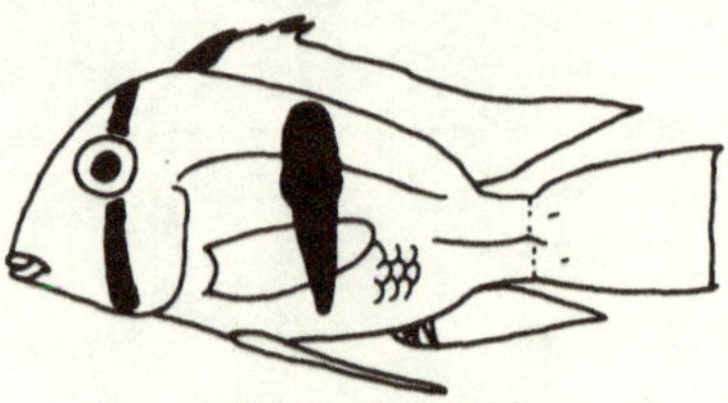

Guianacara owroewefi

Guianacara sphenozona

Cichlasoma bimaculatum

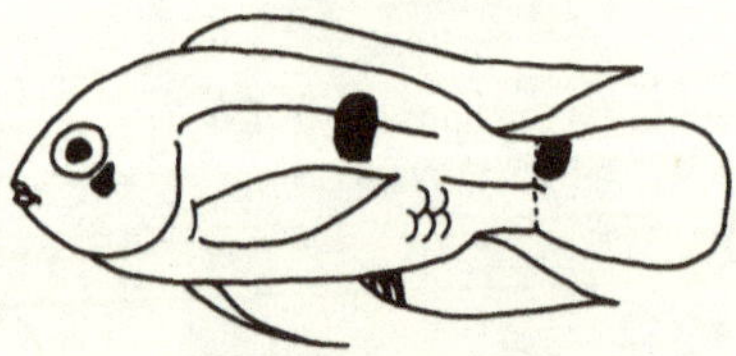

Aequidens tetramerus

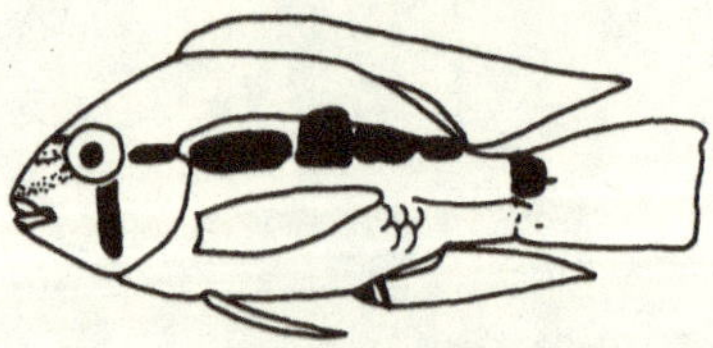

Krobia guianensis

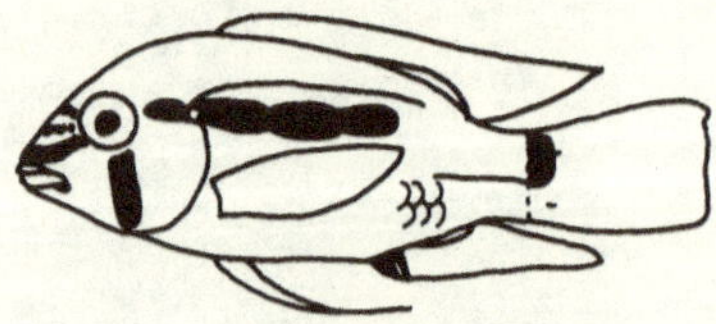

Krobia itanyi

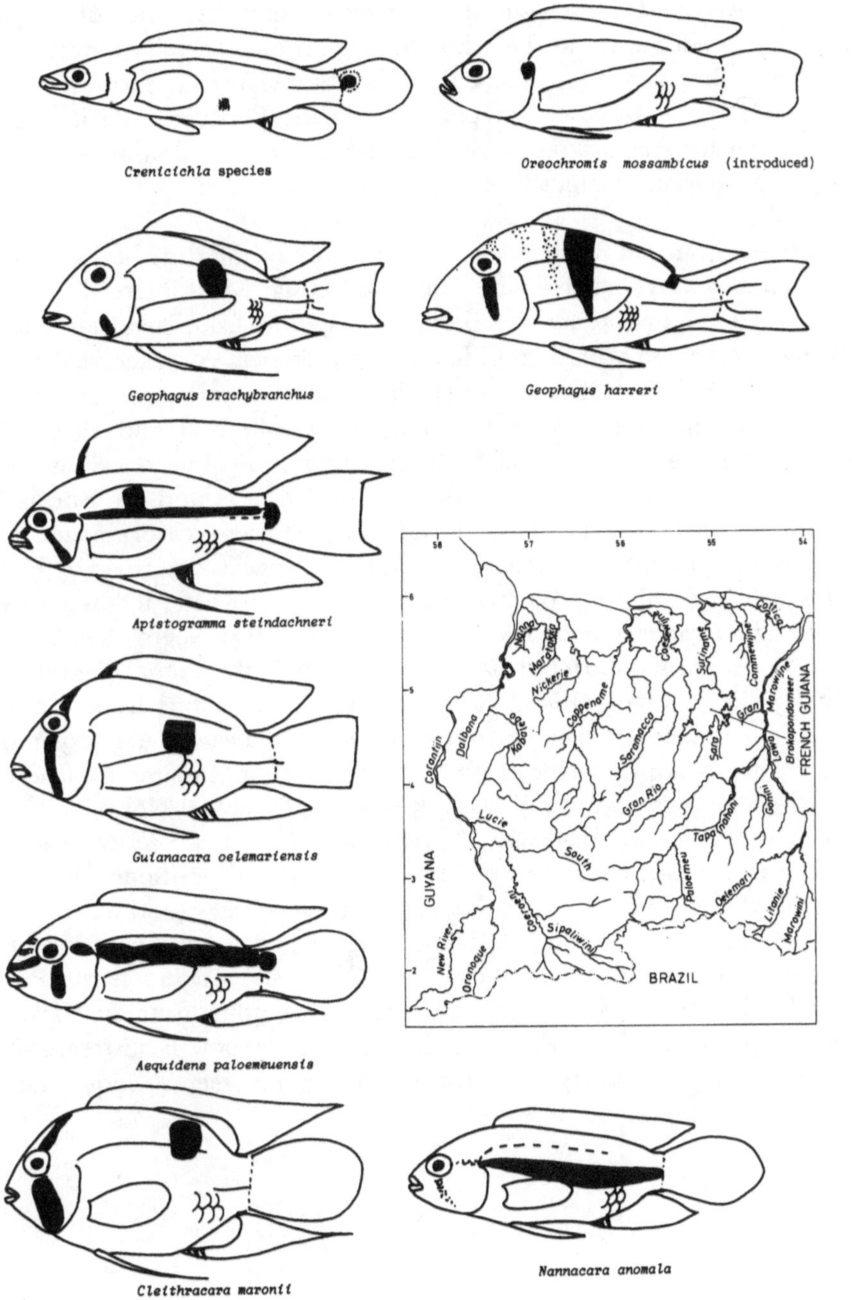

Fig. 6. Diagrammatic representation of Surinamese cichlids recognized in this paper, with principal dark markings, lateral line course, anal fin spines, squamation of vertical fins and body scale size indicated, and map showing area covered by this paper. Do not use this scheme as replacement for the key.

Apistogramma and *Satanoperca*, makes it seem likely that much of the pattern was modelled in the Tertiary. In support of a Tertiary model is the Guianan-Brazilian shield track that likely antedates the forming of the present-day Amazon basin topography in the Pliocene, and it will be more worthwhile to analyze Surinamese cichlid distribution when the backing of a geographically more extensive taxonomic survey becomes available.

'Absences'. Some negative data are notable in the distribution maps, though may turn positive with future collecting. Most interesting is the absence of *Apistogramma* and *Satanoperca* species from the Marowijne. These are genera with a considerable geographical coverage in the rest of South America. The relatively large Coppename and Saramacca drainages have been subject to very little collecting and should be expected to have more species. It may be permissible to assume that species occurring in the Corantijn and/or Nickerie and the Suriname drainage, also occur in the Coppename and Saramacca drainages.

Crenicichla distribution. Among the *Crenicichla* species described by Ploeg (1987), the distribution patterns outlined above are reflected as follows: *Crenicichla lugubris* may be restricted to the lower Rio Negro, Rio Branco and western Guiana. In Surinam it occurs only in the Corantijn drainage (Ploeg 1987). Ploeg (1986a, 1987) includes populations from a much wider area of the Amazon basin. *Crenicichla multispinosa* occurs only in the Suriname, Marowijne (Ploeg 1987) and Mana (NRM material); it seems most closely related to the Oyapock endemic *C. ternetzi* Norman. *Crenicichla nickeriensis*, *C. sipaliwini*, *C. coppenamensis*, and *C. albopunctata* are allopatric species having in common a superiorly positioned humeral ocellus that may be a synapomorphy of these species (and extralimital forms); if all are valid, they reflect drainage specificity to a greater degree than the other Surinamese cichlids. *Crenicichla saxatilis* sensu Ploeg (1986a, 1986b, 1987) has a wide distribution extending from the Nickerie to the Oyapock, with lowland localities except for some headstream sites in the Oyapock drainage. Its distribution pattern seems unique and it may be a composite.

INTRODUCED AND ERRONEOUSLY RECORDED SPECIES

Apistogramma cacatuoides *Hoedeman (1951)*

The species was described on specimens said to come from near Paramaribo. The locality information is apparently erroneous, probably resulting from confusion of labels. The species is otherwise collected only in the upper Amazon basin (Kullander 1980a, 1986; this paper p. 83).

Mesonauta surinamensis *Sauvage (1882)*

Pellegrin (1904, p. 221) reexamined the holotype (from 'Surinam') and identified it as a *Polycentrus schomburgkii* Müller & Troschel. Although Sauvage gives the dorsal fin count as XIII.9 (error for XVI.9?), his description leaves little doubt as to the identity of the specimen. *Mesonauta* Günther is a cichlid genus with chiefly Amazonian distribution, but reported from Guyana (e.g., Eigenmann 1912). *Polycentrus* Müller & Troschel is a monotypic nandid genus; the oldest available specific name is *P. punctatus* (Linnaeus) (Smitt 1892; but cf. Fernholm & Wheeler 1983, Kullander 1983a). Richter & Nijssen (1980) listed *P. schomburgkii* as a cichlid.

Oreochromis mossambicus *(Peters 1852)* (Fig. 7)

Oreochromis mossambicus is an East African oral brooding cichlid species that was introduced in Surinam for brackish water pond culture in 1955 (Lijding 1959b). To which extent escaped fish from this and possible later introductions have established populations is still unknown. The species is also introduced in Guyana (Trewavas 1983). We have examined samples from ditches around Wageningen (RMNH uncat., leg. Boeseman, March 1971) and from near Paramaribo (ZMA 105.806, leg. Nijssen, March 1967) (Fig. 16), confirming the identity of the species. The species can be distinguished from naturally occurring cichlids by, e.g., the cycloid scales and bi- or tri-cuspid jaw teeth. Naturally occurring Surinamese cichlids have ctenoid side scales and caniniform teeth.

Sarotherodon melanotheron *Rüppell (1852)*

Labrus melagaster Bloch (1792) was described from two specimens said to come from Surinam. Müller & Troschel (in Schomburgk 1849, p. 626)

Fig. 7. *Oreochromis mossambicus*, 54.8 mm SL, ZMA 105.806, from near Paramaribo; an introduced African cichlid species.

remarked that the 'Original-Exemplar' of *Labrus melanogaste* (sic) 'is nothing but *Chromis niloticus* Cuv.' (sic), and considered the black abdomen of the figured specimen (Bloch 1792, Pl. 296) to be a mere invention. Reexamination of one of the syntypes (ZMB 2807, 86.5 mm SL) that is still preserved in ZMB confirms that it is an African cichlid, *Sarotherodon melanotheron* Rüppell, 1852 (Trewavas 1987).

KEY TO THE CICHLIDS OF SURINAM

(see also Fig. 6)

1. Lips of 'African' type (Fig. 9); penultimate dorsal fin spine much shorter than antepenultimate and ultimate, so that the fin margin appears deeply notched just anterior to the soft portion .... *Cichla ocellaris*
   - Lips of 'American' type (Fig. 9); posterior dorsal fin spines of approximately equal length...2

2. Gill rakers externally on first ceratobranchial long, slender, numerous (more than 50).............................*Chaetobranchus flavescens*
   - Gill rakers all short, less than 20 externally on first ceratobranchial 3

3. First epibranchial with lobe-like expansion...........................4
   - First epibranchial without lobe-like expansion....................10

4. First ceratobranchial with 0-5 external gill rakers; gill rakers present on lower pharyngeal tooth plate; distinct midbasal caudal spot; lower lip fold continuous...5
   - First ceratobranchial with more than 10 external gill rakers; no gill rakers on lower pharyngeal tooth plate; caudal spot absent or situated on base of dorsal caudal fin lobe; lower lip fold interrupted 6

5. No spot on middle of side; series of spots along abdominal sides; no caudal fin streamers in adult males............*Apistogramma ortmanni*
   - A large spot on middle of side, may be extended dorsad; no distinct spots in series along sides; adult males with prolonged marginal caudal fin rays.......................*Apistogramma steindachneri*

6. Cheek scaled rostrad to lacrimal; vertical fins naked; upper lip fold continuous; first ceratobranchial gill rakers situated over gill filament bases; spot dorsally on caudal fin base; no midlateral spot .... *Satanoperca leucosticta*
   - Anterior half of cheek naked; vertical fins scaled in adults; upper lip fold discontinuous; gill rakers of first ceratobranchial attaching dorsal to gill filament bases; no spot dorsally on caudal fin base; midlateral spot or intense stripe across middle of side prominent......7

7. Black spot on back close to end of dorsal fin base; black stripe across middle of side or spot near middle of dorsal fin; predorsal midline partially naked .... *Geophagus harreri*
   - No black spot on back at end of dorsal fin; black spot on middle of side chiefly below upper lateral line; predorsal midline covered by scales 8

8. Gill filaments with only narrow basal skin cover .... *Geophagus brokopondo*
   - Gill filaments with wide basal skin cover 9

9. Lateral spot on and below 11th-13th lateral line scales; no dark markings on head sides in adults .... *Geophagus surinamensis*
   - Lateral spot on and below 11th-15th lateral line scales; stripe over preopercular corner from 50 mm SL .... *Geophagus brachybranchus*

10. Uni- or tri-serial predorsal squamation; 4 dentary lateralis pores 11
    - Stochastic predorsal squamation; 5 dentary lateralis pores 17

11. Principal caudal fin rays 14; 2, rarely 3, cheek scale series; 3 preopercular scales .... *Nannacara anomala*
    - Principal caudal fin rays 16; 3-4 cheek scale series; 0-2 preopercular scales 12

12. Soft dorsal and anal fins extensively scaled 13
    - Soft dorsal and anal fins with a few basal scales or naked 14

13. No caudal spot; 2 scales on preoperculum; strong bar from nape to interoperculum through eye; caudal fin immaculate; 3 anal fin spines .... *Cleithracara maronii*
    - Superior caudal base spot; preoperculum naked or, rarely, with one scale; suborbital stripe in young; cheek spot in adults; caudal fin dotted; 4, rarely 3 anal fin spines .... *Cichlasoma bimaculatum*

14. Triserial predorsal squamation; cheek spot in adults .... *Aequidens tetramerus*
    - Uniserial predorsal squamation; suborbital stripe retained by adults 15

15. Dorsal and anal fins naked; no accessory lateral lines on caudal fin; lateral band running to caudal fin base........ *Aequidens paloemeuensis*
    - Dorsal and anal fins with a few basal scales; accessory lateral lines on caudal fin present or absent; lateral band running to end of dorsal fin base 16

16. Body depth 40.0-45.3% of SL; lateral band ending well in advance of end of soft dorsal fin base; three facial stripes clearly visible .... *Krobia itanyi*
- Body depth 41.3-50.8% of SL; lateral band continued to end of soft dorsal fin base; two posterior of three facial stripes indistinct .... *Krobia guianensis*

17. Moderately deep; 24-26 scales in a lateral series; pectoral fin extending to over anal fin base 18
- Terete; over 50 scales in a lateral series; pectoral fin not reaching nearly to anal fin base .... *Crenicichla* (20)

18. Large midlateral spot; no produced dorsal fin lappets; single supraneural .... *Guianacara oelemariensis*
- Stripe across side, except in some large specimens in which the stripe reduced to a dorsal spot; anterior dorsal fin lappets produced; two supraneurals 19

19. Flank stripe strongest on and below upper lateral line; dorsal fin membranes between the anterior 3 dorsal fin spines black .... *Guianacara owroewefi*
- Flank stripe wedge shaped, strongest on and above upper lateral line; anterior 3 dorsal fin lappets black in young only .... *Guianacara sphenozona*

20. 100-130 scales in horizontal row below lateral lines................21
- 52-72 scales in horizontal row below lateral lines.................22

21. 112-130 scales below lateral lines; dark spot just posterior to pectoral fin base; nostrils nearer to postlabial skinfold than to orbit......... .... *Crenicichla lugubris*
- 100-115 scales below lateral lines; no dark spot posterior to pectoral fin; nostrils almost halfway between postlabial skinfold and orbit.. *Crenicichla multispinosa*

22. Series of dark blotches along lateral lines in males; humeral spot much longer than deep in females............... *Crenicichla nickeriensis*
- No series of dark blotches along lateral lines in males; humeral spot in females not or only slightly longer than deep..............23

23. Silvery dots present on operculum and cheek in both sexes .... *Crenicichla albopunctata*
- Silvery dots absent 24

24. Maxilla not reaching orbit or anterior margin of orbit; ground colour of body grey .... *Crenicichla sipaliwini*
- Maxilla reaching well below orbit; ground colour of body tan 25

25. Humeral spot round with a white ring; males with yellowish dots .... *Crenicichla coppenamensis*
- Humeral spot with a dorsoanterior notch; males with silvery-white dots.. *Crenicichla saxatilis*

CICHLA Schneider

Cichla Schneider 1801, p. 336 (type species *C. ocellaris* Schneider by subsequent designation in Eigenmann & Bray 1894, p. 611).

Synonymy

Acharnes Müller & Troschel, in Schomburgk 1849, p. 622 (type species *A. speciosus* Müller & Troschel by monotypy).

Cichla is easily distinguished from all other South American cichlids by the shape of the dorsal fin: the spines increase in length to about the 5th, after which there is a gradual decrease to a very short penultimate spine. The soft fin is again about as high as the anterior spinous part. The mouth is large, with prominent jaws, the lower jaw prognathous and the maxilla well exposed. The triacanth anal fin is small and densely scaled in adults. From about 100 mm SL a prominent ocellus is formed on the caudal fin base; otherwise the colouration is very variable ontogenetically and between individuals, although at larger sizes it is species specific.

The bilateral elements of the lower pharyngeal jaw are separate posteriorly and extensively covered by fine teeth. A series of small tooth plates on the 4th ceratobranchial. On the first gill arch there are 7 or 8 epibranchial gill rakers, one in the angle, and 17 to 19 cerato- and hypobranchial rakers, all strongly denticulate, those caudally on the ceratobranchial long, gradually shorter towards arch articulations, anterior 3-5 becoming plate like in large specimens. Microbranchiospines, with a few spines on the exposed face, are present on both sides of all gill arches in adults, only externally in young. Two supraneurals, 35-36 vertebrae (usually 1-2 more abdominal than caudal).

The lateralis canal system of the head includes 7 preopercular and 5 dentary foramina; the suborbital series (Fig. 8) includes 7 bones, the lacrimal and the plate like first infraorbital forming a unit with together 5 lateralis foramina, remaining infraorbitals tubular. The flank lateral line is commonly continuous; tubed lateral line sequences on caudal fin are long, positioned between rays D3 and D4, and V4 and V5. The lip folds are discontinuous symphysially, the upper and lower lips narrowly connected ventrally on the maxilla ('African type' lips, Fig. 9). The preoperculum is entire. Most information on maximum size is imprecise; the largest specimen reliably recorded is Machado's (1971) 62 cm SL *C. orinocensis*. See Stiassny (1982, 1987) and Kullander (1986, 1988) for discussions on generic characters and phylogenetic relation-

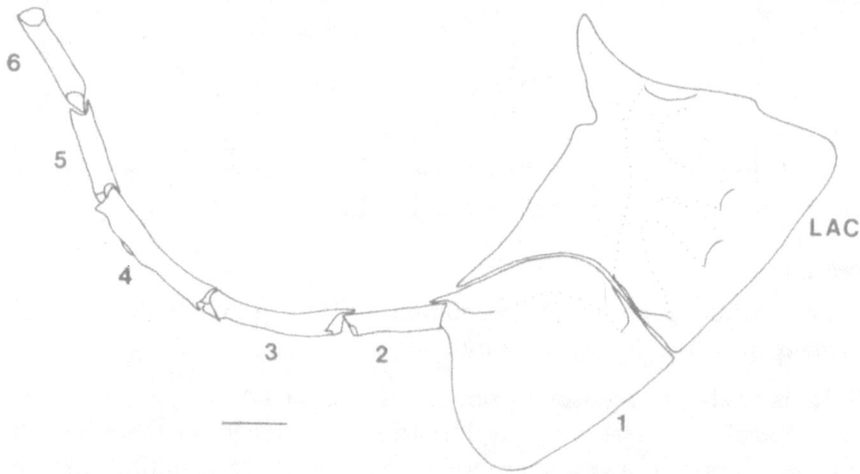

Fig. 8. *Cichla ocellaris*. Suborbital series. Scale 1 mm. NRM A82/1966353.3426, 71 mm SL, alizarin red-alcian blue preparation. *LAC* lacrimal, *1-6* infraorbitals; outline of lacrimal + infraorbital 1 canal dotted.

ships. The taxonomy of the synonym *Acharnes* is discussed in Kullander (1981b).

Species of the genus have been recorded from a wide area including most of the Amazon basin, the Orinoco basin and Guianan rivers. *Cichla* species are among the most important food and game fishes in South America.

Of the 15 nominal species referrable to the genus, five are recognized as valid following the review by Kullander (1986): *C. ocellaris*, *C. orinocensis* Humboldt, *C. temensis* Humboldt, *C. intermedia* Machado-Allison, and *C. monoculus* Spix. Studies in progress suggest, however, that *Cichla* has at least 11 species. *Cichla ocellaris* is generally regarded as a very variable species with wide distribution, but non-Guianan records are mostly for *C. monoculus* (Kullander 1986) or *C. orinocensis* (as in Machado 1971). *Cichla temensis*, *C. orinocensis* (as *C. ocellaris*) and *C. intermedia* were recently described in some detail by Machado (1971), *C. monoculus* by Kullander (1986).

The Surinamese *Cichla* species is *C. ocellaris*. *Cichla* material available from the Oyapock system belongs to the widespread Amazonian species *C. monoculus* (Kullander 1986). The large-scaled *Cichla* in Guyana, described by Eigenmann (1912) and Lowe-McConnell (1969) as *C. ocellaris*, appears to be true *C. ocellaris*. *Cichla temensis* was reported by Lowe-McConnell (1969) from the R. Branco drainage in Guyana. It does not seem to occur naturally in the Guianas, but two species of *Cichla*,

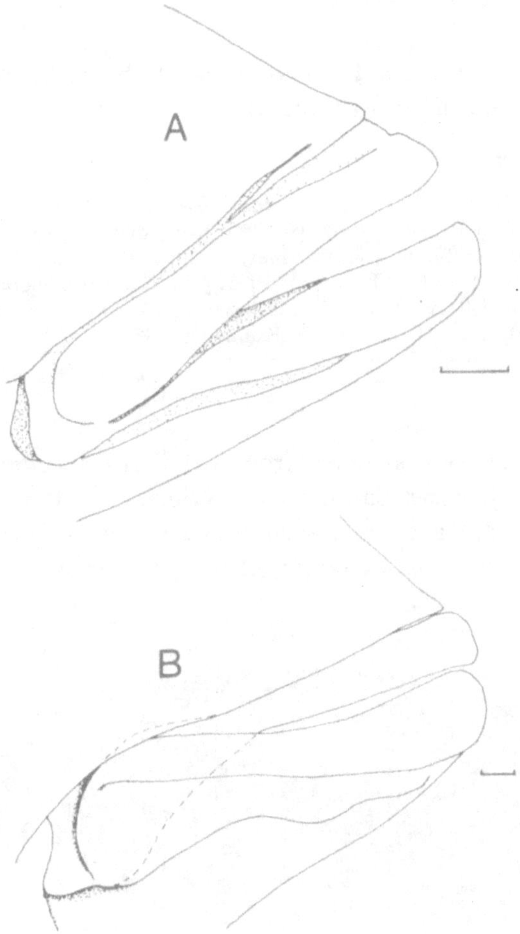

Fig. 9. Comparison of lip configuration in *A Cichla ocellaris* (NRM A82/1966343.3425, 50 mm SL) with 'African type lips', and *B Chaetobranchus flavescens* (RMNH 18166, 93.5 mm SL), with 'American type lips'; in B the outline of the upper lip covered by snout or lower lip indicated by dashed line. Scale 1 mm.

imported from Brazil, were tried in fish culture in Guyana (Lowe-McConnell 1969) and if released into open waters there may now be more than one *Cichla* species in Guyanan Atlantic rivers.

Cichla ocellaris Schneider

(Figs 8-16)

Cichla ocellaris Schneider 1801, p. 340 (India Orientali), pl.66 (colour drawing).

Synonymy

Acharnes speciosus Müller & Troschel, in Schomburgk 1849, p. 622 (Küste und in der Mündung des *Essequibo*).

Surinamese records

Cichla ocellaris; Kappler 1887, p. 149 (local name; distr.; size; life colours; food; gastronomy). — (*pt.*) Pellegrin 1904, p. 184 (descr.; material from Maroni). — Popta 1914-1917, p. 205 (brief descr.; local names). — Puyo 1949, p. 242, fig. 127 (bibliogr.; descr.; Haut Maroni). — (*pt.*) Leentvaar 1973, p. 2 (Lake Brokopondo; local name). — Richter & Nijssen 1980, pp. 124, 125, 129 (listed; Brokopondo reservoir; biomass; local name; [ZMA 107.575, 107.598]). — Kullander 1986, p. 58 (comparison with *C. monoculus*).

Material

Descriptive data were sampled from 101 ZMA specimens, 44.6-414.0 mm SL, from Nickerie, Saramacca and Suriname R. systems; a total of 247 specimens are available, including also series from the Corantijn drainage, as well as the holotype, of unknown provenance.

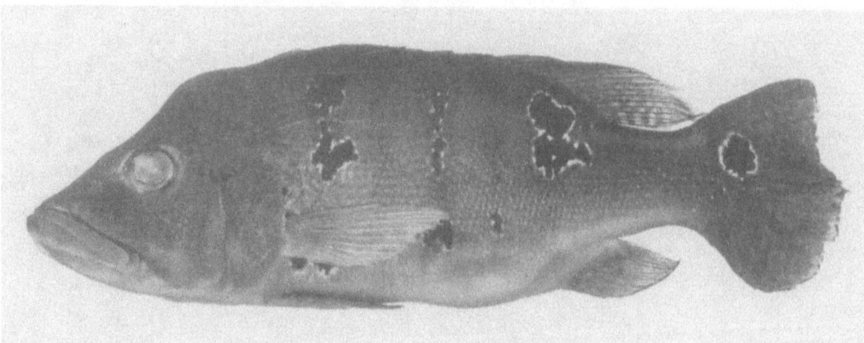

Fig. 10. *Cichla ocellaris*, adult male, 298.6 mm SL, ZMA 105.832, from Brokopondo Meer.

Description

From ZMA 107.832, 298.6 mm (Fig. 10), an adult male, followed by notes on variation in the Surinam material, and on the holotype; observations made on other material specifically indicated. Counts from 25 measured specimens.

Shape: body moderately deep and laterally compressed; little broader anteriorly than posteriorly. In anterior view with flat chest, relatively vertical sides and narrow nape; lateral compression gradually stronger dorsad on back, but abdominal region flattened anteriorly, posteriorly ventral midline rounded. Caudal peduncle tapering caudad, with convex dorsal and ventral margins. Dorsal fin base contour straight horizontal or feebly rising caudad along spinous portion; gradually descending posteriorly. Chest contour straight, nearly horizontal; abdominal contour almost straight horizontal; anal fin base almost straight, slanting upwards.

Head: moderately long, little compressed, with long snout. Predorsal contour ascending, straight save for minor convexity above nostril and a compressed hard elevation in front of dorsal fin. Eye slightly elliptic horizontally, situated below forehead contour, all in upper and most in anterior half of head. Nostril closer to orbit than to snout tip. Interorbital area convex. Mouth large, terminal, low in position, caudally wider than rest of head. Ascending processes of premaxilla extending to behind nostril but not reaching anterior margin of orbit. Caudal part of maxilla well-exposed, extending to nearly middle of orbit. Lower jaw slightly prognathous; articulation little behind vertical from middle of orbit.

Upper lip fold interrupted symphysially, but with a thickened connection with opposite side fold; caudad a simple fold, not curled inwards. Lower lip fold bilaterally from near symphysis, widest medially, much reduced before attachment caudally to maxilla and dorsally to premaxillary tip. Postlabial skin fold with short incision just before lacrimal (cf. Fig. 9).

Opercular bones entire. Preoperculum with uneven straight vertical free margin, rounded ventral margin. Operculum short, not reaching ventrad to interoperculum, which extends little caudal of preoperculum; length about 1.5 times the depth, exposed surface similar in area to that of suboperculum; a wide skin flap along caudodorsal edge, continued around weak caudodorsal blunt subopercular projection.

Scales: ctenoid except where otherwise noted. Cheek naked below a line continuing labial margin of lacrimal caudad; above that, several irregular series of scales, those dorsally and rostrally of about flank scale size, ventrally smaller; a patch of cycloid scales ventrally; dorsal cheek scale series continued onto first infraorbital to level of anterior margin of orbit; 4 vertical series behind orbit; behind upper margin of orbit a narrow naked zone caudad to preoperculum. Preoperculum naked. Operculum, suboperculum save for narrow naked free margin, and interoperculum posteriorly, densely scaled.

Body scales rather small; slightly larger on flanks and abdomen than along dorsal trunk margins and the anterior prepelvic scales. Wide naked area above orbits; predorsal squamation reaching rostrad to not quite anterior margin of orbit; nuchal protuberance naked medially. Prepelvic scales very small save for posteromedial which of about flank scale size. Pectoral axilla and pectoral fin naked. About 32 circumpeduncular scale series.

Lateral line continuous; anteriorly uneven, but more or less horizontal, strongly downturned over 3 horizontal scale series a little caudal to origin of soft dorsal fin, caudad nearly horizontal, to caudal fin base; continued on caudal fin base by 4 canals in ventrad sloping series and a long sequence on dorsal lobe and a short sequence on ventral lobe (which is regenerated distally). Lateral line scales on body slightly smaller than those adjacent.

Fin scales: spinous dorsal fin naked. Along bases of 9 anterior soft rays a series of ctenoid scales, larger than those on fin proper. From behind last spine to behind 11th ray small cycloid scales are arranged in series close behind preceding spine or ray (not filling up space to next ray), double series behind last spine and first ray, otherwise simple series; commencing slightly distad to fin base, anterior series reaching to near tips of rays, posteriorly successively shorter, the last to 1/4 of ray length; interradial scales also on membranes between major ray branches, to near fin edge.

Anal fin with heavy basal sheath of ctenoid scales; interradial scales cycloid, proximally in dense sheath, distad narrower; no scales between ray branches; medially the scale sheath is a little wider than half length of longest rays.

Caudal fin middle membrane and that above naked along middle from ocellus to fin margin; otherwise a wide convex proximal layer of larger scales and densely packed series of small interradial scales that obscure fin rays; distally a narrow zone of cycloid scales and a narrow naked fin edge.

Medial aspect of pelvic fin with basal layer of scales and double series of cycloid interradial scales, extending anteriorly to spine tip, but posteriorly shorter; on lateral side double interradial series of cycloid scales on first two membranes to level of spine tips.

Fins: dorsal fin origin little posterior to vertical from hind margin of operculum; first and last spine of about equal length; some of longer spines apparently with broken tips, but shape otherwise conforming to that described from intact specimens below. Soft fin with narrowly rounded tip, 12th and 13th rays longest, not reaching to caudal fin base. Anal fin short, spines increasing in length to 3rd; soft fin rounded, not extending to caudal fin base. Caudal fin apparently with regenerated lower lobe; dorsal lobe truncate, ventral roundish; ray count impossible without removing dense squamation. Pectoral fin subacuminate, with 4th ray longest, not extending to vent, but halfway to anal fin base end. Pelvic fin short, broad, 2nd ray longest, distal edge straight.

Jaw teeth: teeth small, slightly recurved, in narrow bands in both jaws, symphysial areas edentulous. Mouth and gill cover cannot be much opened without risking damage to specimen, so for details of dentition and gill arches, see notes below from other specimens.

Variation: a representative young specimen (ZMA 105.083, 47.6 mm SL, Fig. 13) elongate with strikingly large head and mouth; outline tapering caudad from head. In frontal view nape broadly rounded, head sides vertical, chest medially flat but narrower than in large adults. Head elongate, deeper than wide, with slowly ascending evenly curved dorsal contour and little sloping prepelvic contour. Interorbital area flattened medially, eye nearly touching forehead contour. Large, oblique, ventrally placed mouth wider posteriorly than interorbital but narrower than occipital region. Tip of maxilla well exposed, reaching to nearly middle of orbit; ascending processes of premaxilla reaching a point just short of orbit; lower jaw barely prognathous, its articulation below middle of orbit. Lips as in adults; operculars like those of adults except that suboperculum not so deep.

Major shape changes correlated with increasing size include a rise of dorsum, straightening of predorsal outline, lowering of eyes relative to predorsal contour and lower jaw prognathy.

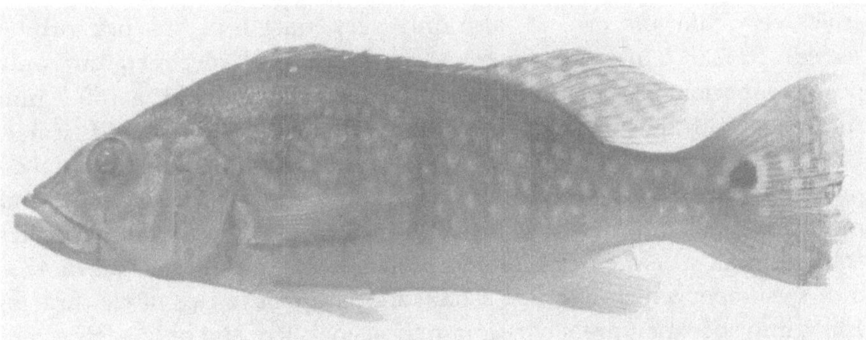

Fig. 11. *Cichla ocellaris*, 141.3 mm SL, ZMA 105.058, from Brokopondo Meer.

A young specimen 138.6 mm (ZMA 105.058) mostly like adults. Elongate, with entire dorsum elevated, spinous dorsal fin base horizontal. Frontal outline rising straight to dorsal fin origin save for minor elevation just in advance of orbit. In frontal view nape keeled, chest flat. Lower jaw distinctly prognathous, articulation below about posterior margin of orbit.

Five males besides 298.6 mm specimen have a nuchal hump. One of these has appearance of a stunted fish and is only 199.7 mm SL (ZMA 107.598). The others are 235.6, 267.8, 268.0, and 414.0 mm SL (ZMA 105.086, two; 105.054; 105.194).

Scales in horizontal series above that containing lower lateral line section from end of caudal peduncle forwards to head 70(1), 71(1), 72(5), 73(1), 74(3), 75(3), 76(5), 77(5), 79(1).

Fig. 12. *Cichla ocellaris*, 86.9 mm SL, ZMA 105.059, from Brokopondo Meer.

Scales on flanks ctenoid at all sizes available. In 47.6 mm specimen cheek and gill cover squamation as in 298.6 mm specimen except that scales very thin and cycloid, also only very margin of first infraorbital scaled. In 138.6 mm specimen as in 298.6 mm specimen but only posterosuperior opercular and subopercular scales ctenoid; a 160.7 mm specimen with posterior/superior half of cheek with ctenoid scales. Predorsal scales cycloid in 47.6 mm specimen; predorsal midline naked medially. Extension of naked midline reduced in larger specimens, and most predorsal scales ctenoid in 138.6 mm specimen. Abdominal scales ctenoid in all specimens; chest and prepelvic scales thin, cycloid in 47.6 mm specimen, which also has a naked area opposite tips of cleithra; in 138.6 mm specimen prepelvic squamation complete and only a few scales remaining cycloid rostrally.

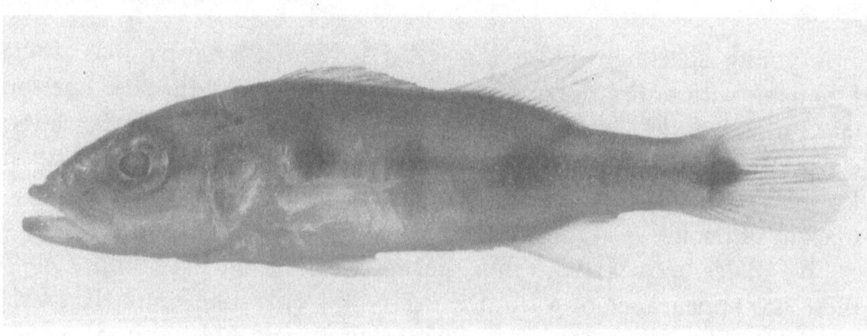

Fig. 13. *Cichla ocellaris*, 47.6 mm SL, ZMA 105.083, from Brokopondo Meer.

Lateral line either continuous or in two sections. When continuous the course is as in the 298.6 mm specimen, comprising 71(1), 73(3), 74(1), 75(4), 76(8), 77(1), 78(2), 79(1) scales, all tube-bearing. In measured specimens line discontinuous on right side in four specimens with 42/33, 44/31, 45/33, and 46/32 scales. In ZMA material of 101 specimens line continuous except in four (including 47.6 mm specimen) with lateral line discontinuous on left side, 12 specimens with it discontinuous on right side, and three with it discontinuous on both sides. Of 8 large USNM Corantijn specimens checked (98.7-236.4 mm), the lateral line is discontinuous in all, however. Photos of four Marowijne specimens show the lateral line continuous. In two old Maroni specimens (MNHN 95-133-134) in very bad condition, a few scales remaining and skin impressions suggest bilaterally continuous lateral line although a continuous lateral line is actually observable on only one side of one specimen. The lateral line is continuous on three of the Guyanan specimens figured by Eigenmann (1912, pl. LXIX), not evident on the fourth. The holotype of *Acharnes speciosus* has discontinuous lateral line on both sides, but the sections are nearly in contact on the left side. When the lateral line is discontinuous, the downwards running section missing; however, when there is a connection, either of the horizontal sections may proceed a few scales caudad or rostrad respectively of that connection, and occasionally also scales preceding the connecting section may overlie it and obscure canals, so that the lateral line may appear discontinuous more often than it actually is; it also occurs that terminal curvatures of anterior and posterior sections are displaced, resulting in discontinuity.

Lateral line continued on caudal fin by 3 or 4 scales but these canals not easily seen in very large specimens. In smallest specimens, about 5 cm, 4-6 canals, in those 6-7 cm, 10-15 canals in a series running between rays V4 and V5, in still larger specimens a long series of tube bearing scales all the way out to near end of caudal fin; a dorsal lobe canal series is present from at least 55 mm SL, then of 2-4 canals, still only of about 10 scales in specimens 80-90 mm, in larger fish extending to near end of fin between rays D3 and D4; usually those accessory lateral line sequences continue across basal caudal fin squamation towards median continuation of lateral line of body, which is usually reached by the ventral series but not by the dorsal. These series also become obsolete in very large specimens. Occasionally there are additional series of lateral line scales on the caudal fin.

Fins except caudal fin naked in 47.6 mm specimen; dorsal fin remaining naked save for basal scales in 160.7 mm specimen (ZMA 105.058); in next larger (170 mm, ZMA 105.566) almost fully scaled as in 298.6 mm and other larger specimens, though only to behind 9th ray. Basal anal fin squamation developing from 50 mm, interradial squamation present proximally on anterior part at 55 mm; 138.6 mm specimen has anal fin only little less densely scaled than reference specimen and other large specimens.

Pectoral axilla and pectoral fin remain scaleless at all sizes. The 138.6 mm specimen has a basal squamation but few interradial scales on medial side of pelvic fin and interradial scales proximally on lateral side; in still larger fish pelvic fin squamation gradually approaches condition in 298.6 mm specimen.

Caudal fin scales cycloid in 47.6 mm specimen; basally a convex sheath of larger scales, beyond marginally thin interradial scales to almost half length of fin; hind margin of scaled area concave. In 138.6 mm specimen caudal fin squamation as in 298.6 mm specimen, only not quite as dense, and membranes between rays D3 to V3 naked along middle, more proximad on median membranes; squamation symmetrical in intact specimens but a regenerated lower lobe is frequent and then scale arrangement on it irregular.

D. XIV.17(1), XV.17(20), XV.18(4); last ray may be actually two, but when base of last major branch/ray could not be distinguished from that of preceding ray/branch it was not counted as separate. In smallest specimens available first spine inserted behind a vertical from hind edge of operculum, but relative position shifts forward with increasing size. First spine about as long as last, and relative spine length increasing until 4th to about 6th which are subequal in length, then decreasing gradually to penultimate, which is about half length of 5th and shorter than ultimate; rays much longer than posterior spines. Soft fin rounded posteriorly, extending to middle of caudal peduncle in smallest, little longer in large specimens, but not reaching to caudal fin base. Two Maroni specimens (MNHN 95-133-134) both have D. XVI.16, a count that is not found in Surinamese material.

A. III.10(6), III.11(19). Insertion of first, minute, spine opposite about 5th to 6th dorsal fin ray; spine length increasing caudad; in large specimens dense squamation prevents individual erection of spines, but also unlike the dorsal fin, spines follow very closely upon each other, so that on cursory inspection at least the first is not easily located. Soft fin has a feebly rounded hind edge in 47.6 mm specimen, with 2nd ray the longest, though median rays reach further caudad; in 138.6 mm specimen fin with a rounded tip with rays 5-6 longest; in smallest specimens reaching backwards to middle of caudal peduncle, in large fish further, but never to caudal fin base.

Caudal fin with 16 principal rays as readily observed in young; in fish over 100 mm rays obscured by scales and excessively branched terminally. Hind margin slightly emarginate in smallest specimens; shape often indeterminable due to damage in large specimens but when determinable truncate with squared upper corner and rounded lower corner.

P. 13(1), 14(16), 15(8). Short, rounded in young, becoming rather acuminate in large specimens; asymmetrical, with 4th ray longest.

V. I.5, spine inserted in advance of a vertical from pectoral axilla; tip rounded; hind edge straight in young, becoming slightly concave in large specimens; 1st and 2nd rays subequal in extension.

Jaw teeth in a specimen ca 245 mm SL (ZMA 105.133, with wide open mouth) small, recurved, simple, pointed, densely arranged in well-defined bands that widen symphysiad, teeth absent symphysially; upper jaw band wider anteriorly than lower jaw band. Teeth anteriorly less than 1 mm long, posteriorly gradual size decrease.

On first gill arch in 245 mm specimen, and a 92.4 mm specimen (NRM 11310) figured (Fig. 14), externally 7 epibranchial, one in angle (or rather epibranchial) and 15-16 ceratobranchial rakers; rostrally (on first hypobranchial) 3 weakly denticulated dermal plates. Length of rakers increasing to uppermost ceratobranchial, ca 8 mm long in the larger fish, then again decreasing, last 3 epibranchial rakers small to rudimentary. Longer rakers rostrocaudally compressed and gradually becoming more slender distad; all strongly denticulate along medial edge. Inner rakers short, denticulate apically. A soft dorsal ridge along ceratobranchial, but no transverse soft ridges.

Seven or 8 epibranchial rakers, one in epi/ceratobranchial joint; 14(1), 15(5), 17(4), 18(10), 19(4) -(1) rakers on lower limb in measured specimens. The number of lower limb rakers decreases with increasing SL apparently because lower rakers, especially the anterior 3, which lie on first hypobranchial, tend to become plate like and are not easily seen or felt on well-preserved specimens. Actual number therefore is apparently 18 or 19, including 3 hypobranchial.

A radiographed specimen (NRM 11310) has 19+16 vertebrae. Cleared and stained specimen with 18+17 vertebrae and hemal arches below 3 last abdominal centra.

Colouration

Colouration extremely variable, due, apparently, mostly to ontogenetic changes and state of preservation. Brightly coloured large specimens presumably present a special breeding colouration. Bar numbering below is from young specimens (*q.v.*).

Male, 298.6 mm, Fig. 10 (right side): Underside yellowish white from caudal fin base to throat and in a zone along ventral sides. Back dark grey, with narrow black scale edges, flanks lighter, brownish. Nape in

Fig. 14. *Cichla ocellaris*. Lateral aspect of first gill arch (shorter scale, 1 mm) and, enlarged (longer scale, 1 mm) second from posteriormost ceratobranchial gill raker. From NRM 11310, 92.4 mm SL.

advance of a line from the gillcleft obliquely rostrad and dorsad across nuchal hump, blackish. Forehead and snout dark grey to blackish. Preorbital brownish grey. Cheek brownish, dirty greyish white on naked ventral part. Preoperculum brownish grey, except edge towards interoperculum, which greyish white, and dull silvery vertical margin. Operculum greyish brown, suboperculum lighter ventrad. Interoperculum whitish save for brownish scaled area. Lower jaw tip white. Upper lip grey, darker symphysiad, lighter caudad; maxillary skin ablabially blackish. Tip of maxilla and lower edge of lower lip whitish; upper edge and tip of lower lip grey. Lower jaw whitish, greyish adlabially. Branchiostegal membrane and intermandibular area greyish white.

Scale centers on back and sides light; on back tending to opalescent; on sides faint opalescent spot basally on most scales, especially below upper lateral line section. Scattered silvery spots on caudal peduncle sides.

Markings: Small black spots irregularly behind orbit, two on cheek, one preopercular, one anterior and two faint posterior opercular, one on subopercular posterior process; more or less ocellated by opalescence. A blackish stripe over mouth angle to preoperculum. On flanks shadows of Bars 1a and 2a; Bars 1, 2 and 3 expressed chiefly as large irregularly shaped blotches dorsally (extended vertically to level of lower lateral line section); ringed with white to silvery or opalescent bright spots, most prominent around Bar 3 spot. Inside of pectoral axilla brown; just anterior to axilla a blackish brown spot with light to silvery hind margin. Under adpressed pectoral fin a large uneven blackish brown spot ringed with silvery white spots, and succeeded caudad and slightly dorsad by 3 similar, minute spots, one chiefly in Bar 2, the last in Bar 2a. Shape and precise positions of postorbital head spots, dorsal blotches and spots anterior and posterior to pectoral axilla slightly different on the both sides of the fish.

Dorsal fin with anterior 8 interradial membranes blackish brown, every 2nd membrane base with a pale to opalescent spot; posteriorly mottled dark brown and light. Soft fin brownish, paler towards distal margin, covered by large light spots, especially prominent posteriorly. Anal fin mottled brownish, edge narrowly dark. Caudal fin light brownish with large, indistinct light spots on dorsal lobe and along middle; close to base, above lower lateral line section level a black spot with white to silvery spots contained in a pale ring around, diameter little less than that of eye. Pelvic fin brownish outwardly, lighter inwardly.

The 298.6 mm specimen is the only specimen available with three large dorsal flank ocelli.

Young fish (from ZMA 105.083, 47.6 mm, Fig. 13) yellowish white on underside and lower part of head, trunk and caudal peduncle, up to pectoral axilla level. Dorsum brown grey, sides lighter. Nape and forehead brownish grey. Lower lip white posteriorly, grey anteriorly. Upper lip grey dorsally and along margin, white caudally and ventrally. Preorbital greyish. Cheek silvery dorsally; operculum, preoperculum and suboperculum greyish with some silver especially anteriorly on operculum and suboperculum. No pre- or post-orbital dark stripes. A pale narrow stripe obliquely dorsad-rostrad across nape from gill cleft.

Along middle of sides 3 dark spots, (1) dark brown, roundish, just posterior to pectoral axilla level, below upper lateral line section, with faint narrow dorsal extension; (2) faint, brown, above vent, with fainter extensions dorsad and ventrad; (3) dark brown, horizontally extended, above soft anal fin, with fainter extension along caudal peduncle middle to caudal base spot.

Spinous dorsal fin smoky, soft fin clear with dark spot on each ray base. Anal fin clear, base narrowly dusky. Pelvic fin clear. Caudal fin faintly smoky, especially medianly; midbasally a roundish spot, slightly extended caudad; large whitish spot dorsocaudal and ventrocaudal to base spot.

This colour pattern is found in specimens up to about 55 mm; then short vertical bars develop, the first (Bar 1) with the first spot in center, the second (Bar 2) replacing the second spot.

Specimens ca 60-70 mm develop large light spots on the back, two horizontal dark bands outwardly in the soft dorsal fin; occasionally narrow dark vertical shades behind Bar 1 (= Bar 1a), and behind Bar 2 (= Bar 2a); a bar (Bar 3) may also show through the anterior part of the 3rd flank spot; light spots preceding caudal spot still indistinct.

Specimens ca 70-80 mm have light spots also on sides (or only on the sides), well-developed vertical bars (1, 1a, 2, 2a, 3; but a-bars not always apparent); the 3rd flank spot tends to become absorbed by Bar 3, the caudal base spot tends to be isolated from caudal peduncle pigment, and lighter spots around it brighten.

At 90-100 mm (Fig. 12), no flank spots remain, the soft dorsal fin becomes light-spotted and the isolated caudal base spot tends to have a more dorsal position than in smaller fish. Specimens ca 130 mm have light spots on the back, side spots whitish or silvery, the caudal base spot is ocellated and lies dorsal to the lateral line, a-bars always distinguishable (Fig. 11).

Specimens 161 mm and larger develop an ocellated spot dorsally in Bar 3 and spots behind pectoral axilla appear. Specimens over 200 mm have lost the light back spots and on the sides they also disappear, Bar 3 is lost at larger sizes, but Bars 1 and 2 have dark spots dorsally and all are more or less spotted behind the pectoral axilla.

Large adults (> 200 mm) are more or less brownish with light underside. The vertical bars are pointed ventrally. The Bar 3 ocellus usually is roughly boomerang shaped with apex directed rostroventrad. There is commonly a second caudal ocellus below the lower lateral line level, contiguous or not with the major ocellus. A spot or bar is more or less evident from the gill cleft, obliquely rostrad and dorsad to the nape sides (in very young, the posterior light margin is more conspicuous, in the 298.6 mm specimen contiguous with dark anterior nape). Small silvery scale center spots are found in about half of the specimens over 200 mm, more or less neatly arranged in horizontal series on flanks and caudal peduncle; in those without, the sides are indistinctly light/dark marbled.

Holotype. The holotype (Fig. 15) is preserved in a relatively good condition, but somewhat soft and slightly discolored. It is a female, 183.4 mm SL, 219.7 mm TL, agreeing well with Surinamese *Cichla* of the same size. The shape and remains of colour pattern are apparent from Fig. 15. It departs from other Surinamese *Cichla* studied only in the scaled pectoral fin base, and is notable for having the right lateral line interrupted (as pointed out by Müller & Troschel, in Schomburgk 1849). It has 7 or 8 series of teeth in the upper jaw, about 5 in the lower jaw, not a single series as stated in the original description, apparently an observation lapse (already subjected to Heckel's [1840] sarcasm).

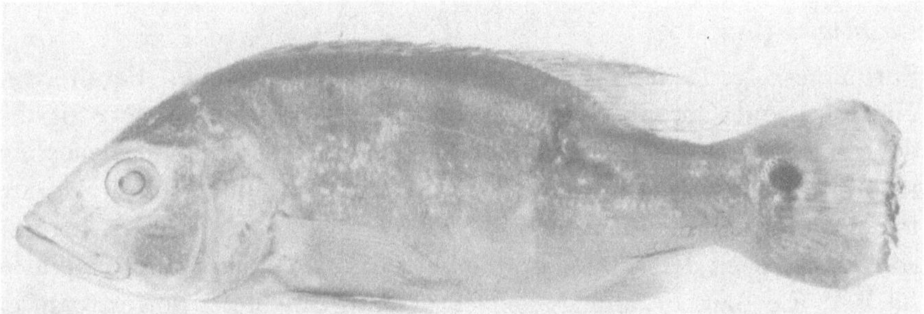

Fig. 15. *Cichla ocellaris*, holotype, 183.4 mm SL, ZMB 2839, origin unknown.

In shape it departs from the 298.6 mm specimen described above chiefly in lacking a nuchal protuberance. Squ. long. 71. First infraorbital scaled adbuccally; most of predorsal midline naked. Pectoral fin base, and on left side also adjacent pectoral axilla, scaled between rays 1 to 10. Pelvic fin scaled anteriorly on both sides. Dorsal fin scales running in single series on soft fin membranes except last four; series gradually shorter posteriorly on fin. Anal fin heavily scaled anteriorly, to edge of fin, squamation gradually restricted to base posteriorly, last membrane naked. Left side lateral line smoothly continuous, on right side interrupted, with two horizontal scale series separating the sections (41/34); four tubed scales on caudal fin base, continuing lower lateral line, and long tube series, to distal edge of fin squamation, on membranes D3-D4, V4-V5 on right side, D3-D4, V3-V4, V4-V5 on left side. D. XV.18, A. III.11, P. 14. Soft portions of dorsal and anal fins with rounded tips, not reaching to base of caudal fin. Caudal fin lower lobe regenerated, hind edge damaged, otherwise appearing truncate. Rakers 7 + 1 + 18. Vertebrae 19 + 17; only one epural (two in other specimens of the species).

Back olivaceous, middle flanks somewhat lighter, below level of pectoral fin base yellowish, abdomen greyish. Traces of narrow, ventrally narrowed dark Bars 1 and 2 from halfway between lateral line and dorsal fin down to lower lateral line level. Angled but irregular brown blotch above lower lateral line, anteriormost portion with a few silvery dots around it; scattered silvery dots also anteriorly on flanks associated with Bars 1 and 2. Soft dorsal fin dusky with large round pale yellowish spots. Anal fin immaculate. Caudal fin with dorsal dark brown ocellus, with bright white semicircle anteriorly, paler white semicircle posteriorly; trace of another, contiguous spot on level of lower lateral line; large light spots on naked portion.

Distribution (Fig. 16)

Surinamese collections available from the Suriname, Saramacca, Nickerie, and Corantijn systems do not give a clear picture of the distribution. The species is probably more widespread than the collecting record indicates. Two upper Marowijne *Cichla* specimens examined are too small for definite identification, but Marowijne *Cichla* referable to *C. ocellaris* has been determined on the basis of fine colour photos, supplied by P.Y. Le Bail, of adults collected at Saut Singatetei downstream of Grand Santi, and two old Maroni specimens (MNHN 95-133-134).

The *Cichla* material reported from Guyana by Eigenmann (1912) and Lowe-McConnell (1969) has not been checked, but probably represents *C. ocellaris*, judging from Eigenmann's figures. Puyo (1949) reported '*C. ocellaris*' from the upper Mana, the Oyapock at Saut Cafesoca, and the upper Maroni, and Pellegrin (1904) listed material from Cayenne (used by Popta (1901) for gill arch study) and Maroni (MNHN 95-133-134).

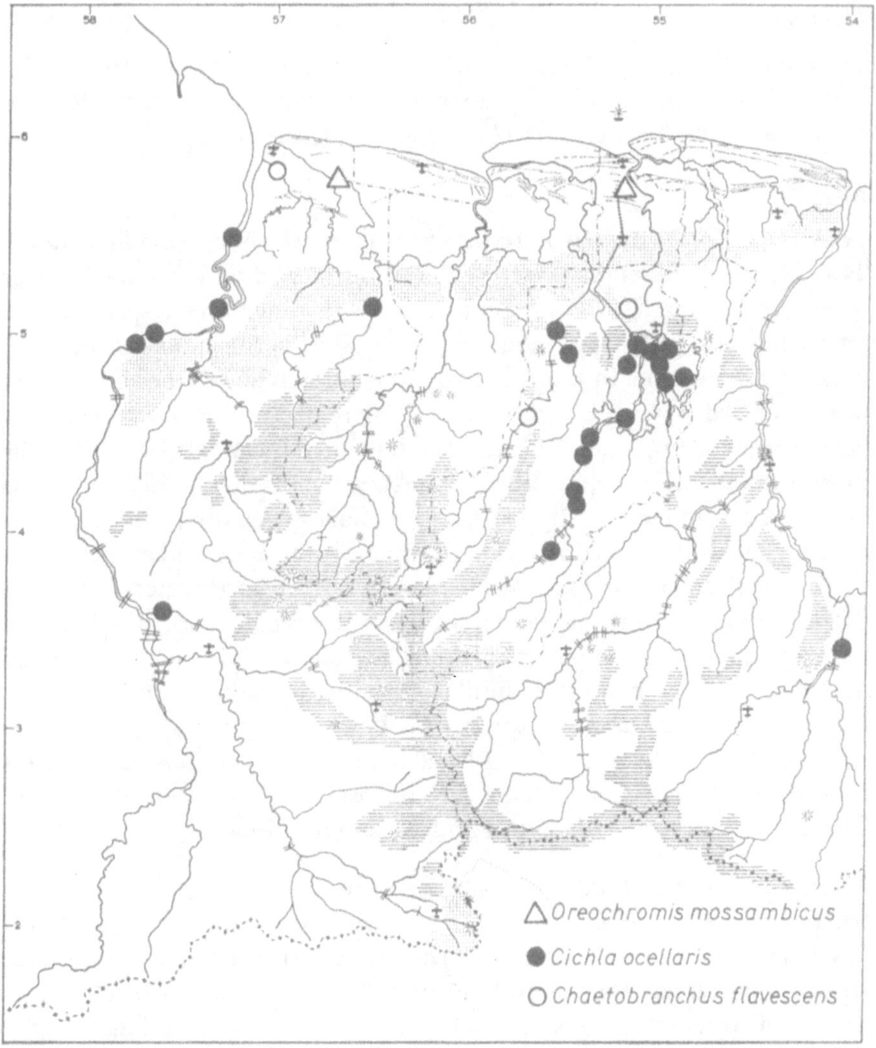

Fig 16. Collecting localities of *Cichla ocellaris*, *Chaetobranchus flavescens* and *Oreochromis mossambicus* in Surinam.

The Cayenne specimen (MNHN A. 9484) is a fine specimen, referable to *C. ocellaris*, with bilaterally continuous lateral line and characteristic colour pattern. The locality is open to doubt, as it may have been merely shipped from Cayenne, but collected elsewhere.

Cichla ocellaris appears to be restricted in distribution to the Guianas with its eastern limit in the Marowijne, the western limit still to be investigated.

Local names

Toekoenali (Lijding 1959a), Zonnevisch, Toekanale (Popta 1914-1917), Tukanale (Kappler 1887), Toekoenari (Surinam, Saramaccan; Richter & Nijssen 1980, Leentvaar 1973, Brokopondomeer).

Notes

Diagnostics: Cichla ocellaris is recognized as relatively deepbodied (depth 24.8-34.1% of SL in size range 50-414 mm), and with relatively large scales (squ. long. 70-79) in comparison with other *Cichla* species. The lateral line is usually continuous at all sizes. The vertical bars are narrow, in large adults spots in the first two bars situated above the lateral line, ocellus of 3rd vertical bar prominent. There are no horizontal series of yellow spots along sides and no dark bar on the caudal peduncle. The most similar species may be *C. monoculus* which has a discontinuous lateral line, ontogenetically earlier lower jaw prognathy, and lacks the a-bars characteristic of *C. ocellaris*. Bars 1-3 are much wider than in *C. ocellaris* and extend ventrad from the dorsal fin base instead of being slightly removed from the dorsal fin base; in breeding fish, the bars are reduced in length and appear as three large contrasted black spots dorsal to the lateral line. Young are similar in shape, but adult *C. monoculus* are more slender than adult *C. ocellaris*, as best evident in caudal peduncle depth: 12.0-13.2% of SL, $\bar{x} = 12.5 \pm 0.12$ in 11 *C. ocellaris* 138.6-324.1 mm SL, 10.3-12.3%, $\bar{x} = 11.2 \pm 0.15$ in 14 *C. monoculus* 138.0-324.5 mm SL, using data from this paper and Kullander (1986).

Holotype. Schneider (1801) published a very fine plate (no. 66) of *Cichla ocellaris* besides the description: 'Cichla with large gape, oblique, lower jaw long, pointed, with small teeth in single series, two dark transverse stripes, large black spot at end of dorsal fin, black ocellus bright white at base of rounded and scaled caudal fin, lateral line descending towards anal fin, whereafter straight, dorsal fin with middle hollowed, basally scaled, scales small, vent far back. Branchiostegal rays 5. P. 15. V. 1/6, A. 3/14, C. 20, D. 15/32. Habitat as the preceding [India orientali].' (Translation from the Latin.) That part of the description is probably written by Bloch; a note is appended by Schneider: 'No pseudobranch, neither gill jaws, but two rough bones at the end of the roof of the mouth. Schn.' (Translation from the Latin.)

The specimen has since been studied only by Müller & Troschel (in Schomburgk 1849) who merely remarked on the dextrally interrupted, sinistrally continuous lateral line.

The stated type locality, 'India orientali', is certainly an error, either from incorrect original locality data or a lapse in the compilation of the manuscript. It is very likely that the fish came from Surinam as this country is a common source of 18th century South American natural history objects.

Synonymy: Acharnes speciosus was synonymized with *C. ocellaris* by Peters (1866) and Eigenmann (1912), and the synonymy is tentatively verified by examination of the holotype of *A. speciosus*. It is a small specimen in very bad condition which does not offer justification for recognition of the species.

CHAETOBRANCHUS Heckel

Chaetobranchus Heckel 1840, p. 401 (type species *Chaetobranchus flavescens* Heckel, by subsequent designation in Eigenmann 1910, p. 469).

This genus is recognized for large (to ca 250 mm SL) South American cichlids distinguished in particular by specializations apparently associated with planktivory: gill rakers externally on first gill arch long, slender, close set (gradual size decrease towards skull and hypobranchial), 20-35 on epibranchial, ca 50-90 on cerato- and hypo-branchial, other rakers shorter and more spaced, but still relatively long and numerous. The mouth is large; jaw teeth are small, simple, pointed, in narrow bands in both jaws, running along entire upper jaw rim in adults. On the roof of the branchial chamber there is a narrow skin ridge above each first ceratobranchial. The lower pharyngeal tooth plate is long and slender, with deeply incised hind margin, and covered by numerous small pointed teeth. Both lip folds are continuous; the lower lip attaches to the hind edge of the upper lip fold well above distal tips of premaxilla and maxilla ('American type lips', Fig. 9).

There are no gill rakers on the lower pharyngeal tooth plate, but there is a series of rudimentary microbranchiospines along the lateral edges. Microbranchiospines present also externally on the 2nd-4th arches and internally on the 4th, plate like, with rounded top and truncate bottom, and with many small teeth on exposed surface and some along lateral margins ventrally. Basibranchial 3 is usually toothed. Ceratobranchial 4 is edentulous.

The lateralis system includes 7 preopercular and 5 dentary lateralis canal openings; suborbital series consisting of lacrimal, 4 tube like infraorbitals (2nd with middle foramen); and several series of tubed scales on caudal fin: one long between rays D1-D2; on lower lobe either a few scales continuing lower lateral line plus long sequence between rays V2-3 or V3-4, or 2-3 long sequences between rays V1-2, V2-3 (variably present) and V3-4 respectively.

The predorsal squamation is irregular; anteroventrally on the cheek is a group of scales narrowly separate from the rest of the cheek squamation.

The position of the caudal fin lateral lines is unusual among cichlids. Basibranchial tooth plates are not known from any other cichlid. They need more study with regard to possible homology with those of other percoids (e.g. in Greenwood 1976). A feature of uncertain interpretation is the shape of the microbranchiospines, shared with *Astronotus* and *Chaetobranchopsis*. The presence of microbranchiospines on the lower pharyngeal tooth plate is a feature unique to *Chaetobranchus*.

Included species are *Chaetobranchus flavescens* Heckel (widely distributed in the Amazon basin, Guianas, and Orinoco basin) and *C. semifasciatus* Steindachner (central Amazon basin). These differ considerably, e.g. in colouration and squamation. *Chaetobranchus semifasciatus* has vertical bars and caudal ocellus, smaller scales (e.g. 20 instead of 16 circumpeduncular scale series) and narrowly scaled instead of naked vertical fins.

Chaetobranchopsis Steindachner is similar to *Chaetobranchus* but has extensively scaled vertical fins, 5-6 (occasionally 4) instead of 3 anal fin spines, and lacks basibranchial tooth plate.

Chaetobranchus flavescens Heckel

(Figs 9, 16-17)

Chaetobranchus flavescens Heckel 1840, p. 402 (Flüss Guaporé ... Ausflüsse des Rio-negro).

Synonymy

Chaetobranchus bruneus Heckel 1840, p. 405 (Rio-negro).
Centrarchus ? cyanopterus Jardine 1843, p. 165 (Essequibo).
Chromys ucayalensis Castelnau 1855, p. 15 (Sarayacu (Pérou)).
Chaetobranchus robustus Günther 1862, p. 310 (Guiana).
Geophagus badiipinnis Cope 1872, p. 251 (Ambyiacu River).

Surinamese records

Chaetobranchus flavescens; Boeseman, 1952, p. 197 (listed; Nanni creek, Doksen savanna [pt., RMNH 18802]; Coropina creek, Republiek [RMNH 18166]; Upper (?) Saramacca River [RMNH 18323]; local names).

Material

Only four specimens, 86.4-161.8 mm SL, none in good condition, are available, from Saramacca R. (RMNH 18323), Para R. near Republiek (RMNH 18166) and Nanni Kreek (RMNH 18802).

Description

All available specimens were measured and counted, but the qualitative data are principally from the largest specimen (RMNH 18802; Fig. 17). Proportions in the text are given in sequence from the smallest to the largest specimen.

Fig. 17. *Chaetobranchus flavescens*, 161.8 mm SL, RMNH 18802, from Doksipan, Nanni Kreek.

Shape: moderately elongate, laterally compressed, depth —, 47.1, —, 42.1% of SL; in facial aspect sides straight, nape and chest rounded. Caudal peduncle deeper than long (depth 16.3, 16.7, 16.5, 16.1%, length 9.5, 8.4, 8.6, 11.0% of SL).

Head: moderately elongate (length 38.8, 37.8, 37.4, 35.9% of SL), width uniform in dorsal view (behind orbit 18.9, 19.3, 19.1, 18.9% of SL); predorsal contour rising straight to anterior 3rd of orbit, then elevated to slightly anterior to dorsal fin origin where it levels out; prepelvic contour with strongly inclined straight lower jaw and straight sloping posterior section. Snout broad, broadly rounded anteriorly in dorsal view, in lateral aspect a wide triangle with rounded tip, length 11.7, 11.4, 10.8, 10.4% of SL. Mouth wide, but narrower than interorbital; jaws about equal anteriorly, lower very slightly prognathous; lower jaw articulation below middle of orbit; ascending premaxillary processes nearly as long as lower jaw, extending to little behind anterior margin of orbit; maxillary tip well exposed, reaching slightly behind anterior margin of orbit. Orbit touching predorsal contour; diameter 12.7, 12.1, 12.2, 11.2% of SL; interorbital width 13.1, 14.1, 13.5, 14.7% of SL; preorbital depth 7.1, 7.9, 7.7, 8.3% of SL.

Scales: Squ. long. 26(3), -(1); circumpeduncular scale series 16. Predorsal scales irregularly arranged, smaller than flank scales; cycloid in advance of dorsal fin and above gill cover. Flank scales finely ctenoid; small scales along dorsal and anal fin bases, posteriormost caudal peduncle scales, scales under interoperculum and a few rostral prepelvic scales, and all cheek and gill cover scales cycloid. Prepelvic scales posteriorly little larger than half a flank scale, smaller anteriorly in a median series of about 19 scales and 2 parallel series in front of each pelvic fin; small numerous scales on throat sides under interoperculum.

Cheek scales irregularly arranged, in about 4 series above, and 2 below a narrow naked area extending from anterior margin of lacrimal towards (but not to) preopercular angle; one vertical series behind orbit. Two horizontal series of scales over dorsal edge of operculum, on bone proper a ring of large scales around a central patch of small scales, hind margin naked. Suboperculum with 2 series of scales and 2 (left side) or 3 (right side) scales on suspensoriad process; wide naked free margin. Interoperculum with 5 scales in a single series, lower margin widely naked. All head scales smaller than flank scales.

Upper lateral line in the 3rd scale series above the lower, at 3 1/2 (initially) to 3 (at middle) to 2 (plus marginal vestigial scale; terminally) scales distance from dorsal fin, extending to below 3rd dorsal fin ray, overlapping lower line by 3 scales. Lower line extending forwards to above 3rd anal spine, at 5 scales distance from anal fin. Lateral line counts 16/11(1), 16/12(1), 17/10(1), 18/12(1). Fins, except caudal fin, naked. Caudal fin scales with few cteni or cycloid; a narrow basal sheath continued distad by small interradial scales (some between major ray branches), dorsally not reaching half of fin length, medially still less, ventrally to over half of fin length; medially 2 tubed scales continuing lower lateral line and between rays D1-D2, and V3-V4 long series (up to 30 scales) of tubed scales reaching to near end of fin, and tending to converge basally on fin. RMNH 18323 with caudal fin lateral lines between rays V2-V3 as well as V3-V4 on right side.

Fins: D. XII.14(1), XIII.13(3); first spine inserted in advance of vertical from hind margin of operculum, 2/7 length of last, which 22.1, 22.6, 21.8, 20.3% of SL; relative length increasing to last, but moderately from 4th; lappets free, moderately beyond spine tips; soft fin lost in largest specimen, in the others pointed, reaching beyond middle of caudal fin. A. III.11(3), III.11.i(1); first spine inserted opposite last dorsal fin spine; soft fin pointed, 5th ray longest, slightly produced, to 2/3 of caudal fin. Caudal fin with 3 procurrent and 8 principal rays in each lobe; damaged except in 93.2 mm specimen in which truncate. Pectoral fin with 14(1), 15(2), 16(1) rays; long (length 45.3, 44.6, 43.6, 43.6% of SL), pointed, 4th ray longest, to above 5th anal fin ray. Pelvic fin pointed, first ray longest, produced, to 6th anal fin ray, hind edge otherwise straight (pelvic fin slightly shorter in smaller specimens; length 37.6, 41.3, 40.5, 41.8% of SL).

Jaw teeth: relatively small, simple, subconical, slightly recurved, in narrow bands, relatively spaced. In upper jaw about 40 teeth in outer hemiseries, a single inner series near symphysis, posteriorly 2, but for most of length posteriorly, to near jaw tip, only one inner series; outer teeth conspicuously larger than inner. In lower jaw 38 teeth in outer hemiseries, inner teeth only slightly smaller, anteriorly in 3 series, posteriorly wanting.

Gill rakers: all long, slender, especially those externally on first arch (length of uppermost ceratobranchial rakers about equal to orbital diameter), close set, with paired short projections along internal edge (on external rakers) or on external edge (on internal rakers) edges; 20-23 epibranchial and 60-62 lower limb rakers externally on first arch in 3 smaller specimens. Microbranchiospines externally on 2nd-4th arches, internally on 4th arch, and, rudimentary, along edge of lower pharyngeal tooth plate.

Tooth plates: 2-4 teeth on pharyngobranchial 2. A minimum of 7(2), 11(1) and 25(1) teeth on 3rd basibranchial tooth plate.

Vertebrae: one specimen radiographed (RMNH 18802), with 13 + 13 vertebrae and 2 supraneurals.

Colouration

The poor state of preservation has probably affected the colouration to a great extent. In general appearance countershaded, dark above, sides brassy, ventrally light; two dark stripes backwards from the mouth; a large dark midlateral spot; ventral fins prominently spotted; no caudal fin base spot.

Largest specimen brown on forehead, nape and back; back scales with an illdefined central brassy spot, which successively increases in size on scales further ventrad, reducing brown scale area to a distal marginal spot. Preventrally and abdominally greyish white. Pectoral axilla greyish brown on inside. Cheek light brown over brassy sheen, operculum similar, ventral operculars shading to brownish yellow. Lip folds brownish, lower lip otherwise light. Intermandibular area brown; branchiostegal membrane light. Brown stripe over hiatus in cheek squamation and on to subopercular corner; another along adlabial margin of mandible, which light ventrally. A very indistinct dark spot below 10th to 12th upper lateral line canals including scales of horizontal series below.

Dorsal fin pale grey, soft portion lighter distad, with large brown spots on soft part and posterior 2/3 of spinous part (spots gradually less distinct anteriorwards). Soft anal fin with large brown spots with narrow whitish to hyaline interspaces. Caudal fin brownish, with indication of a few small light spots on scaleless portion. Pelvic fin outwardly brownish, inwardly nearly hyaline.

Smaller specimens, 86.4-93.5 mm, lighter, tending more to olivaceous than brownish. One lacks post-oral dark stripes but is dark intermandibularly; the other two have post-oral stripes indicated, but intermandibular area light. Midlateral spot distinct, though edges not sharp, below 10th to 12th lateral line tubes, 1-1 1/2 scales deep. Fin spotting more contrasted than in the large specimen, with about 10 vertical series of dark spots on soft and posterior spinous dorsal fin, anal fin similar but with light interspaces much narrower; caudal fin base dark, posteriorly greyish to brownish with 7-8 vertical series of small light spots.

Distribution (Fig. 16)

Outside Surinam, *C. flavescens* has been collected in the R. Guaporé, upper R. Madeira, along the R. Ucayali, Peruvian R. Amazonas, R. Solimões and Brazilian R. Amazonas, R. Negro, R. Branco, R. Oyapock, mouth of R. Approuague, Essequibo system (pers. obs.), and the Orinoco drainage (Fernandez-Yépez 1951).

Very few specimens of *C. flavescens* have been collected in Surinam. In the Amazon basin, *C. flavescens* is a floodplain fish, chiefly collected in whitewater regions. The species is not expected to be widespread in Surinam, but a continuous range may extend along the coast connecting Guyanan and French Guianan localities.

Local names

Kunapari, Kamara paru, Ouwi weifi (Boeseman 1952).

Notes

Chaetobranchus flavescens is readily distinguished from all other Surinamese cichlids by the long setiform gill rakers.

Peruvian *C. flavescens* were recently described by Kullander (1986), which see for discussion on synonyms. Eigenmann (1912) described Guyanan *C. flavescens*. Lowe-McConnell (1969) studied *C. flavescens* in the Rupununi area (Guyana) and gave data on breeding, growth, and life colours. The morphology of the branchial apparatus indicates a plankton feeder, but stomachs so far examined (Lowe-McConnell 1969, Knöppel 1970) were empty or rotten.

No critical study has ever been made of *C. flavescens* material from all over the extensive range of the species. Compared to Peruvian and French Guianan material at hand, the Surinamese specimens appear to have the midlateral spot 1-2 scales more posteriorly, and on average to be more slender than Peruvian specimens. The basibranchial 3 tooth patch seems better developed in Surinamese specimens.

GEOPHAGUS Heckel

Geophagus Heckel 1840, p. 383 (type species *Geophagus altifrons* Heckel, by subsequent designation in Eigenmann & Bray 1894, p. 621).

This genus includes medium sized to moderately large (to 230 mm SL) South American cichlids with an extensive laminar ventral expansion of the first epibranchial margined by connective tissue, a so-called epibranchial lobe (Figs 20, 24, 27, 30).

In Gosse's (1976) revision, *Geophagus* includes 10 species with a nearly pan-South American range. Kullander (1986) proposed a more restrictive diagnosis, but also considered a larger number of species in *Geophagus* s.str. Some species of *Geophagus* sensu Gosse were then referred to the resurrected *Satanoperca* (this paper, p. 66), and still others are left *incertae sedis*.

Four Surinamese cichlid species belong to *Geophagus* s.str., distinguished from similar genera (with epibranchial lobe) by having more caudal than abdominal vertebrae (14-15 + 17-19) and epihemal ribs continuing the pleural rib series on the anterior 6-12 hemal spines over long paired swimbladder diverticula. Other characters helpful for distinguishing the genus from one or more other geophagine groups include: Epibranchial 1 as deep as long; pharyngobranchial 1 with widened epibranchiad end; minute interarcual cartilage; ceratobranchial rakers on first gill arch (9-14 externally) attaching to the ceratobranchial and not to the skin fold over gill filaments; 3-5 series of jaw teeth, in adults upper jaw outer tooth series widely separated from inner series; no rakers on lower pharyngeal tooth plate; 3-6 tooth plates on ceratobranchial 4; single supraneural; scaled dorsal fin, anal fin scaled or naked; caudal fin marginally scaled to near distal end; small scales ((31) 32-34 in squ. long. series and 18-22 around caudal peduncle); anterior half of cheek naked; a large dark spot on middle of side, no spot on caudal fin base; 4 infraorbitals (second with middle foramen); microbranchiospines absent, or present externally on 2nd to 4th arches; accessory cartilage between hypurals 2 and 3.

Characteristic is also smooth opercular and pectoral girdle bones. The lip folds are interrupted anteriorly; lip attachment of 'American type'. Anal fin with three spines; dorsal fin long, with 15-19 spines and 10-13 rays, dorsal fin spines equal or subequal in length from 5th to 7th, last again a little longer; caudal fin emarginate, with 8 principal and 5-7 procurrent rays in each lobe. Predorsal (well over ten scales along midline), chest and prepelvic scales small; naked pit opposite tip of cleithra. Long lateral lines (18-24/9-18, plus 2-3 on caudal fin base), upper well removed from dorsal fin base (by 7-9 1/2 scales initially, terminally 1 1/2 scale); accessory lateral line sequences on caudal fin between rays D3-D4 and V4-V5, dorsal isolated (sometimes proximally downturned in *G. harreri*), ventral usually curved towards lower lateral line proximally.

Other characters—not repeated in the descriptions below—common to the four Surinamese species include (in adults): sides strongly compressed; dorsal outline much more arched than ventral; in facial aspect laterally compressed, wider ventrally, with keeled nape and flattened chest; ventral margin of caudal peduncle longer than dorsal margin; maxilla very little or not exposed; ascending premaxillary processes of about length of lower jaw, not reaching to orbit; flank scales ctenoid; 18 circumpeduncular scale series (7 above, 9 below lateral line) except in *G. harreri* (with 7-8 above, 11 below lateral line); upper lateral line in 3rd scale series above that containing lower; first dorsal fin spine above opercular opening; pectoral fin long, asymmetrical, with dorsal tip rounded, 4th ray longest; pelvic fin pointed, first ray longest.

The genus is widely distributed in the Amazon and Orinoco basins, and the Guianas. The only other Guianan species aside from those reported here, is *G. camopiensis* Pellegrin, in the Oyapock and Approuague systems. The central and western Amazonian *G. proximus* (Castelnau) was redescribed by Kullander (1986). The two remaining nominal species of the genus are *G. altifrons* Heckel (Central Amazonia) and *G. megasema* Heckel (Bolivian Amazonia). Many species remain to be described.

Geophagus surinamensis (Bloch)

(Figs 18-22)

Sparus surinamensis Bloch 1791, p. 112, pl. 277, fig. 2 (Surinam).

Surinamese records

Geophagus surinamensis; (?; *pt.*) Regan 1906a, p. 55 (descr.; Surinam [not reexamined]). — (*pt.*) Gosse 1976, p. 25, fig. 3 (descr; only: ZMA 105.687, 105.747, 106.197).

Material

364 specimens, including holotype, from the Saramacca, Suriname and Marowijne R. drainages.

Description

The principal study material is composed of ZMA 105.687 and ZMA 105.745 from the Suriname R. system, and ZMA 106.197 from the Marowijne R. system. Notes on the holotype are given separately. Description based on a 143.1 mm male (ZMA 105.745, Fig. 18). Counts from 25 specimens. Largest specimens examined 148.0 mm SL.

Fig. 18. *Geophagus surinamensis*, 143.1 mm SL, ZMA 105.745, from Parwapa Kreek, Suriname R. drainage.

Shape: moderately elongate; dorsal head contour rising straight to in front of orbit, where slightly concave, continued ascent gradually more straight; dorsal fin base sloping, gradually steeper caudad. Ventral outline feebly sloping, straight save for projecting lower jaw end, from snout to pelvic fin base, abdomen straight, horizontal, anal fin base and ventral caudal peduncle margin sloping moderately upwards.

Head: moderately deep. Snout slightly extended, rounded anteriorly. Orbit below forehead contour, lying all in dorsal and almost all in caudal halves of head. Mouth low, wider than interorbital; maxilla not extending to orbit; lower jaw articulation little behind (in others it may be at) vertical from anterior margin of orbit; jaws equal anteriorly. Lips moderately wide, slightly wider lower lip fold than in *G. brokopondo*. Operculum depth:length ratio 5:2.

Scales: Squ. long. 32(2), 33(21), 34(2). Dorsal head scales rapidly becoming smaller forwards from extrascapular canal series region; along predorsal midline about 16 scales, posteriorly some minute median cycloid scales, larger scales about half size of anterior flank scales. Chest scales laterally 2/3 size of anterior flank scales; posteriormost prepelvic scales ctenoid, rest cycloid, small (posterior 1/3 size of anterior flank scales, smaller anteriorly), 25 along middle (not naked opposite tip of cleithrum, but normally so in other specimens), naked under interopercula; bases of medial pelvic fin rays covered by scales.

Cheek scales mostly cycloid (posterosuperior ctenoid), small (larger are half the size of anterior flank scales), in about 5 oblique series (5-7 in entire material counted). Operculum with a group of 9 scales dorsally, below hinge 29 of varying sizes, larger 3/4 size of anterior flank scales, mostly ctenoid. Suboperculum with 3 series of mostly ctenoid scales, the larger half the size of flank scales. Single small cycloid scale posteriorly on interoperculum.

Lateral lines 20/14(1), 21/13(1), 21/15(1), 21/16(1), 22/12(1), 22/14(7), 22/15(4), 22/16(5), 23/13(1), 23/14(1), 23/15(2). Upper lateral line at 7 (initially) to 1 1/2 (terminally) scales distance from dorsal fin, rising for about 5 scales, posteriorly straight, to below 4th dorsal fin ray; overlapping lower by 5 scales. Lower lateral line continued on caudal fin by 2 scales.

Fin scales: dorsal fin squamation as in *G. brachybranchus*, last two membranes naked, interradial scales from behind 2nd spine, up to 13 along a membrane, depth of scaled portion at most 1/3 of length of last spine, distal scales cycloid. Largest specimen without dorsal fin interradial scales 100 mm, smallest with scales 76 mm. Anal fin with a series of small cycloid scales along base anteriorly; no interradial scales in the entire material studied. Pectoral fin naked. Caudal fin densely scaled (in other large specimens with intact fin, to one-fourth along middle, mostly naked between rays D3 and V3, along fin margins scaled almost to hind edge, distalmost scales cycloid).

Fins: first dorsal fin spine length 1/3 that of last, relative spine length increasing to about 5th from which spines are subequal, last a little longer; anteriorly little curved (very little curved also in others, even in small specimens), posteriorly straight; soft fin damaged. D. -.-(1), XVI.13(1), XVII.12(14), XVII.13(2), XVIII.11(5), XVIII.12(1), XIX.10(1). Soft fin frequently injured, otherwise pointed, reaching slightly behind caudal fin base, 4th ray longest; longest, slightly produced, in a male, 81 mm.

First anal fin spine opposite last dorsal fin spine; soft part pointed, reaching to 1/3 of caudal fin though damaged. A. III.7(2), III.8(23). Soft anal fin like soft dorsal fin, only a little shorter.

Caudal fin damaged; in others emarginate, in specimens over 90 mm usually with a long point formed by rays D6 and D7. Pectoral fin reaching to above 2nd (or 3rd) anal fin spine. P. 14(1), 15(24). Pelvic fin with first ray produced, to behind anal fin base, 2nd ray not much shorter; in others reaching to somewhere along soft anal fin base.

Fig. 19. *Geophagus surinamensis*, 62.2 mm SL, ZMA 105.745, from Parwapa Kreek, Suriname R. drainage.

Jaw teeth: simple, conical, slightly recurved, outer erect, inner distinctly orad-pointing in upper, nearly erect in lower jaw. Anterior 1/3 of each lower jaw half toothed, an outer series of larger teeth, and anteriorly about 4 inner series of small teeth in a band that narrows to 1 series posteriorly. Most of upper jaw rim toothed; an outer series of larger teeth, well separated from inner band of small teeth in 4 series anteriorly, narrower posteriorly, a wide gap at the symphysis. In other specimens none or one inner series posteriorly in lower jaw. Specimens less than 100 mm with band of uniform sized teeth in upper jaw, outer series not separated from other. In 13 specimens 56.9-143.1 mm, 17-30/17-25 teeth in upper/lower jaw outer hemiseries, 2-4 inner series anteriorly.

Gill rakers: on first gill arch 7-11 epibranchial and 11(4), 12(17), 13(4) on lower limb. First gill arch in a 90 mm specimen (ZMA 105.745, Fig. 20): 6 conical medial epibranchial rakers, laterally 3 blunt somewhat flattened, with slightly expanded tips, and 1, next to ceratobranchial, conical, all reaching beyond margin of lobe; a raker in the angle and 6 upper ceratobranchial rakers more or less conical, but short, blunt tipped, 4 lower short, knob like rakers and 2 anterior rudiments; upper 4 ceratobranchial connect with median ledge by swellings, 6 below rather by ridges, rudiments free. Very little of gill filaments exposed, especially ventrally. Inner rakers conical, 5 + 1 + 15, basally connected with narrow swelling to median ledge; all ceratobranchial rakers attached over ceratobranchial.

Epibranchial 1 deeper than long; ventrally a connective tissue pad with mediad process; pharyngobranchial 1 expanded ventrally. Microbranchiospines observed only in 103.1 mm Saut Gostou specimen, medially on external face of 3rd and 4th arches.

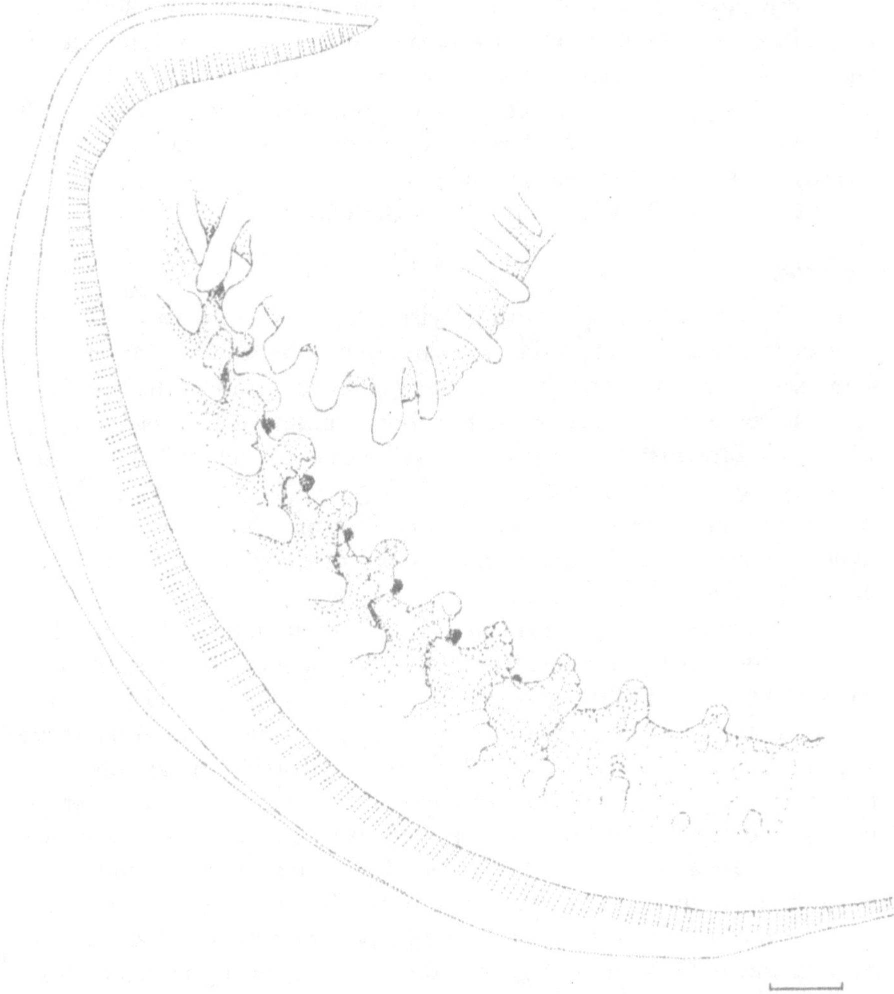

Fig. 20. *Geophagus surinamensis*. First gill arch in lateral aspect. Scale 1 mm. From ZMA 105.745, 90 mm SL.

Tooth-plates (90 mm specimen): ceratobranchial 4 with 6 small tooth plates with 2, 5, 5, 5, 4 and 2 teeth respectively. Lower pharyngeal dentition as in *G. brachybranchus*.

Young (especially 41 mm specimen) elongate, with nearly equal, little arched dorsal and ventral contours. Orbit touching dorsal profile, in about middle of head length; operculum about twice as deep as long; maxilla reaching a little behind vertical from anterior margin of orbit. Scales cycloid on head anterior to extrascapular canals, on cheek and gill cover, and anterior to pelvic fin bases. Fins naked except caudal fin, which, however, is not as densely scaled as in adults. Pectoral fin extending to above first anal fin spine; pelvic fin pointed, not produced, reaching to vent; caudal fin emarginate, but without elongations; soft dorsal and anal fins rounded, former almost reaching caudal fin base, latter only to middle of caudal peduncle.

Vertebrae: 14 + 18(1), 15 + 18(6); caudal ribs 9.

Colouration

Male, 143.1 mm (Fig. 18), with greyish white ground colour on chest, anterior to pelvic fins and along belly midline; yellowish area anterior to pelvic fin bases, anterior to that duskied. Preorbital, lower lip and jaw, inter- and pre-opercula pale greyish, suboperculum greyish yellow, operculum pale brownish, cheek ventrally yellowish, dorsally pale brownish. Upper lip and forehead grey. Nape dark grey. No markings on head. Dark grey branchiostegal membrane and intermandibular area. Faintly silvery on chest. Inside and pocket of pectoral axilla brown dorsally. Sides pale yellowish.

Vertical bars showing as brownish traces, except anterior 2 which dorsally greyish brown. First bar from grey nape down to posterior margin of orbit; 2nd from midline of nape behind extrascapular canal series and below 4 anterior dorsal fin spines, down to operculum and a little above pectoral fin base, nearly divided by narrow lighter vertical zone; 3rd from below 6th to 8th dorsal fin spines, very indistinct across sides, pigmentation on squ. long. scales slightly stronger, forwards to first squ. long. scale; 4th across 11th to 13th lateral line scales, not below squ. long. series. Behind 4th bar no distinct bars, but rather dorsal sides and caudal peduncle darker, ventral sides and caudal peduncle lighter. Bars slightly inclined and wider than interspaces. Along sides below upper lateral line, but not on caudal peduncle, faint pale brownish scale junctions forming faint horizontal stripes. Midlateral blotch in 4th bar on 11th to 13th lateral line scales and down to include squ. long. scales, but scale centers light below lateral line. Dorsal margin of caudal peduncle dark brown.

Dorsal fin pale greyish, scaled part slightly darker, lappets brownish; upper margin of soft fin narrowly brownish. Faint traces of light spots on inner half of soft part. Anal fin clear along hind margin, for the rest brownish with large light spots that form about 4 light stripes paralleling body axis. Caudal fin pale brownish with large ligh spots across entire fin. Pelvic fin with white line along outer margin of first ray, rest brownish grey, paler inwardly.

Most specimens, like the one described above, appear to be slightly faded; those well-preserved differ chiefly in showing dorsal fin spotting better. Chest commonly silvery, operculum and suboperculum silvery chiefly in ZMA 105.687 and 106.197 series. No horizontal striping in specimens 95 mm and smaller. Second bar usually more or less distinctly divided in specimens over 90 mm; no distinct bars behind the 4th, but in young especially, 3 faint spots below dorsal fin and a spot distally on caudal peduncle; occasionally 4th bar tending to split below midlateral blotch. In well preserved specimens, dorsal fin membrane base light, on soft portion 3 horizontal stripes of light spots, 2 inner of which continued

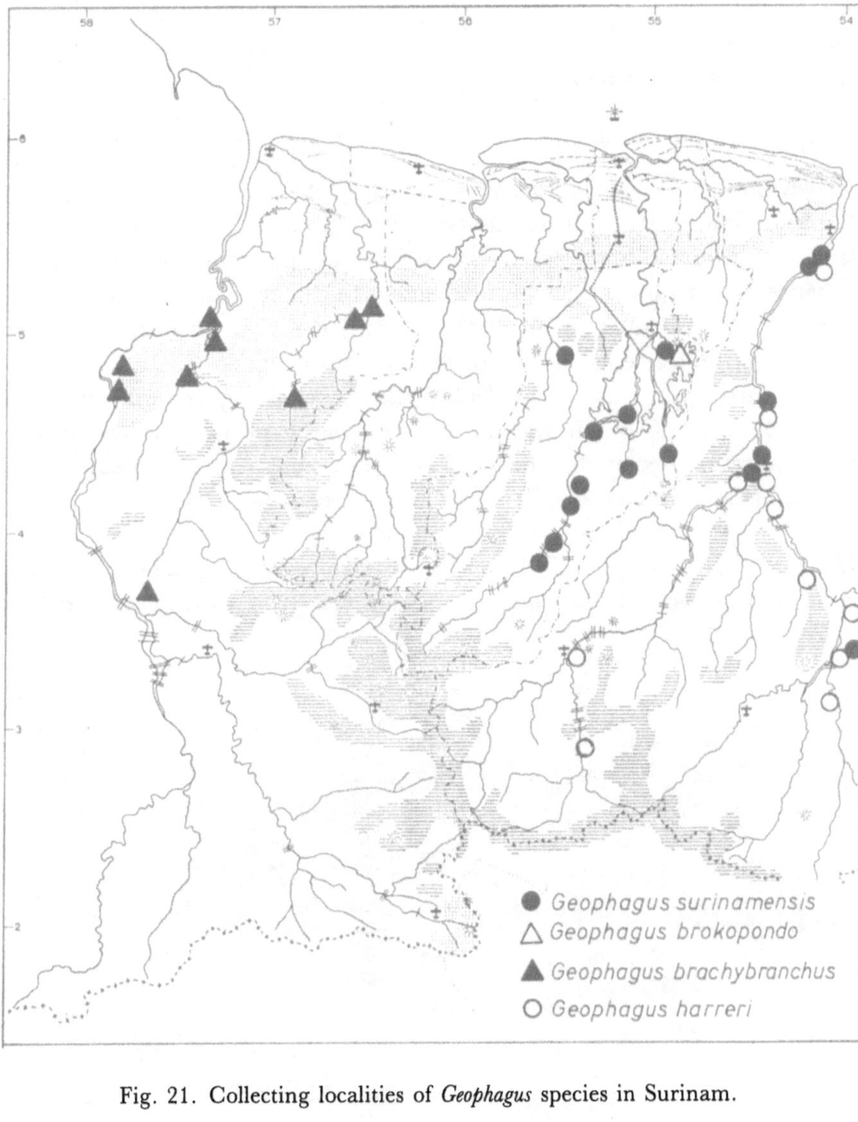

Fig. 21. Collecting localities of *Geophagus* species in Surinam.

on posterior part of spinous portion. Caudal fin spotted in specimens over ca 90 mm. In those smaller rather with about 4 vertical stripes of spots. Branchiostegal membrane either whitish or, in large specimens, dusky to blackish.

Distribution (Fig. 21)

Geophagus surinamensis has been collected in the Suriname R. drainage upstream from Afobaka, in the Marowijne from the coast to right bank upland tributaries. There is only one Saramacca locality. The species is occasionally sympatric with *Geophagus harreri*.

Local names

Agankoi (Lijding 1959a).

Notes

Diagnostics: Geophagus surinamensis is distinguished by its long head (29.9-32.9% of SL); gill filaments on first gill arch with wide skin cover (Fig. 20); interradial dorsal fin scales, in large specimens, in single series; absence of dark markings on head sides; midlateral flank blotch extending from lateral line scales (11th to 13th) down to include squ. long. scales; and absence of dark spot below end of dorsal fin base.

It is distinguished from *G. brokopondo* especially by the gill arch morphology, from *G. harreri* and *G. brachybranchus* especially by the absence of dark markings on the head. Those characters in combination also distinguish *G. surinamensis* from all other *Geophagus* species.

Holotype. The holotype (ZMB 2825, 78.5 mm SL, Fig. 22) is in a poor condition, firm, but discoloured, scales lost to a considerable extent,

Fig. 22. *Geophagus surinamensis*, holotype, 78.5 mm SL, ZMB 2825.

eviscerated. The jaws and lips are damaged and many teeth lost. The gills are stuck together by deposits. The posterior dorsal fin spines are decalcified and all soft fins badly broken.

Shape moderately elongate, with dorsal contour much more arched than ventral. Predorsal contour straight ascending. Snout somewhat extended. In frontal aspect compressed, with keeled nape and flattened chest. Tip of maxilla exposed, not reaching to vertical from anterior margin of orbit. Width of exposed gill blade on first gill arch about equal to width of arch plus proximal skin cover. Teeth in bands anteriorly in jaws. Anterior half of cheek naked; few opercular scales remaining; 3 scale series on interoperculum; 2 scales posteriorly on interoperculum. Fins naked. Nape and chest scales small and numerous (largely covered by deposits). Circumpeduncular scale series 18. Anterior dorsal fin spines distinctly curved backwards. Deep skin pigment remaining, but the only marking readily recognizable is a roundish blotch on and, chiefly, below about 11th to 12th lateral line scales (estimated).

Measurements: SL 78.5 mm; head length 25.4 mm; snout length 9.0 mm; body depth 30.0 mm; orbital diameter 8.8 mm; head width 11.4 mm; interorbital width 6.1 mm; preorbital depth 7.5 mm; caudal peduncle depth 9.4 mm; caudal peduncle length 12.9 mm; pectoral and pelvic fins badly damaged; last dorsal fin spine ca 13.2 mm. Counts: D. XVII.11; A. III.-; squ. long. approximately 32-33; lateral lines ca 23/14; cheek scale series 7; gill rakers on first gill arch 9 + 1 + 12; no microbranchiospines.

Figures are given by Gosse (1976, fig. 3; inaccurate in some details) and Bloch (1791, pl. 277; expressing beauty and not subscribing to dull reality).

Gosse (1976) gave some notes on the specimen. As he remarked, it has 17 dorsal spines, instead of 15 as stated in the original description and shown in the figure (Bloch 1791). The posterior spines are very soft and may have been mistaken for rays.

Bloch described and figured the specimen as having three black spots on the side; one just behind the gill cover, one on the middle of the side, and one on the end of the caudal peduncle. The middle spot can still be traced, but at present there is no indication of the other two. Traces remain, however, of the vertical bars shown in Bloch's figure, but we doubt, like Gosse (1976), that they were ever so red as shown in Bloch's illustration.

In his generic revision, Gosse (1976, p. 94 especially) defined as *G. surinamensis* a species with interrupted lower lip fold, 17 or fewer gill rakers, 17 or more scales (scale series) around the caudal peduncle, absence of a spot on back below dorsal fin end, D. XVI-XIX.10-13, and slightly higher (but overlapping) scale counts than *G. camopiensis* and *G. crassilabris*. The geographical range of Gosse's material included most of the Amazon basin and the Guianas from the Essequibo system eastwards to the Marowijne system. Synonyms included *G. altifrons*, *G. megasema* and *Chromys proxima* Castelnau.

Gosse compared measurements and counts from Guianan and Amazonian material statistically, and concluded that they were not part of the same population; further, heterogeneity in the Amazonian material suggested the existence of geographical races.

Revision of part of Gosse's material as well as additional material including the relocated types of *G. altifrons* and *G. megasema*, suggests that Gosse's *G. surinamensis* is actually a large group of species, each readily distinguishable by colouration and other characters which were not studied by Gosse.

In Suriname, there are three species that were included in *G. surinamensis* by Gosse. One deep-bodied, with long gill filaments, and two relatively elongate species, with gill filaments largely covered by skin.

The poor condition of the holotype of *G. surinamensis* renders the identification of any fresh material as belonging to this species difficult. However, the relatively long snout and skin covered gills, as well as what remains of the colouration points to either the species recognized in the Marowijne + Suriname drainages or the Nickerie + Corantijn species. Fresh material of these species is easily identified on the basis of differences in colouration, but colour traits are of limited value in comparison with the nearly colourless type of *G. surinamensis*.

Measurements and counts are not particularly helpful; although a very short caudal peduncle is notable for the holotype (16.4% of SL, but 17.0-20.5% of SL in fresh material available). However, the dorsal spines are decidedly curved, as in the Marowijne + Suriname species, but not in the Nickerie + Corantijn species, and the remains of the midlateral blotch are more like the Marowijne + Suriname species, i.e., a vertical from the hind margin of the blotch passes through or anterior to the anal orifice rather than through the genital papilla as in the Nickerie + Corantijn species. Furthermore, there is no trace of a preopercular spot or stripe in the *G. surinamensis* type, although pigment still remains on the head elsewhere, and a preopercular marking is not mentioned or shown on the figure in the original description (which is, however, thought to be unreliable, see above).

Bloch was not detailed about the origin of his fish ('Der ihm beigelegte Nahmen zeigt den Ort seines Aufenhalts an') and did not always have reliable localities for his specimens (especially in describing the South American *Hoplias malabaricus* as coming from India). Comparison with non-Surinamese *Geophagus* fails to suggest any more likely candidate for the name *G. surinamensis* than the Marowijne + Suriname R. species.

Conclusively, there is evidence for identifying the Marowijne + Suriname R. *Geophagus* species as *G. surinamensis*, and we consequently describe the other two similar Surinamese *Geophagus* species as new.

Geophagus surinamensis, as here conceived, has a very restricted distribution, collected in only three Surinamese rivers. There is a vast literature on '*G. surinamensis*' (cf., e.g., the bibliography in Gosse 1976), but almost all of it pertains to other species, which will not be discussed further here.

In its specialized gill arch morphology, *G. surinamensis* is most similar to *G. brachybranchus*, which also has a similar body shape and, compared to *G. brokopondo*, reduced dorsal fin squamation. The colour pattern is, however, more like that of *G. brokopondo*, especially in the lack of suborbital markings.

Geophagus brokopondo n. sp.

(Figs 21, 23-25)

Surinamese records

Geophagus surinamensis; (*pt.*) Gosse 1976, p. 25 (description of Surinamese material; only ZMA 105.480, 105.081).

Holotype (Fig. 23)

ZMA 119.530. An adult female, 107.7 mm SL. Suriname, Distr. Brokopondo, Suriname R. system, Brokopondomeer along N shore at Afobaka. 15 February 1967. Leg. H. Nijssen (HN 103).

Fig. 23. *Geophagus brokopondo*, holotype, 107.7 mm SL, ZMA 119.530, from near Afobaka.

Diagnosis

Head short (27.8-30.4% of SL); gill filaments on first gill arch with only narrow skin cover proximally; interradial dorsal fin scales mostly in 2 or 3 series in large specimens; no dark markings on head sides; midlateral flank spot squarish, extending from lateral line canals (11th to 12th) down to include squ. long. scales.

The wide gill blade (Fig. 24), deeper general aspect and more extensive dorsal fin squamation distinguish *G. brokopondo* from other Surinamese *Geophagus* species. The colour pattern is much like that of *G. surinamensis*, but different from that of *G. brachybranchus* in the smaller midlateral spot and absence of head stripes. These characters in combination distinguish *G. brokopondo* from all other *Geophagus* species.

Material

Holotype and 37 paratypes, 30.7-123.2 mm SL, all from the Brokopondomeer. Many, especially among the smaller specimens, are preserved distorted and with open mouths (all of those in the larger series), hence only a few specimens could be measured. A try was made with 5 specimens with open mouth (55-59 mm).

Description

From holotype (Fig. 23), unless indicated otherwise; counts from 25 specimens.

Shape: moderately elongate. Dorsal head contour rising straight to dorsal fin origin; dorsal fin base contour initially little ascending or horizontal, descending from about 9th spine. Ventral outline with feebly sloping chest, horizontal abdomen and feebly sloping anal fin base and caudal peduncle.

Head: deep with elevated nape. Snout not particularly extended, rounded anteriorly. Orbit below forehead contour, all in dorsal, most in caudal halves of head. Mouth low, wider than interorbital; maxilla not extending to orbit; lower jaw articulation in advance of orbit; lower jaw slightly prognathous. Lips moderately wide, fold of lower restricted to mouth. Operculum about twice as deep as long. Juveniles much more elongate than adults, with orbit more or less touching predorsal contour, and rounded (not keeled) nape; maxillary tip not reaching orbit, but nearer to orbit than in adults; depth:length ratio of operculum 3:2.

Scales: Squ. long. 32(1), 33(20), 34(4). Dorsal head scales rapidly smaller forwards from above first lateral line scales; along predorsal midline about 14 scales plus some irregularly inserted, but no median series, larger scales about 2/3 size of anterior flank scales. Chest scales laterally 2/3 size of anterior flank scales; cycloid and very small (larger 1/3 size of

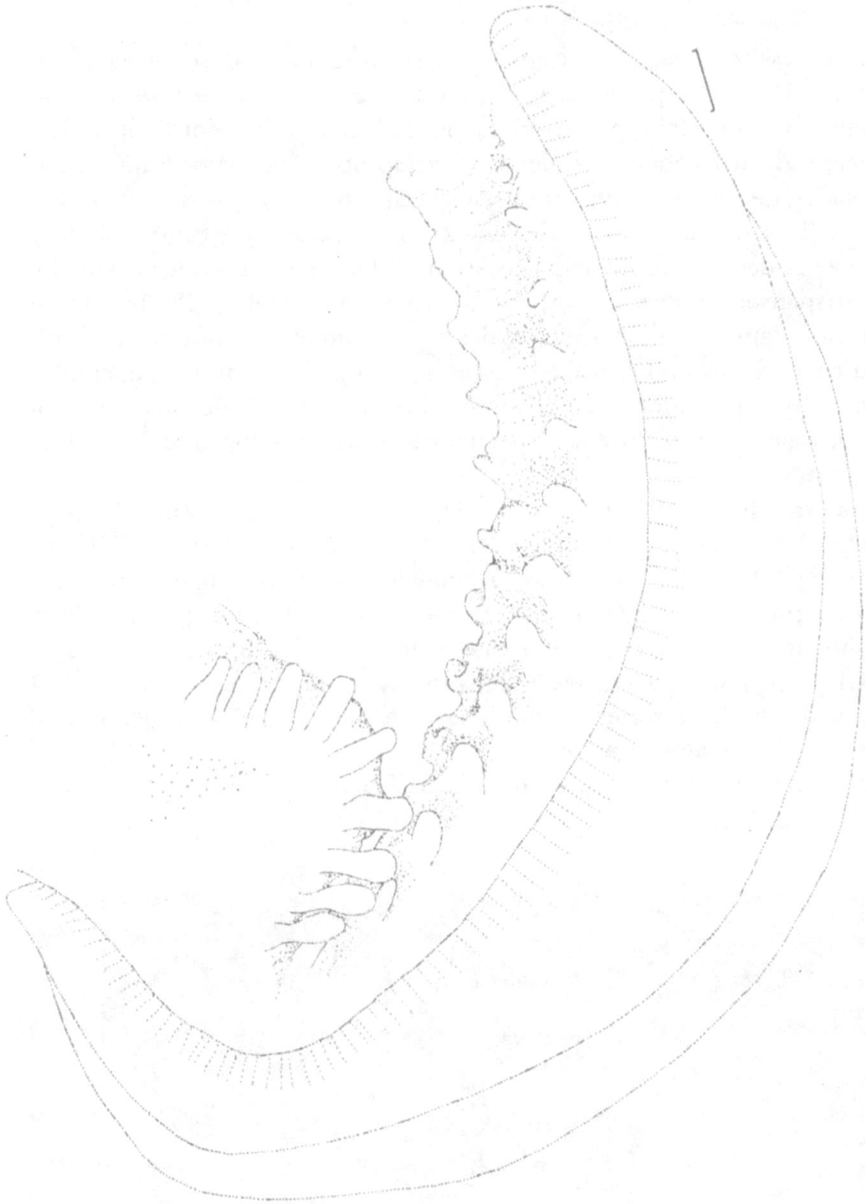

Fig. 24. *Geophagus brokopondo*. First gill arch in lateral aspect. Scale 1 mm. From NRM A87/1967073.3941, 96.6 mm SL.

anterior flank scales) preventrally; bases of medial pelvic fin rays covered by scales; along midline from between pelvic fins forward 17 posterior to and about 6 anterior to naked spot opposite tip of cleithra; narrow naked zone anteriorly extending little laterally.

Cheek with small thin scales not arrranged in clear series (approximately 5 or 6 oblique series in the entire material counted), lower- and anterior-most cycloid, posteriorly about half size of anterior flank scales. Operculum with about 8 scales in a group dorsally, below hinge about 38 scales, dense, not in apparent series, but rather a ring of slightly larger scales (half the size of anterior flank scales) encircling central slightly smaller scales. 20 subopercular scales in 3 series plus one scale proximally on suspensoriad process, largest a little larger than half the size of anterior flank scales. 3 small scales dorsocaudally on interoperculum. Gill cover scales cycloid except some larger opercular and subopercular scales. Specimens less than 70 mm with gill cover scales and most of cheek scales cycloid; 30.7 mm specimen also with predorsal scales cycloid along midline.

Lateral lines 19/14(1), 20/11(1), 20/14(1), 20/16(2), 21/11(1), 21/14(1), 21/15(3), 21/18(1), 22/12(1), 22/13(1), 22/14(2), 22/15(5), 22/16(3), 22/17(2); lower counts in smaller specimens. Upper lateral line at 7 (initially) to 1 1/2 (terminally) scales distance from dorsal fin, rising for about 6 scales, posteriorly straight, to below 3rd or 4th dorsal rays; overlapping lower by 5 scales (6-7 in others). Lower lateral line continued on caudal fin by 2 scales; dorsal lobe with sequence of 18 tubed scales, ventral of 25 scales in holotype.

Fin scales: dorsal fin with a band of small ctenoid scales along base, but 5 last membranes naked and with ray muscles exposed distally; from

Fig. 25. *Geophagus brokopondo*, paratype, 58.7 mm SL, ZMA 105.081, from Brokopondo Meer.

behind 4th spine cycloid interradial scales, usually in double or triple series, but leaving a naked area close to next spine or ray, last 3 membranes naked, depth of scaled zone at most about half length of posterior spines. Specimens smaller than 70 mm without interradial dorsal fin scales, those 70.0 and 70.8 mm with only a few interradial scales between posterior spines, larger specimens (from 96.6 mm) like holotype. Anal fin with small cycloid scales along base, spinous portion basally covered by body scales, 5 last membranes naked. Two largest specimens, 118.6 and 123.2 mm, with 1-6 interradial scales between each soft fin membrane except on last or last two; basal sheath in 96.6 mm specimen, anal fin naked in those smaller. Caudal fin squamation ctenoid, dense, but with deeply concave hind margin, only 1/4 of fin scaled medianly, marginally 3/4.

Fins: first dorsal fin spine length little more than 1/3 that of last, relative spine length increasing to 6th from which equal; 2nd to 11th distinctly backwards curved, remainder less so or straight, soft fin damaged. D. XVI.12(1), XVII.12(16), XVII.13(2), XVIII.10(1), XVIII.11(2), XVIII.12(3). Largest specimens with soft dorsal fin pointed, 5th ray produced, reaching to near end of caudal fin; other large specimens similar; 70.0-70.8 mm specimens with rounded-subacuminate fin reaching to caudal fin base, specimens ca 55 mm with rounded fin, reaching caudal fin base, smaller specimens with shorter fin. Specimens 70.0-70.8 mm and smaller with rather straight dorsal fin spines; curved in those larger.

First anal fin spine opposite antepenultimate dorsal fin spine; soft part pointed, 3rd ray slightly produced, to little beyond middle of caudal fin. A. III.7(5), III.8(18), III.9(2). In others, soft anal fin shape much like that of soft dorsal fin, only a little shorter.

Caudal fin injured, emarginate with slightly produced dorsal hind corner (rays D6 and D7), ventral hind corner shape uncertain. In smallest specimens also emarginate; largest specimens generally with injured caudal fin, but when intact, rays D6-7 and V6 form slightly produced points. Pectoral fin to above 2nd anal fin spine. P. 14(8), 15(17). Pelvic fin much produced, outer branch of first ray reaching to middle of caudal peduncle. Juveniles with pelvic fin pointed, but not produced, not reaching to anal fin base, only three large specimens ($>$ 100 mm) with long filaments to caudal peduncle.

Jaw teeth: simple, conical, slightly recurved, loosely attached, outer erect, inner pointing more or less orad. Posterior half of each lower jaw half naked, anteriorly rather few teeth, comprising an outer series distinctly forwards curved anteriorly, and an inner band of adsymphysially about 3 series, posteriorly one series; outer not much larger than inner but gradual size decrease ablabially. Most of upper jaw rim toothed; outer series teeth little larger than those of inner, which form band of about 3 series anteriorly, not extending so far backwards; widely edentulous at symphysis. In 123 mm specimen a distinct differentiation in upper jaw of an outer series of larger teeth, increasing in size anteriorwards and a well-separated inner band of smaller teeth, other specimens like holotype but small fish with fewer teeth. In five large specimens 70.0-123.2 mm, 21-30/20-24 teeth in outer hemiseries in upper/lower jaw, 2 or 3 inner series.

Gill rakers: 9 or 10 epibranchial and 11(1), 12(17), 13(4), 14(3) lower limb rakers on first gill arch. First gill arch in 96.6 mm specimen (Fig. 24): Epibranchial about as deep as long, with ventral expansion with extensive connective tissue margin on which inserted 9 fingerlike rakers near and reaching beyond margin, 3 rakers next to ceratobranchial with somewhat expanded tips; one in the angle and uppermost ceratobranchial rakers subconical, remaining ceratobranchial rakers flattened and round-tipped, short, save for lowermost, rudimentary, also a rudiment on hypobranchial. All rakers, also 4+1+15 internal rakers, insert on gill arch, rather than on skin covering gill filament bases. Ceratobranchial rakers connected with dorsal skin crest by swellings which more prominent posteriorly; internal ceratobranchial rakers which subconical, erect, connected basally by similar swellings. Pharyngobranchial 1 subtriangular, expanded ventrally. Ceratobranchial 4 with 5 tooth plates with 3, 4, 3, 3, 3 pointed teeth respectively. Microbranchiospines usually wanting; in 14 specimens checked, 43.2-123.2 mm, present externally on 2nd-4th arches in 3 (57.6-70.0 mm),in 96.6 mm specimen a few only on 4th arch. Lower pharyngeal dentition as in *G. brachybranchus*.

Vertebrae: 14 + 19(1), 15 + 18(4); 10 caudal ribs.

Colouration.

Chest and prepelvic region whitish, abdomen ventrally yellowish. Underside of head greyish white; intermandibularly and on branchiostegal membrane whitish, but duskied. Operculum, suboperculum, cleithrum and scaled part of cheek silvery, rest of head greyish brown, darker on snout and forehead. No dark markings on sides of head.

Sides shiny brassy; faint opalescent lines over scale junctions along sides below upper lateral line. Faint brown vertical bars: (1) narrow, from posterior part of orbit up to nape midline, (2) from below first to 4th spine of dorsal fin and adjacent nape, wide, slightly narrowing ventrally, obliquely backwards to anteriormost upper lateral line scales, (3) below 6th to 8th dorsal spines, straight vertically to far down on abdominal sides, (4) below 9th to 11th dorsal fin spines, slightly curved caudad, fading out on abdominal sides, (5) below end of spinous dorsal fin, not distinctly delimited from the next bar, reaching ventrally to about level of lower margin of caudal peduncle, (6) below anterior 3 dorsal fin rays, extension as preceding bar, (7) very wide, below posterior half of soft dorsal fin, not distinctly delimited posteriorly, as caudal peduncle of about same colour, save for dark brown dorsal margin and greyish yellow lower margin. Bars slightly darker dorsally than over flanks, and slightly wider than their interspaces. Inside of pectoral axilla brownish, pocket dorsally dark brown.

In 4th bar a prominent brown blotch, about squarish in form, below 10th to 12th lateral line scales, reaching up to 11th to 12th lateral line canals and down to include squ. long. scales.

Dorsal fin greyish, lappets darker; a few light spots indicated on posterior half of soft portion. Caudal fin greyish to pale brownish, with whitish round spots especially on median (asquamous) portion, forming about 4 vertical series. Anal fin smoky, with indications of pale spots on soft portion. Outer margin of first pelvic fin ray white, rest of fin greyish.

Other adults (96.6-123.2 mm) similarly lack head markings and have very feebly spotted fins; midlateral blotch limited to 4th bar, between lateral line canals and squ. long. scales. However, vertical bars on sides (2nd to 7th) more or less split vertically, each specimen different, 4th bar divided below midlateral blotch. Branchiostegal membrane and intermandibular area faintly pigmented.

Two smaller topotypes, 70.0-70.8 mm, distinguished chiefly by a slight darkening on posterior part of caudal peduncle, and by light spots on soft dorsal fin, irregular or in 2 horizontal series.

Young specimens (30.7-58.7 mm; ZMA 105.081, Fig. 25), especially the smaller ones, have chest, prepelvic area, gill cover, cheek and part of preorbital silvery. Vertical bars not as distinct as in adults, rather forming spots along dorsal fin base and along body midaxis, one posteriorly on caudal peduncle distinct though faint. Midlateral blotch roundish. Two horizontal series of light spots, one along base, one along middle, along most of dorsal fin, and one additional series on soft dorsal fin, but fin virtually immaculate in smallest fish. Usually a distinct brownish spot midbasally on caudal fin, lost with increasing size, rest of fin light grey to smoky with about 4 light vertical stripes or series of spots. Pelvic fin white, darkening with increasing size.

Distribution (Fig. 21)

Available collections are from close to Afobaka in the Suriname R. drainage.

Etymology

The species epithet refers to the district Brokopondo, as well as to the Brokopondomeer (officially, the Blommensteinmeer), an artificial lake along the shores of which the type material was collected.

Local name

Songeh (Saramaccan).

Notes

Geophagus brokopondo appears to be the most generalized *Geophagus* species in Surinam, being rather deep and with relatively well-scaled fins and having a 'normal' gill arch morphology. Alone the absence of head markings is notable, though shared with *G. surinamensis*.

Geophagus brachybranchus n. sp.

(Figs 21, 26-28)

Surinamese records

Geophagus surinamensis; (*pt.*) Gosse 1976, p. 25 (description of Surinamese material; only ZMA 106.193).

Holotype (Fig. 26)

RMNH 30974. A female, 107.4 mm SL. Surinam, Distr. Nickerie, Nickerie R. system, rocky pool just above Blanche Marie Vallen. 13 January 1971. Leg. M. Boeseman.

Fig. 26. *Geophagus brachybranchus*, holotype, 107.4 mm SL, RMNH 30974, from Nickerie R. near Blanche Marie Vallen.

Diagnosis

Head long (31.0-34.0% of SL); gill filaments on first gill arch largely covered by skin; interradial dorsal fin scales in single series in large specimens; from 50 mm SL with dark stripe in angle of preoperculum; midlateral flank spot round, from 11th to 15th lateral line canals down to include most of squ. long. scales below.

The colour pattern alone distinguishes *G. brachybranchus* from other Surinamese *Geophagus* species; from extralimital species with similarly large flank blotch and preopercular stripe, the extensive skin cover over the gills provides a useful distinguishing character.

Material

The description is based on study of the holotype and 174 paratypes, 40-138 mm SL, from the Corantijn and Nickerie river systems.

Description

From the holotype (Fig. 26), unless otherwise indicated; counts from 25 specimens measured.

Shape: moderately elongate. Dorsal head contour rising straight to dorsal fin origin, save for concavity above anterior half of orbit (in others usually straight, without concavity, ascending processes of premaxilla often slightly projecting from forehead outline); dorsal fin base contour initially horizontal, descending straightly from 4th spine. Ventral outline with feebly sloping, straight chest and head contour; about straight horizontal abdomen and rising anal fin base and ventral caudal peducle contour.

Head: moderately deep; snout deep, but little produced, its ventral contour nearly horizontal, dorsal contour much steeper; rounded anteriorly. Orbit below forehead contour, all in dorsal, almost all in caudal halves of head. Mouth low, wider than interorbital; maxilla not extending to orbit; lower jaw articulation below anteriormost part of orbit; jaws equal anteriorly. Lips moderately wide, but wider than in *G. brokopondo*, fold of lower slightly expanded (but generally restricted to mouth in small specimens). Operculum a little more than twice as deep as long.

Scales: Squ. long. 32(11), 33(14). Dorsal head scales rapidly smaller from above anteriormost lateral lines scales; along predorsal midline about 17 scales, but arrangement irregular, sizes varying, larger about half flank scale size, many smaller. Chest scales of about half size of anterior flank scales size. Prepelvic scales ctenoid close to pelvic fin bases, otherwise cycloid and small, larger 1/4 size of anterior flank scales; along midline about 16 posterior to and about 8 anterior to naked spot opposite tip of cleithra; narrow naked zone under interoperculum.

Cheek scales in 5 or 6 (occasionally 4 or 7) not well defined oblique series of mostly ctenoid, anteriorly and ventrally cycloid scales, larger scales about 2/3 size of anterior flank scales, all cycloid in small specimens, ca 50 mm. Operculum with a group of 9 scales dorsally; below hinge about 43 of mixed sizes, cycloid and ctenoid, larger about half flank scale size like larger of suboperculum (with 3 series of mixed cycloid and ctenoid scales); a small posterosuperior cycloid interopercular scale.

Lateral lines 21/14(1), 21/15(1), 22/-(1), 22/14(2), 22/15(6), 23/-(2), 23/14(5), 23/15(3), 23/16(2), 24/13(1), 24/15(1). Upper lateral line at 7 (initially) to 1 1/2 (terminally) scales distance from dorsal fin, rising for about 5 scales, posteriorly straight, to below 6th dorsal ray; overlapping lower by 6 (to 8) scales. Lower lateral line continued on caudal fin by 2 scales, dorsal branch of 22 scales, ventral branch of 27 scales in holotype.

Fin scales: dorsal fin with a series of small cycloid and ctenoid scales along base, but last 4 membranes naked, with ray muscles exposed distally; from behind 3rd spine cycloid interradial scales, in single series (usually a proximal caudad directed short sequence, distal sequence along preceding spine or ray), with up to 19 scales, last 4 membranes naked, depth of scaled zone at most half, usually only 1/3 length of last dorsal fin spine; more interradial scales in large than in small specimens, smallest specimen with is 50 mm, largest specimen without is 72 mm. Base of spinous anal fin covered by scales; a series of small scales along base, but last 2 membranes naked; no interradial scales in any specimen examined. Four medial pelvic fin rays covered by scales basally. Pectoral fin naked; in 110, 116, and 138 mm specimens a few small cycloid scales on pectoral fin base. Caudal fin scales mostly ctenoid, distally and marginally very small and cycloid; 2/7 scaled medianly, chiefly naked between rays D3 and V3, marginally scaled to near hind edge.

Fins: first dorsal fin spine length about 1/3 that of last, relative spine length increasing to 5th from which subequal, last again a little longer; anterior spines very little curved, posteriorly straight. Soft fin damaged. Lappets free, pointed, reaching well beyond spine tips. D. XVII.10(1), XVII.11(3), XVII.12(16), XVII.13(1), XVIII.11(4). Soft fin short and rounded in small specimens ca 50 mm, to little behind caudal fin base, in large specimens (over 70 mm) with broad point, not much beyond caudal fin base; slightly produced 4th ray only in large males 101 and 110 mm, to 1/3 to middle of caudal fin. First anal fin spine opposite penultimate dorsal fin spine; soft part pointed, 3rd ray longest, not produced, reaching slightly behind caudal fin base. A. III.8(22), III.9(3).

Soft fin rounded, slightly shorter than soft dorsal fin in small specimens (ca 50 mm), not reaching to caudal fin base; in large specimens (over 70 mm) rounded or, usually, pointed, but produced only in 110 mm male, to middle of caudal fin.

Caudal fin emarginate; corners damaged; in others short prolongations of corners from rays D6-D7 and V6; an 89 mm female exceptional with dorsal prolongations nearly half length of rest of caudal fin. Pectoral fin reaching to above 3rd anal fin spine (to at most above 2nd anal fin ray in others). P. 14(11), 15(14). Pelvic fin first ray outer branch produced, reaching to middle of caudal peduncle; generally first ray not very long, in 50 mm specimens with short prolongation, to 3rd anal spine, in larger fish only exceptionally beyond anal fin base, to caudal fin base in 110 mm male which also with long dorsal and anal fins.

Young: three smallest specimens available, 38-40 mm (ZMA 106.193) more elongate than adults, with more depressed head (though nape keeled), orbit touching predorsal contour, maxilla almost reaching to orbit, and operculum twice as deep as long. Cheek, gill cover, and all prepelvic scales cycloid. Caudal fin not so extensively scaled, and hence with fewer lateral line canals. Soft dorsal and anal fins rounded, dorsal fin reaching almost to, anal fin by far not to caudal fin base. Pectoral fin as in adults. Pelvic fin pointed, not produced, reaching to anal fin origin. Caudal fin emarginate.

Jaw teeth: teeth simple, conical, recurved, loosely attached, inner pointing more or less orad. Posterior 1/4 of each upper jaw half naked, anteriorly a narrow band of small teeth, widely interrupted symphysially. Anterior 1/3 of each lower jaw half with a single (commonly two in other specimens) series of teeth anteriorly forwards curved to leave space for adsymphysial patch of slightly smaller teeth, symphysis edentulous. Larger specimens, especially distinct in 138.4 mm specimen, with enlarged anteriormost lower jaw teeth and outer row teeth of upper jaw larger than and spaced from wide band of small inner teeth. 24-34/20-30 teeth marginally along upper/lower jaw half in 18 specimens 65.6-107.4 mm; 2-4 inner series anteriorly.

Gill rakers: usually 9, occasionally 8 or 10 epibranchial and 11(3), 12(18), or 13(4) lower limb rakers on first gill arch. Gill arches dissected from 82 mm specimen (ZMA 106.193; Fig. 27): on first arch externally, 5 medial epibranchial rakers conical, 4 lateral more compressed with slightly expanded tips, all reaching beyond margin of lobe; a raker in angle conical; 8 upper ceratobranchial rakers flattened, with somewhat expanded tips, 2 lower knob like, anteriormost 2 rudimentary. Upper ceratobranchial rakers connect with median ridge by swellings which become ridges further down the arch, 3 anteriormost rakers free. Very

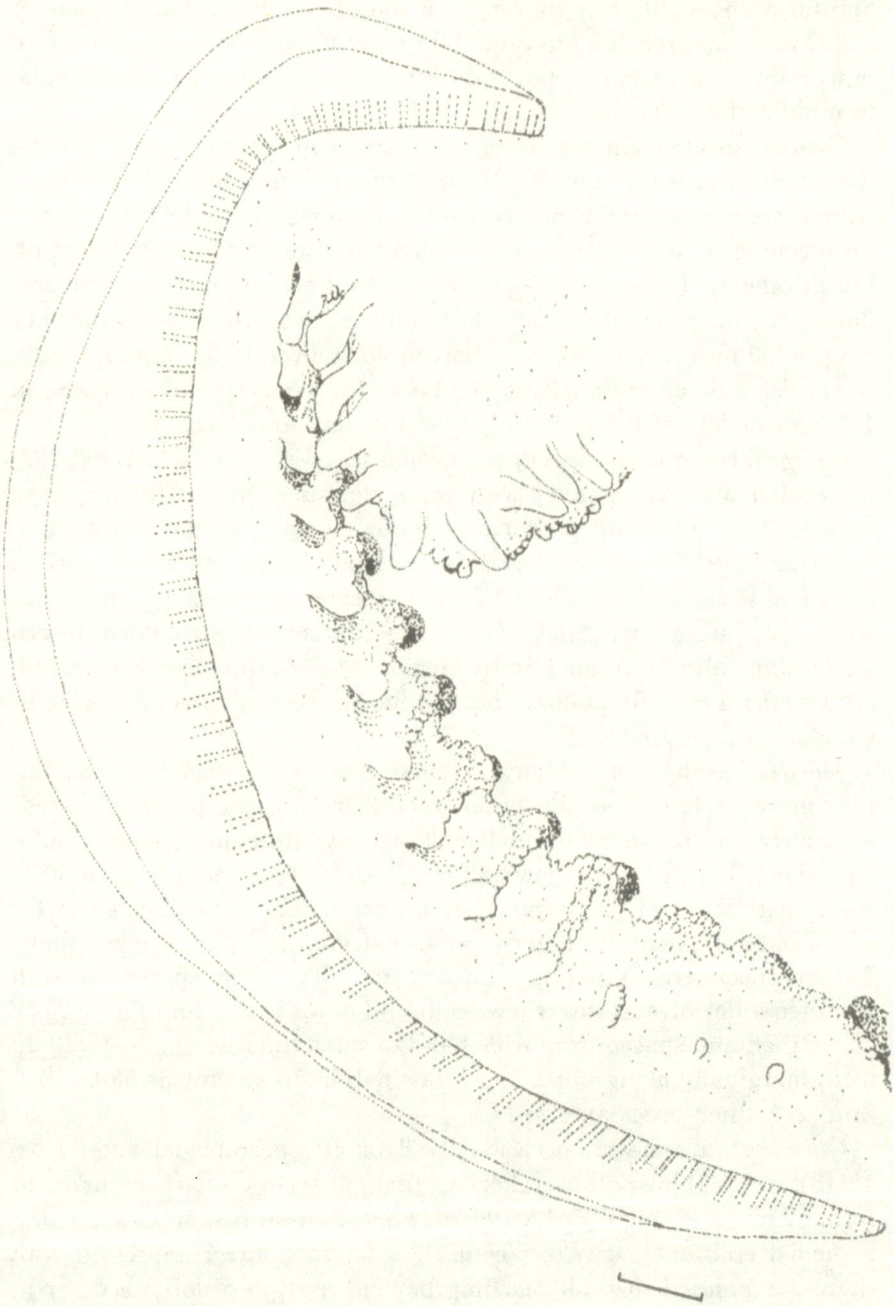

Fig. 27. *Geophagus brachybranchus*. First gill arch in lateral aspect. Scale 1 mm. From ZMA 106.193, 82 mm SL.

narrowly exposed gill filaments. Internal rakers 4 + 1 + 14, conical, connected with medial ridge by swelling basally. All ceratobranchial rakers attached to ceratobranchial. Microbranchiospines usually present all along 2nd-4th ceratobranchial; in 15 specimens checked, 65.6-138.4 mm, missing completely in 2, from 1-2 arches in 5. Epibranchial 1 deeper than long; ventrally a connective tissue pad extended mediad; pharyngobranchial 1 expanded ventrally.

Tooth plates: lower pharyngeal tooth plate (Fig. 28) with a few anterior teeth simple; posteromedian rounded in occlusal aspect, with flattened apex with central blunt projection, and small cusp on rostral edge; remaining slender, with antrorse posterior cusp and small projection on rostral edge. Five tooth plates with 13, 8, 8, 8, and 6 teeth respectively, on ceratobranchial 4. Ceratobranchial 4 tooth plates 6 in cleared and stained specimen.

Vertebrae: radiographs of 5 specimens show 14 + 19 (1), 15 + 17 (1), 15 + 18 (3) vertebrae, single supraneural, 10-11 caudal ribs, 6 dorsal and

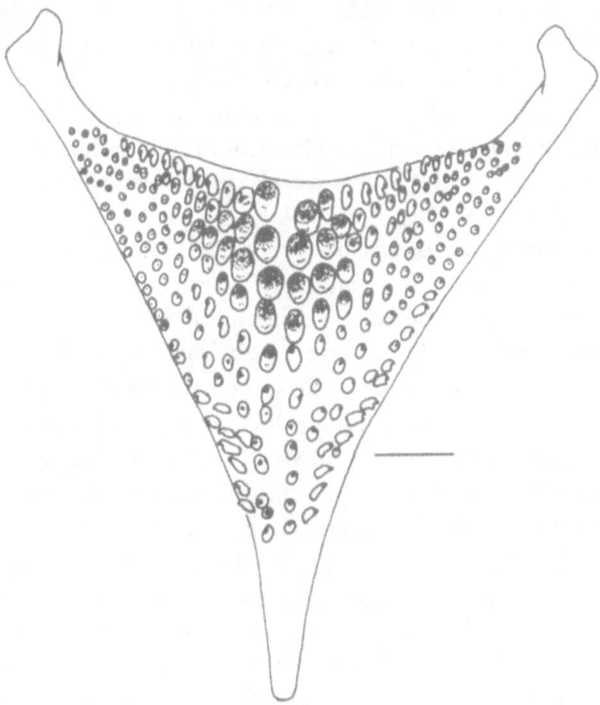

Fig. 28. *Geophagus brachybranchus*. Lower pharyngeal tooth plate in occlusal view. Scale 1 mm. From ZMA 106.193, 82 mm SL.

5-6 ventral procurrent caudal fin rays. Cleared and stained specimen with 32 vertebrae, 15th articulating with anal fin pterygiophore, like 13th and 14th with hemal arch, hemal spine from 16th; epipleural ribs to 10th vertebra, pleural ribs on 3rd to 24th vertebra; long unpaired hypapophysis on 3rd vertebra; parhypural spine reaching middle of lowermost hypural; cartilage piece between 2nd and 3rd hypurals.

Colouration

Whitish along belly midline, rostral to pelvic fins, on chest and pectoral axilla. Underside of head yellowish (interoperculum, lower limb of preoperculum, lower jaw, lower lip). Intermandibularly and on exposed part of branchiostegal membrane blackish. Operculum dorsally brownish grey, major part, like suboperculum, silvery. Snout, forehead, preorbital, and upper lip greyish. Cheek pale brownish, posterosuperiorly with silvery sheen. Preoperculum with faint blackish stripe between inner and outer margins of corner, continued by spot on adjacent interoperculum.

Flanks pale brownish yellowish, darker dorsally. Nape brownish grey. No vertical bars. Along back scale bases light, but major portion of each scale dark brown; on sides fainter dark scale part tends to be restricted to upper and lower margins, forming, especially ventrad, ill-defined dark lines along side. Dorsal margin of caudal peduncle brown. Inside of pectoral axilla and dorsal portion of pectoral axilla pocket dark brown. A large round blackish brown blotch on middle of side, on upper lateral line scales 11-15 (extending above canals) and down to over most of squ. long. scales below.

Dorsal fin lappet tips blackish, rest of fin with large light (opaque) spots with narrow grey interspaces, arranged in 3 uneven horizontal series, irregular only on posterior half of soft fin. Caudal fin brownish grey with large, irregularly arranged light (opaque to clear) spots over entire fin. Anal fin with 3 narrow dark lines paralleling body axis, outer along fin margin, interspaces light. Outer margin of pelvic fin whitish, otherwise fin outwardly greyish, inwardly gradually lighter.

Juveniles 38-40 mm similar, but notably lacking preopercular stripe and dark branchiostegal-intermandibular pigment; sides with pale brown indistinct vertical bars or spots: (1) on head, preceding light (ground colour) zone marking extrascapular lateralis canal from posterior margin of orbit up to midline of nape; (2) below anterior 3 dorsal fin spines and immediately preceding nape down to slightly below upper lateral line; (3) below 5th to 7th dorsal fin spines, confluent with (2) below body midline; (4) a spot close to dorsal fin base just behind (3) or reaching down to midlateral blotch; (5) another spot, little behind (4); (6) from below anterior portion of soft dorsal fin to lower lateral line; (7) from below posterior portion of soft dorsal fin to lower lateral line; (8) a spot on lower lateral line posteriorly on caudal peduncle.

Anterior bars wider than interspaces, strongest close to dorsal fin; notably, 4th and 5th bars appear to be restricted to give space to prominent midlateral blotch, 6th bar hardly distinguishable from succeeding. Midlateral blotch not extending above lateral line canals. Dorsal fin clear, lappets blackish; anterior about 9 membranes pale brownish; posteriorly 2 brownish stripes paralleling fin base, outer on distal half of spinous fin, inner on proximal half, but both continued narrower on soft portion, outer margin of which lightly pigmented. Anal fin lightly pigmented, especially marginally. Caudal fin clear; base brownish, margins duskied, over middle portion about 3 narrow vertical brownish stripes, curved towards fin base, and distal edge with diffuse dark pigment. Pelvic fin dusky.

Largest specimens (over 100 mm) similar to holotype, but usually narrow dark lines along sides distinct (from dark uppermost and lowermost parts of otherwise light scales).

Midlateral blotch always prominent, but tending to remain below lateral line in small specimens, usually reaching little above it in large specimens. Colour pattern rather diffuse in largest specimens, perhaps due to preservation; several fine specimens ca 70-90 mm, however, showing vertical bars relatively distinct and lacking have narrow horizontal stripes; first 4 bars as in juveniles, 4th including midlateral blotch, posterior bars diffuse, 5th and 6th confluent. In well-preserved specimens silvery sheen to cheek, gill cover, chest and pectoral girdle prominent; also bar interspaces with faint metallic, but rather golden sheen.

Branchiostegal membrane and intermandibular area more or less blackish in specimens over 70 mm, but may be pigmented also in smaller specimens. Preopercular stripe very variably expressed, from 50 mm SL, in many fine specimens continued some distance up along anterior margin of preoperculum. Juvenile caudal fin pattern (of dark lines) replaced by adult pattern (of large spots) at ca 70 mm; fin spots tending to be clear rather than opaque.

Distribution (Fig. 21)

Surinamese collections are from the middle-lower Corantijn drainage, and the Nickerie drainage. The species probably ranges westward to the Essequibo drainage.

Etymology

From the Greek *brachys*, short, and *branchia*, gills, a noun suggested by the only narrowly exposed gill filaments.

Notes

The Surinamese *G. brachybranchus* apparently represent the easternmost populations of the species, as Eigenmann's (1912) *G. surinamensis* from the Demerara and lower Essequibo system in Guyana are apparently also *G. brachybranchus*, possibly also Lowe-McConnell's (1969) *G. surinamensis* from the Rupununi (Essequibo and Branco drainages).

Geophagus harreri Gosse

(Figs 21, 29-33)

Geophagus harreri Gosse 1976, p. 88, fig. 21 (rivière Ouaqui à Saut Bali).

Surinamese records

Original description with many Marowijne localities.

Material

144 specimens were examined, all from the Marowijne system. The holotype, described and figured by Gosse (1976), was only checked for identity with material studied.

Fig. 29. *Geophagus harreri*, 135.5 mm SL, ISNB 17569, from the R. Marouini, downstream of Epoia.

Description

Based on 135.5 mm female (ISNB 17569; Fig. 29); measurements and counts from 25 specimens. Largest specimen examined 173.9 mm SL (ZMA 107.751).

Shape: elongate. Head moderately deep, with extended snout. Predorsal contour about straight ascending (feebly concave on snout) to above anterior margin of orbit, posteriorly gradually less steep; prepelvic contour little sloping, straight save for slightly projecting lower jaw articulation.

Head: snout long, rounded anteriorly. Mouth of about interorbital width, terminal but near ventral body margin. Jaws equal anteriorly; lower jaw articulation in advance of orbit; maxilla reaching to halfway between orbit and nostril. Lips thick, lower lip fold wide, extending over lower jaw anteriorly. Preoperculum and pectoral girdle bones entire. Operculum more than twice as deep as long. Eye in posterior and upper halves of head, well below forehead contour, interorbital strongly convex.

Scales: Squ. long. 31(1), 32(18), 33(4), -(2); circumpeduncular scale series 20 (7 + 1 + 11 + 1; usually with 8 series dorsal to lateral lines in large specimens). Predorsal scales about half anterior flank scale size (distinct size reduction below dorsal fin origin forwards compared to scales behind); about 18 ctenoid scales along midline, but anterior half of midline naked, behind small midline and overlapping cycloid scales. Chest scales ctenoid, of about 1/3 size of anterior flank scales. Anterior to pelvic fins numerous small embedded cycloid scales (ctenoid between pelvic fins), 1/4 size of anterior flank scales and smaller, a group also on median pelvic fin ray bases.

Cheek scales small, ctenoid posterosuperiorly and in series up behind orbit, about half flank scale size, embedded, smaller and cycloid ventrally, in about 5 to 8, not regular series. Operculum with ctenoid (posteriorly) and cycloid (anteriorly) scales along dorsal, posterior and anterior margins, small and thin. Suboperculum with 2 scale series, cycloid and ctenoid, hind margin naked. Interoperculum naked (in all specimens).

Upper lateral line at 9 1/2 to 1 1/2 scales distance from dorsal fin, overlapping lower by 8 scales, ending little before caudal peduncle. Lower line extending forwards to above 2nd or 3rd anal fin spine; continued by 2 scales on caudal fin. On right side about 30 tubed scales between caudal fin rays D2-D3 and V4-V5; on left side 2 dorsal lateral line sequences, between rays D2-D3 and D3-D4 (typical); rostrally curved towards middle sequence or not. Lateral line counts in specimens 29-142 mm, 18-24/9-17, increasing with increasing SL; 2 or 3 canals continuing lower on caudal fin; either or both of dorsal and ventral lobe canal series usually curved towards middle series proximally.

Fin scales: dorsal and anal fins with few scales. Dorsal fin on left side with 1 to 6 interradial scales in single series close to fin base, on membranes between 1st and 8th rays; basal scales between 4th and 6th rays.

On right side scattered interradial scales basally between 1st and 6th rays. In other specimens interradial scales only basally on anterior soft dorsal fin membranes, one to 11 along a membrane, in single series, in specimens 57-142 mm; dorsal fin naked in smaller specimens. Anal fin on left side without interradial scales, on right side basally 5 interradial scales on membrane behind 3rd ray and one behind 4th ray. In other specimens those 127 mm and larger similarly with few anal fin scales, those 109 mm and smaller with naked anal fin. Caudal fin densely scaled, less than 1/3 along middle (membranes between rays D2 and V3 largely naked), to near fin end marginally; all scales ctenoid. Pectoral fin naked.

Fins: first dorsal fin spine 2/7 length of last, spines equal from 7th, last 2 again slightly increasing. Anterior 4 lappets with rounded tips, posteriorly truncate, with posterior point; free, moderately long. Soft dorsal fin pointed, 4th ray longest, produced, reaching little behind hind edge of caudal fin, 3rd to 5th rays distally filamentous. D. XV.13(3), XVI.12(39), XVI.13(11), XVII.11(1). Soft fin similar in other specimens over 120 mm, shorter in smaller specimens, without filament. Anal fin origin in 3rd quarter of standard length. Soft fin with long rounded tip, 3rd and 4th rays longest, to 1/3 of caudal fin (never produced, to 1/4 to 1/3 of caudal fin in large specimens. A. III.8(51), III.9(3). Caudal fin lower lobe injured, dorsal lobe with ray D7 slightly prolonged; emarginate, with rays V5 and D6 slightly prolonged in (the few) large specimens with undamaged fin. Pectoral fin tip above first anal fin ray (or 3rd spine). P. 15(1), 16(24). Pelvic fin first ray little produced, to base of 3rd soft ray of anal fin.

Jaw teeth: lower jaw with an outer series of stronger, feebly recurved blunt tipped teeth along anterior half, symphysis widely naked; near symphysis a few much smaller teeth in about one series lingual to outer series. In upper jaw an outer series of stronger, feebly recurved teeth with abraded tips, along anterior half of each jaw half; inner dense narrow band of much smaller pointed teeth, linguad pointing, well separated from outer series, about as far back as outer series. Outer series teeth in both jaws moderately spaced, uniform in size.

In young upper jaw with a narrow band of teeth, the outer not much larger or separated from the inner series, lower jaw with only a few inner teeth anteriorly. Lower jaw outer dental arcade curves forwards anteriorly in large but not in small specimens and upper jaw dentition clearly differentiated in an outer series of stronger teeth well separated from inner narrow band of small teeth; only a few inner series anteriorly in lower jaw. In 8 specimens 88-142 mm, 14-20/17-21 teeth in upper/lower jaw outer hemiseries, 2-3 inner series in upper, 1 (of about 10 teeth) to 2 (one or a few irregularly close to symphysis) inner series in lower.

Gill rakers: ceratobranchial rakers 11(1), 12(15), 13(8), 14(1). Gill arches and lower pharyngeal tooth plate examined in 88.3 mm specimen (ISNB 17659; Figs 30-31). Epibranchial with 8 medial digitiform long rakers and 4 lateral rakers with slightly expanded tips, lateral reaching beyond, medial to about lobe margin; papilliform raker in corner and 10 short blunt upper ceratobranchial rakers, medially attached to well-developed tuberculate swelling continuous across median thin longitudinal membrane with opposite side swellings, anteriorly 2 caudad directed pointed rakers, the rostral rudimentary, on the rostral edge of the ceratobranchial.

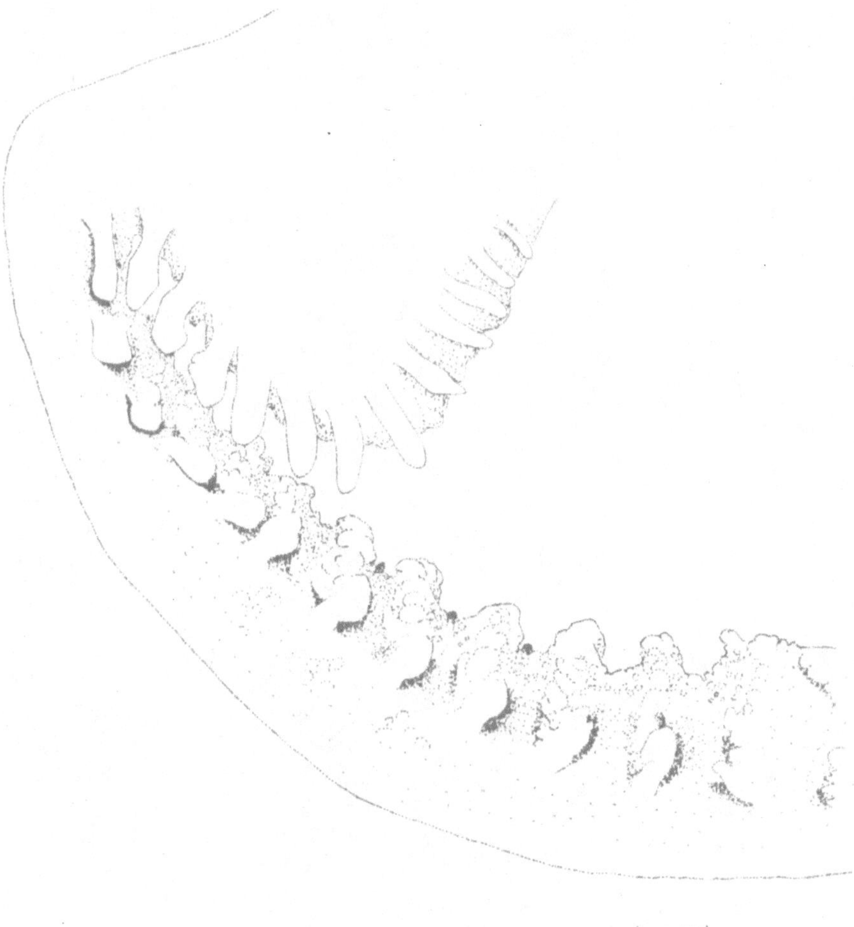

Fig. 30. *Geophagus harreri*. Lateral aspect of first gill arch. Scale 1 mm. From ISNB 17569, 88.3 mm SL.

Skin fold over gill filaments extensive, covering filaments completely rostrally; upper ceratobranchial rakers attach to bony arch, but rostrally a gradual slight ventrad shift, so that ante- and penultimate rakers clearly attached in skin below bony ceratobranchial. 7 + 1 + 13 long inner rakers, 6 lower basally connected to extensive medial swellings.

First epibranchial a little longer than deep, with extensive ventral connective tissue cap with medial projection. Pharyngobranchial 1 broad, approximately triangular with concave sides, 3 arms cartilage tipped. Interarcual cartilage cylinder shaped. Microbranchiospines present or absent; in 15 specimens checked, 43.0-172.9 mm, present externally on upper half of 2nd-4th ceratobranchial in 5, 88.3-172.9 mm, absent in 10, 43.0-141.6 mm. Buccal cavity extensively papillose, tubercles arranged

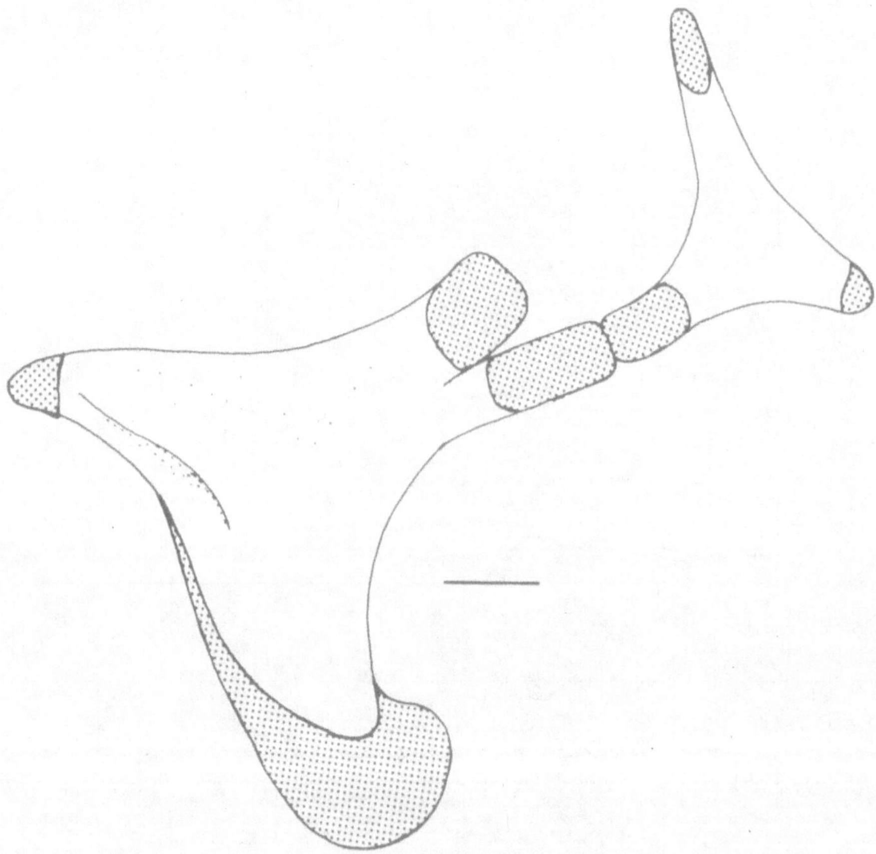

Fig. 31. *Geophagus harreri*. Rostrolateral aspect of epibranchial 1 and pharyngobranchial 1. Epiphyseal cartilage and ventral epibranchial connective tissue pad stippled. Scale 1 mm. From ISNB 17569, 88.3 mm SL.

in longitudinal ridges dorsally (medially smooth), transverse ridges laterally.

Tooth plates (88.3 mm specimen): lower pharyngeal tooth plate similar to that of *G. brachybranchus* in shape and dentition except that the marginal teeth are mostly simple, and the medioposterior teeth are more slender. Ceratobranchial 4 with 3 toothplates with 3, 5, 3, pointed teeth respectively.

Young, ca 29 mm elongate, in frontal aspect with nape and chest rounded; interorbital flat, orbit touching forehead contour. Maxilla reaching back to anterior margin of orbit; lips thick, but fold of lower not particularly wide. Predorsal, chest and lateral head scales cycloid, predorsal midline naked. Soft dorsal and anal fin tips rounded, dorsal reaching to, anal not to caudal fin base; caudal fin hind edge emarginate; pelvic fin pointed, first ray longest, not produced, reaching to vent. 1/3 of caudal fin scaled; only a few tubed scales on dorsal and ventral lobe. Two close-set teeth series in upper jaw, lower jaw dentition uniserial.

Fig. 32. *Geophagus harreri*, 90.8 mm SL, ISNB 20849, from the R. Marouini.

Osteology

Vertebrae: 14 + 18(2); pleural ribs on 6-7 anterior caudal vertebrae, in radiographed specimens. Cleared and stained specimen with 14+18 vertebrae; 1 supraneural; long median hypapophysis on 3rd vertebra; 15th vertebra (1st caudal) with hemal arch, basapophyses, pleural ribs and hemal spine, no hemal arch on 14th; epipleural ribs to 11th vertebra; pleural-epihemal ribs from 3rd to 21st vertebra (from 15th autogenous); 3 tooth plates on 4th ceratobranchial; cartilage between 2nd and 3rd hypural; long parhypural spine to middle of 2nd hypural; infraorbitals: 2, 3 + 4, 5, 6.

Colouration

Ground colour fleshy anterior to pelvic fine and along abdominal midline, becoming grey on head, sides pale yellowish. Snout, forehead, preorbital, lips, lower jaw, branchiostegal membrane, interoperculum, preoperculum, naked cheek grey, darker dorsally, lighter ventrally. Scaled cheek, operculum and suboperculum grey brown. Faint brownish band from orbit straight down to preoperculum posterior to corner. Scattered dark spots on vertical limb of preoperculum, operculum and suboperculum. Inside and pocket of pectoral axilla brown to greyish.

Nape dark brown; a narrow vertical greyish white zone marking extrascapular canal series, and forming anterior limit of vertical bar from nape midline and below first dorsal fin spine, dark brown to well above lateral line, ventrad faint, to pectoral girdle; next vertical bar dark brown below 4th to 6th dorsal fin spines, very faint, from light brown to beige-brownish, ventrad from well above lateral line to abdominal sides where continuous with next bar and beige-brownish chest (area next above anal fin base much lighter than abdomen and chest); next bar across 8th to 10th lateral line scales from below dorsal fin, narrower ventrad to abdominal sides, containing blackish brown blotch between dorsal fin base and lateral line (but not reaching to either); shades of another 2 bars below dorsal fin posteriorly.

Blackish brown spot below last 5 soft dorsal fin rays down to upper lateral line level, continuous with dark brown dorsal margin of caudal peduncle; faint brownish bar down to lower lateral line from spot and dark caudal peduncle margin. Vertical bars wider than interspaces. No horizontal striping along flanks. Scales of dorsal sides with slightly lighter bases.

Dorsal fin greyish, lappets darker. On spinous portion 2, posteriorly 3 series of oblong whitish spots forming stripes paralleling body midaxis, on soft portion irregularly arranged clear round spots (5 to 7 along longer membranes); outer margin of soft part blackish. Caudal fin pale brownish with white narrow line submarginally along dorsal margin, dorsal lobe and upper half of ventral lobe with many whitish spots. Anal fin light brownish with white spots on last 2 membranes. Pelvic fin brownish grey, lighter inwardly, outer margin of first ray white.

In other large specimens, 88.3-141.6 mm (ISNB 17569, Fig. 32), best developed in those over 100 mm, pattern of yellowish whitish scale centers and, on lower sides brownish, on back dark brown scale margins, very distinct. Broad brown vertical stripe over cheek very distinct in those 88.3 to 108.6 mm, fainter in those over 126 mm, more or less evident also on preoperculum. Branchiostegal membrane never darkened.

Blotch below middle and at end of dorsal fin as described above, always prominent, but the bar in which anterior blotch contained may be as strong as blotch, whereby a narrow wedge-shaped stripe across side forms (95.5 mm specimen); also the anterior spot varies in size and position, though always above lateral line. The other verical bars vary in intensity but are not divided; those behind that containing flank blotch forming rather uniform pale brownish dorsal sides. Nape dark. Dark spots on gill cover either very faint or absent. Upper margin of caudal fin with light submarginal stripe or series of dots.

Young (ISNB 17570, 28.9-56.8 mm; Fig. 33) with distinct broad brown band from eye down to preoperculum behind corner or onto preoperculum. Body slightly countershaded, with reticulate pattern of brownish scale margins and yellowish whitish scale centers. Nape dark brown; extrascapular canal series yellowish white, behind variously strong, spot-like or wedge-shaped backwards inclined bar; another bar behind faint; next bar intense blackish brown or dark brown, wedge-shaped, without blotch, extending down onto abdominal sides. Intense dark spots below dorsal fin end; dorsal margin of caudal peduncle dark brown, most intense at posterior end. Caudal fin in smallest specimens, ca 30 mm, immaculate or faintly spotted, without dark spot at base, but from ca 35 mm caudal fin with large whitish spots. Dorsal fin as in adults but spots fewer and not distinct on soft portion; anal fin light, immaculate; pelvic fin whitish.

Distribution (Fig. 21)

Geophagus harreri is a Marowijne drainage endemic with a fairly even distribution throughout collected parts of the Marowijne drainage,

Fig. 33. *Geophagus harreri*, 37.4 mm SL, ISNB 17570, from the R. Marouini, downstream of Epoia.

including highland tributaries and mainstream localities north to the Saint Laurent area.

Notes

Geophagus harreri may be diagnosed by the long head (32.7-37.8% of SL); gill filaments on first gill arch largely covered by skin; adults with few dorsal fin scales, on anterior soft part in single series; dark vertical stripe from eye to preoperculum; prominent dark blotch just below end of soft dorsal fin; young with wedge-shaped stripe over middle of side, in adults fainter but including dark blotch between dorsal fin and lateral line.

In comparison with Surinamese congenerics, the species is notable also for higher pectoral fin count (mode 16; range 14-15 in the others), lower dorsal fin count (mode XVI.12 versus XVII.12 in the others), more ventrally inserted anterior ceratobranchial rakers, fewer inner teeth in lower jaw, and fewer upper jaw outer series teeth, naked interoperculum and embedded rather than superficial lower head and prepelvic scales, and naked anterior nape midline.

Geophagus harreri is a very distinctive species, not only for its unusual, contrast rich colour pattern (Figs 29, 32-33). The almost triangular pharyngobranchial 1 shape (Fig. 31) appears unique among cichlids. The naked nape midline is paralleled among cichlids only by *Cichla*. The absence of abdominal hemal arches, if our specimen is representative, is unusual for a large geophagine cichlid.

The thick lips, small embedded or (on preoperculum) absent scales on lower head and thoracally, and especially the naked anterior head midline, suggest a rheophilic species. The relatively reduced jaw dentition and gill arch morphology tends towards *Satanoperca*, but reminds also of the rheophilic *Retroculus*. We thus tentatively recognize *G. harreri* as a very specialized *Geophagus* species.

The colour pattern bears a striking similarity to that of nominotypical *Guianacara* species (cf. Figs 29, 32-33 with Figs 43-44, 66-68). *Geophagus harreri* and *Guianacara owroewefi* are commonly collected together and a mimicry situation appears fairly evident as the wedge-shaped contrasting dark bar over the side is unique to these fishes among cichlids. The similarity is strongest at small sizes, when the major superficial differences are in the more elongate shape of *G. harreri* and the dark spot below the end of the dorsal fin. Large *G. harreri* have a less contrasting pattern of dark markings.

It is interesting to note that *Guianacara owroewefi* has the least pronounced wedge-shaped flank stripe of the nominotypical *Guianacara* species. That may reflect intergeneric character displacement affecting *G. owroewefi*. The unique spot at the end of the dorsal fin base in *G. harreri* may also have evolved as a mark facilitating species recognition, even though we would have expected a mimicry pair, in which one species probably reproduces at sizes larger than that attained by the other, not to amplify differences. Is it possible even that the aberrant colour pattern of the sympatric *Guianacara oelemariensis* resulted from interaction with *G. harreri*? We suggest tentatively that *G. harreri* is the mimic and *G. owroewefi* (or its ancestral form) the model, because the colour pattern is more widespread (geographically and in taxa) in *Guianacara* than in *Geophagus*, but above all we recommend field work to investigate the actual ecological relationships of these species, that may elucidate their possible coevolutionary history. This is the first strong case of potential mimicry involving particular species in South American cichlids. More general mimicry involving *Cichla* and species with a caudal fin ocellus has been described by Zaret (1977), and aggressive mimicry is known from African cichlids (e.g., Trewavas 1947).

A further point of interest is the range overlap of *G. harreri* and *G. surinamensis*. It is unusual to find congeneric South American cichlids with shared geographical range.

Our study of *G. harreri* involves many more characters than those studied by Gosse (1976) for the original description. Gosse measured and counted 62 specimens, and found D. XVI-XVII.11-13, A. III.7-9 (we have 8-9 rays, but 7 or 9 rays are evidently very rare), 12-16 ceratobranchial rakers (we have 11-14, mode 12, and suspect that Gosse included all rakers not definitely on the epibranchial lobe); scale counts were obtained by different methods, and Gosse's counts appear mostly slightly higher than ours. Gosse's largest specimen is 190 mm SL. He figures (pl. I) a specimen with 14+18 vertebrae, but gives 15 + 17 (in 12 specimens) and 15+18 (in one specimen) as vertebral counts, so apparently there is disagreement in our judgement about which is the first caudal vertebra.

Gosse considered *G. harreri* to be close to *G. surinamensis*. As his *G. surinamensis* is a mixture of several species, and as we found no character uniting *G. harreri* more closely with *G. surinamensis* than with any other species, the relationship question remains open.

SATANOPERCA Günther

Satanoperca Günther 1862, p. 312 (type species *Geophagus daemon* Heckel, by subsequent designation in Jordan & Evermann 1898, p. 1542).

Satanoperca are geophagine cichlids with a well developed epibranchal lobe, uniquely distinguished by having almost all of the 15-22 external first ceratobranchial gill rakers attached to the skin over the bases of the gill filaments. Other characters distinguishing the genus from one or more of other genera of geophagines are: epibranchial 1 about as deep as long; pharyngobranchial 1 and interarcual cartilage rod like; jaw dentition principally uni-or bi-serial; a dark, usually ocellated spot dorsally on base of caudal fin; naked dorsal and anal fins; single supraneural; gill rakers absent from lower pharyngeal tooth plate; more abdominal than caudal vertebrae or equal numbers; no ribs on caudal vertebrae; no median hypural cartilage; 3 infraorbitals; no tooth plates on ceratobranchial 4; cheek scaled rostrad to mouth angle; upper or both lip folds continuous; preoperculum and supracleithrum frequently serrated. The largest species reach about 250 mm SL.

Satanoperca was included in *Geophagus* by Gosse (1976), but rehabilitated by Kullander (1986) who gave a brief diagnosis of the genus and redescribed the widespread Amazonian species *Satanoperca jurupari* (Heckel). *Satanoperca jurupari* may occur in French Guiana; the only other Guianan *Satanoperca* species is *S. leucosticta*, described below. The genus includes also the nominal species *Geophagus daemon* Heckel (upper R. Negro drainage, Orinoco drainage), *G. acuticeps* Heckel (Central Amazonia), *G. mapiritensis* Fernandez-Yépez (Venezuela), and *G. pappaterra* Heckel (Paraguay and Guaporé drainages), besides several undescribed species.

Satanoperca leucosticta (Müller & Troschel)

(Figs 34-37)

Geophagus leucostictus Müller & Troschel, in Schomburgk 1849, p. 625 (See Amucu und in den Sumpfen der Savanne).

Synonymy

Satanoperca macrolepis Günther 1862, p. 314 (Demerara; British Guiana).

Surinamese records

Geophagus jurupari; (*pt.*) Gosse 1976, p. 49 (Kreek van Nickerie River, Surinam [ZMA 105.818, 106.196]; only Guyana and Surinam material in description is *S. leucosticta*). — (?) Boeseman 1952, p. 198 (listed: Lucie River [not found in RMNH]).

Material

Only five specimens, 56.1-98.2 mm available, from the Stondansi Val area (ZMA 105.818, 106.196).

Description

Based on 89.4 mm specimen (ZMA 105.818; Fig. 34); counts from all specimens available. Largest specimen examined 98.2 mm SL (ZMA 106.196).

Fig. 34. *Satanoperca leucosticta*, 89.4 mm SL, ZMA 105.818, from Nickerie R. tributary near Stondansi Val.

Shape: elongate, moderately compressed laterally; dorsal contour more arched than ventral. Ventral margin of caudal peduncle longer than dorsal. In frontal aspect compressed, with keeled nape and flat chest.

Head: elongate, with long snout. Predorsal contour rising straight to just anterior to dorsal fin; prepelvic contour almost horizontal. Laminar operculum twice as deep as long; interoperculum narrow. Orbit in upper and posterior halves of head, touching forehead contour. Mouth low, much wider than flat interorbital. Tip of maxilla well exposed; lower jaw much prognathous; jaws situated entirely rostral to orbit. Lips wide, upper continuous medially, lower interrupted. Supracleithrum serrated in 3 (only on left side in one) specimens, smooth in 2.

Scales: flank scales ctenoid, relatively large, gradually smaller onto median anterior areas; squ. long. 27(4), 28(1); circumpeduncular scale series 16; predorsally about 11 scales along midline, anteriorly median, posteriorly scale pairs overlapping along midline, cycloid, anterior 2/3 size of anterior flank scales, smaller posteriorly. Chest scales ctenoid, except under gill cover, and small scales on pectoral axilla base; largest nearly of flank scale size. From between pelvic fin bases about 10 + 7 (gap opposite tip of cleithrum) forwards smaller scales, posterior 2/3 size of anterior flank scales; cycloid like in 3 parallel series on each side; narrowly naked along gill cavity margin. Abdominal scales ctenoid.

Cheek with about 5 series of scales, naked ventral to line continuing edge of anterior preorbital skin margin caudad; posterosuperior scales largest, 2/3 flank scale size, and ctenoid, cycloid and smaller ventrally; one series of cycloid scales up behind orbit. Operculum largely scaleless; scales dorsally and along rostral margin of laminar part cycloid, largest 2/3 size of anterior flank scales. Eight cycloid subopercular scales in principal series, and 1 or 2 below, largest 2/3 size of anterior flank scales. Two or 3 small cycloid scales posteriorly on interoperculum.

Upper lateral line in third scale series above that containing the lower, at 3 1/2 (anteriorly) to 1 1/4 (terminally) scales distance from dorsal fin, overlapping lower by 1-4 scales, ending below anterior part of soft dorsal fin. Lateral lines: 18/3 pores +7(1), 18/10(1), 19/10(1), 19/12(1), 17/11(1). One or, usually, 2 canals continuing lower line on caudal fin; also a longer canal series between rays V4 and V5, but upwards bent anteriorly; shorter sequence between rays D3 and D4 absent in 2 smaller specimens.

Fins naked except caudal fin, which scaled for 1/3 along middle, marginally for about half length of fin; interradial scales in single series, most cycloid.

Fins: dorsal fin origin above first lateral line scale, first ray about 1/3 length of last; spines increasing to 4th to 9th, or 5th to 7th, or 5th to 8th, which subequal in length; spines behind slightly shorter, either subequal or slightly decreasing to slightly before last, which as long as 5th or little shorter. Soft fin rounded in small specimens,reaching little behind base of caudal fin, in large specimens pointed, 3rd ray longest, to or past middle of caudal fin. D. XV.10(5). Anal fin in posterior 3rd of body, soft fin like soft dorsal fin, but shorter. A. III.7(5). Caudal fin damaged except in 61 mm specimen in which truncate; principal rays 16, procurrent rays 4 in dorsal, 3 in ventral lobe. Pectoral fin subacuminate, 3rd ray longest, reaching to above 2nd or 3rd anal fin spine. P. 14(5). Pelvic fin pointed, first ray longest, slightly produced, reaching to 2nd anal fin spine base (to 2nd anal fin ray in largest specimen).

Jaw teeth: in upper jaw a single tooth series along most of jaw rim, wide edentulous area at symphysis; teeth small, conical, but tips rounded,

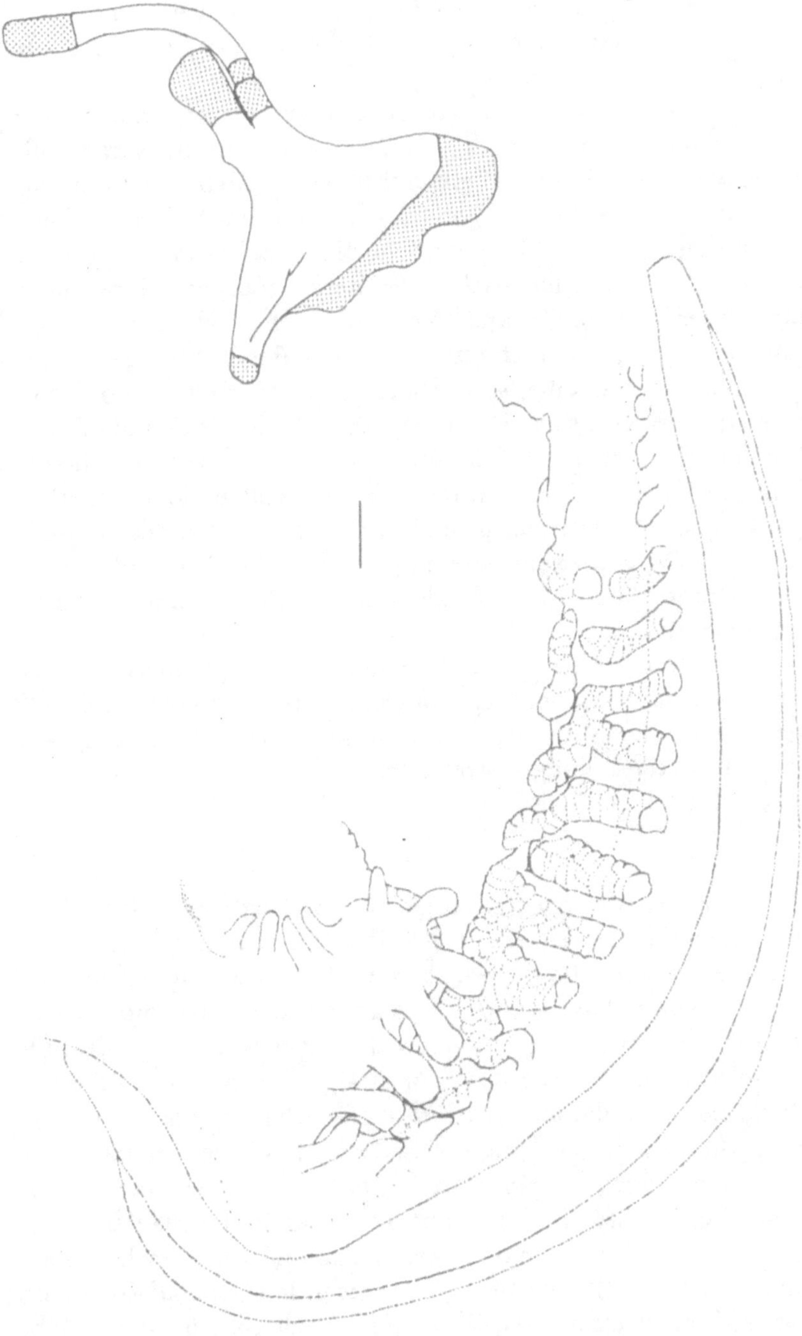

Fig. 35. *Satanoperca leucostica*. Lateral aspect of first gill arch (left) and rostrolateral aspect of epibranchial 1 and pharyngobranchial 1 (right). Epiphyseal cartilage and epibranchial connective tissue pad stippled. Scale 1 mm. From ZMA 105.818, 74.4 mm SL.

20 (in smallest) to 32 (largest specimen). Anterior 1/3 or little more of each lower jaw half with a series of 18-20 small, conical, teeth; wide edentulous area at symphysis; anteriorly a short inner series (2 in 98.2 mm specimen).

Gill rakers: in 5 specimens counted, 10-12 epibranchial, and 16(3) or 17(2) lower limb rakers on first gill arch. Microbranchiospines externally and internally along 2nd and 3rd gill arches, on 4th arch only externally save for short sequence internally in 89.4 mm specimen. First gill arch in 74.4 mm specimen (Fig. 35) with 5 medial papilliform and 5 broad flattened, apically somewhat expanded lateral epibranchial rakers, a papilliform raker in the angle, and 16 ceratobranchial rakers, the uppermost short, attached to the arch and free, remaining attaching to broad skin cover over gill filaments, short, obtuse, and forming end of transverse tuberculate ridges, save for lowermost 4 which free and tending to conical in shape. Inner 5 + 1 + 17 short papilliform rakers attached to arch, connected medially with transverse tuberculate ridges. Ceratobranchial 4 without teeth. Epibranchial 1 (Fig. 35) about as deep as long, medial margin of ventral expansion margined with cartilage and connective tissue; pharyngobranchial 1 slender, curved laterad ventrally. Interarcual cartilage rod like.

Lower pharyngeal tooth plate (Fig. 36) slender, with deeply incised caudal border; marginal and anterior teeth simple, pointed, posteromedial teeth strongest, laterally compressed, with posterior antrorse blunt cusp and one or two projections along rostral edge.

Vertebrae: 14 + 14(5).

Colouration

Ground colour whitish on chest, underside of head, and along belly medianly; yellowish on sides. Snout, preorbital, upper lip, forehead, and gill cover grey; operculum brownish dorsally. Cheek light brownish yellowish. Numerous faint white dots scattered over preorbital, cheek and gill cover. Across forehed 2 narrow dark stripes between eyes, the anterior forwards curved, the posterior straight. Nape brownish grey, light zone in extrascapular area. Brownish on back along dorsal fin base. Very faintly duskied on squ. long. scales, 4 shadows of vertical bars close below dorsal fin, dorsal margin of caudal peduncle brown; sides otherwise warmly light brown. Inside and pocket of pectoral axilla dark grey. Light dot on base of each scale on sides. Dorsal fin light greyish, lappets dark grey, basally on soft portion scattered clear dots. Caudal fin greyish with scattered small clear dots; faint dark small spot on dorsal lobe slightly distal to base. Anal fin light greyish, with small clear dots on last 3 membranes. Pelvic fins smoky, outer margin white.

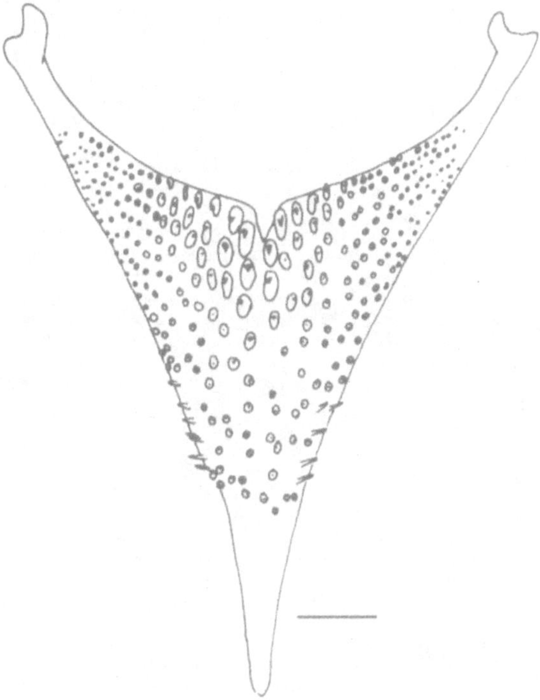

Fig. 36. *Satanoperca leucosticta*. Lower pharyngeal tooth plate in occlusal view. Scale 1 mm. From ZMA 105.818, 74.4 mm SL.

Smaller specimens in the same lot lack head dots, but vertical bars slightly more intense. ZMA 106.196 specimens faded.

Distribution (Fig. 37)

Only two Surinamese samples have been available for examination, both from the Stondansi Val area, Nickerie drainage. The species apparently ranges westwards into the Essequibo drainage and may be expected to occur in the Corantijn.

Notes

The holotype (ZMB 2831, 104.1 mm SL), probably from the Rupununi area, Guyana, has a shorter snout (only 14.1% of SL), more elongate body (depth 36.3% of SL) and very long caudal peduncle (15.3% of SL) compared to Surinamese material. The short snout may be due to shrinkage, as in the fresh material the firm flesh on the snout tip adds to the snout length. It agrees in dentition, gill morphology, and traces

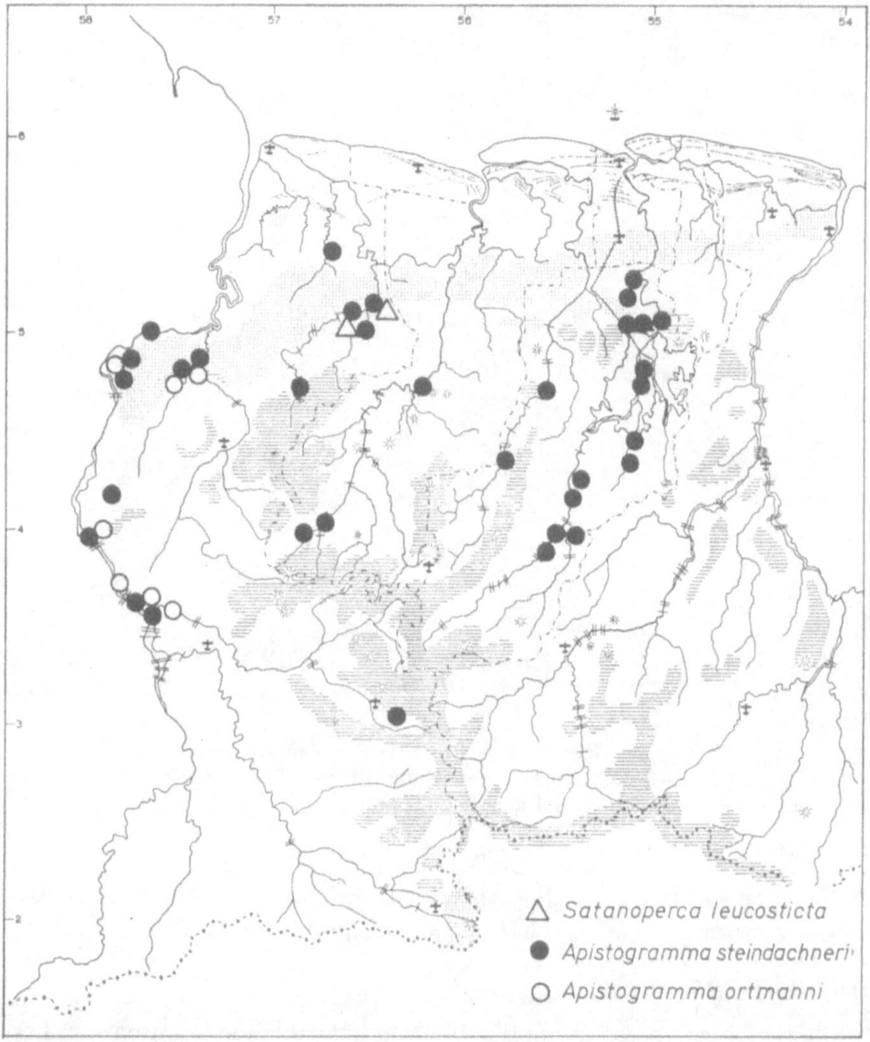

Fig. 37. Collecting localities of *Satanoperca leucostica* and *Apistogramma* species in Surinam.

remaining of colour pattern, so the Surinamese material is identified as *S. leucosticta*.

Satanoperca leucosticta is distinguished from all other *Satanoperca* species by the following characters in combination: absence of dark blotches on the sides; biserial or triserial lower jaw dentition; narrowly exposed gill filaments on first gill arch; preorbital, cheek and gill cover with numerous small whitish dots.

Eigenmann's (1912) material of *Geophagus jurupari* from the Essequibo, Demerara and Mahaica apparently represents *S. leucosticta* as well, to judge from his description and photo and examination of specimens from Lama and Maduni Stop-Off (ZMA 109.560, 109.561). The breeding behaviour of *S. leucosticta* was described in detail by Reid & Atz (1958). Although their material was identified as *Geophagus jurupari*, the colour pattern of the adults shown on their photographs (pl. I, figs 1-2), is rather that of *S. leucosticta*. The species is a biparental mouthbrooder, taking up the eggs after about one day of substrate brooding.

The syntypes of *S. macrolepis* Günther, were not reexamined; the description and localities (Demerara; British Guiana) give little room for doubt about the synonymy with *S. leucosticta*. Limited material of a *Satanoperca* species from the upper R. Negro and upper R. Orinoco can be distinguished by in having the light head spots excluded from the gill cover. *Satanoperca* is represented in collections from French Guiana; the identity of this form, more similar to *S. jurupari* than to *S. leucosticta*, has not been decisively determined.

APISTOGRAMMA Regan

Apistogramma Regan 1913b, p. 382 (replacement name for *Heterogramma* Regan; type species as for *Heterogramma*: *Mesops taeniatus* Günther, 1862, by subsequent designation by Eigenmann 1910, p. 478).

Synonymy

Heterogramma Regan 1906a, p. 60 (type species *Mesops taeniatus* Günther, by subsequent designation by Eigenmann 1910, p. 478).
Pintoichthys Fowler 1954, p. 316 (type species *Biotodoma trifasciatus* Eigenmann & Kennedy, by original designation).

This group is made up of small (largest known specimen 65 mm SL) South American geophagine cichlids, with a single supraneural; small epibranchial lobe with marginal rakers; very long interarcual cartilage; 0-5 external ceratobranchial rakers; anguloarticular either without lateralis canal or with posterior skin canal opening separate from rostral preopercular canal skin opening; and only two infraorbitals.

Apistogramma was treated monographically by Kullander (1980a), and is subject to continuous revision on species level (Kullander 1979a, 1979b, 1980c, 1981a, 1982a, 1982b, 1982c, 1983b, 1986, 1987; Koslowski 1985b). See Kullander (1987) for discussion of generic characters. Forty-six species are currently recognized as valid, and species awaiting description indicate a total of well over 60. Four species occur in the Guianas, viz. *A. gossei* Kullander (1982c) in the Oyapock, Approuague and Araguari (Amapa) river systems; *A. rupununi* Fowler in the Rupununi and Branco river systems; and the two Surinamese forms, *A. steindachneri* and *A. ortmanni*, which are also collected in Guyana and the latter also in Venezuela.

The Surinamese *Apistogramma* species belong to different species groups. *Apistogramma ortmanni* is a member of the *regani* group (Kullander 1980a), with many species in the Amazon basin, and also including *A. gossei*. *Apistogramma steindachneri*, *A. rupununi*, and the Amazonian *A. hippolytae* Kullander form the *steindachneri* group (Kullander 1982a).

Apistogramma steindachneri (Regan)

(Figs 37-40)

Heterogramma steindachneri Regan 1908, p. 370 (Georgetown, Demerara; [syntypes: BMNH 1909.4.30:31-32]), fig. p. 371 (sketch, probably of syntype, habitus, lateral aspect).

Synonymy

Apistogramma ornatipinnis Ahl 1936, p. 141 (Britisch-Guiana).
Apistogramma wickleri Meinken 1960, p. 665 (aquarium specimens).

Surinamese records

Apistogramma steindachneri; Hoedeman 1951, p. 1 (listed; Paramaribo; [ZMA 100.033]).— Boeseman 1952, p. 198 (listed; Lucie River [RMNH 18470]; Surinam, (Upper Corantyne River?) [RMNH 18471]; lengths). — Boeseman 1956, p. 196 (listed: Surinam [RMNH 21402]). — Richter & Nijssen 1980, p. 124 (listed; nascent Brokopondo reservoir; local names). — (*pt.*) Kullander 1980a, pp. 16-19, 24, 25 (listings; incl. *A. rupununi*). — (*pt.*) Kullander 1982a, pp. 191, 192 (data from present paper; incl. *A. rupununi*). — Kullander 1982c, p. 71 (distribution).

Material

About 700 specimens from the Corantijn, Nickerie, Coppename, Saramacca and Suriname R. systems.

Description

From a male, RMNH 30998, 29.7 mm, except where otherwise stated. Quantitative data are from the specimens measured (n = 20 or less), unless otherwise stated. The largest specimen examined is a male, 65.3 mm SL (ZMA 107.008, Fig. 38).

Fig. 38. *Apistogramma steindachneri*, largest known specimen, adult male 65.3 mm SL, from Dabikwen Kreek, Suriname R. System.

Shape: body moderately elongate, broadest and deepest just behind gill cover. Dorsal contour straight sloping from origin to end of spinous dorsal fin, caudally gradually descending, rather steeply close to end of soft dorsal fin; meeting about straight, horizontal caudal peduncle contour at an angle. Chest contour almost straight, sloping; abdominal contour about straight, slightly sloping, deeper caudally, horizontal just in advance of anal fin origin; anal fin base contour straight, upwards sloping, meeting straight sloping caudal peduncle contour at an angle. Head moderately elongate, longer than deep, slightly shorter than body depth.

Predorsal contour straight ascending, slightly steeper on snout and distinctly levelled out behind level of posterior margin of orbit. Prepelvic contour evenly curved, sloping.

Head: snout very little extended, profiles about equal, straight or dorsal slightly convex. Very large males more robust with rather broad head, occasionally with indication of swelling of nape; snout may be conspicuously extended or well rounded. Ascending processes of premaxilla reaching to about 1/4 of orbit, angle ca 50°. Angle of maxilla ca 55-60°; tip exposed, reaching to anterior margin of orbit. Lower jaw longer than upper, but jaws equal anteriorly. Orbit diameter 32.3-41.0 ($\bar{x}$ = 37.0 ± 0.47) per cent of head length; situated chiefly in rostral and dorsal halves of head, touching predorsal contour. Preopercular serrations occur, but are very weak (RMNH 31006, 22.4 mm; RMNH 19334; RMNH 30992; cf. fig. 1 in Kullander 1980a).

Scales: depth of exposed part of the ctenoid body scales greatest on middle of flanks, gradually lesser posteriorly and marginally (reduced to about one-half on caudal peduncle). A 7.9 mm specimen has cycloid instead of ctenoid body scales, and predorsal scales absent; a 10.9 mm specimen has ctenoid flank scales, and cycloid predorsal scales. Predorsal scales cycloid above forwards curved line between preopercular tip and dorsal fin origin. Chest not naked anteriorly. Cheek completely scaled except for rostroventral corner, scales cycloid except those behind posterior margin of orbit. Most of opercular scales ctenoid. Subopercular scales cycloid, in one series. Interopercular scales cycloid. Circumpeduncular scale series 16. Upper lateral line to below 2nd soft dorsal fin ray, overlapping lower line by 2 scales; 1-3 subserial pores in 18 of 20 specimens; lower line forwards to above 3rd soft anal fin ray, continued on caudal fin by a canal or pore in 13 of 20 specimens. Dorsal and anal fins naked. Scales on base of caudal fin with ctenoid proximally, cycloid distally; very large males have half of caudal fin scaled, in young only 1/3 is scaled.

Fins: dorsal fin spines subequal or but slightly increasing in length from 6th or 7th, last spine longest. Lappets pointed, not produced, but reaching beyond spine tips; short, obtuse in females. Soft part pointed, not produced, 2nd ray reaching 1/3 of caudal fin. Soft dorsal fin produced in large males, point reaching middle or end of caudal fin (without extensions); in females rounded, reaching to little behind caudal fin base.

Soft anal fin similar, 3rd ray reaching 1/4 of caudal fin; in large males produced, similar to soft dorsal fin; in females rounded, reaching to little behind caudal fin base. Last dorsal fin spine longer in males than in females of comparable SL (18.2-22.1% vs. 16.1-18.9% of SL in the SL range 26.1-33.5 mm).

Pectoral fin rounded, 4th ray longest; 2 uppermost and 4 lowermost rays simple, remaining rays branched; reaching to above genital papilla. Pelvic fin acuminate, first ray slightly produced, reaching to beginning of soft anal fin. In large males first pelvic fin ray produced, reaching to middle of caudal peduncle; in females always short, but pointed, reaching to about spinous anal fin base. Large males in material measured have long pelvic fins (52.7-60.3% of 46.3-61.9 mm) but males and females of comparable lengths have comparably long pelvic fins. Caudal fin subtruncate; 2 marginal rays in each lobe simple, inner rays with branched. Caudal fin of females roundish to subtruncate. In males, from ca 30 mm, rays D4 and D5 in upper lobe and corresponding rays in lower lobe more or less produced to form short streamers; commonly even in rather large specimens only either streamer present, or both absent, fin then long and more or less truncate.

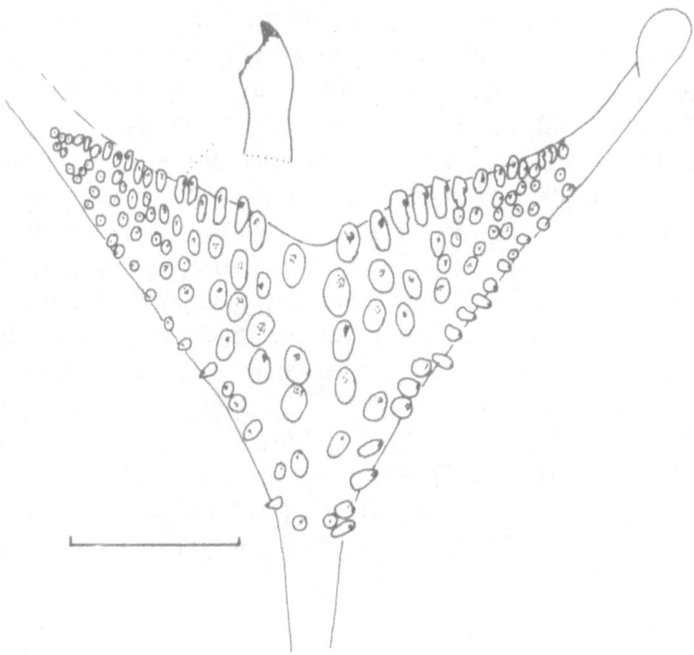

Fig. 39. *Apistogramma steindachneri*. Lower pharyngeal tooth plate in occlusal view and posterior tooth in medial view. Scale 1 mm for tooth plate, 0.5 mm for tooth. From RMNH 31016, 46.3 mm SL.

Jaw teeth: simple, conical, moderately (outer adsymphysial may be strongly) recurved. Outer and anterior teeth larger compared to inner and posterior. Two series in upper jaw, outer series alone continued laterally; 3 series in lower jaw, only outer series continued laterally. In the remaining material occasionally a few teeth in upper jaw forming a 3rd series in symphysial region, and occasionally only 2 series in lower jaw; one inner series in upper jaw may be extended laterally about half as far as outer series. Teeth on one side of upper/lower jaw counted in 6 specimens 40.5-61.9 mm: 31/26, 25/29, 26/29, 27/28, 25/31, 25/30.

Gill rakers: on first gill arch (in RMNH 30998, 29.7 mm) 4 short rakers on margin of epibranchial lobe, one in epi-/ceratobranchial angle, and 3 short but distinct ceratobranchial rakers; generally 2 or 3 upper ceratobranchial rakers distinct, additional rakers hardly distinguishable.

Tooth plates: lower pharyngeal bone as in Fig. 39 (from RMNH 31016, 46.3 mm). Anterior and lateral teeth simple, slightly broader subapically than near base. Posterior and medial teeth with rostroventrally extended base, bicuspid, with posterior slightly forwards curved cusp, and short obtuse anterior lower cusp. In posterior edge series, close to middle, teeth usually more or less tricuspid, with two small cusps on anterior edge and strong posterior cusp.

Counts. From 268 RMNH and USNM specimens: D. XIV.7(3), XIV.7.i(1), XV.-(5), XV.6(7), XV.6.i(11), XV.7(211), XV.7.i(3), XV.8(5), XVI.6(11), XVI.6.i(2), XVI.7(8), XVII.6(1). A -.-(1), III.-(2), III.6(144), III.6.i(43), III.7(77), III.7.i(1). Squ. long. 21(5), 22(227), 23(9), -(27). Cheek scale series -(9), 0(9; juvs), 1(4; juvs), 2(8; juvs), 3(150), 4(88). Gill rakers -(27), 1(4), 2(32), 3(159), 4(37), 5(8), 6(1). Lower pharyngeal tooth plate rakers -(86), 11(6), 12(19), 13(35), 14(55), 15(49), 16(13), 17(5). From 20 measured specimens: P. 11(2), 12(16), 13(2). Upper lateral line canals 11(2), 12(7), 13(5), 14(3), 15(2), -(1); upper lateral line total 14(4), 15(13), 16(2), -(1); lower lateral line canals 3(1), 4(6), 5(6), 6(5), 7(2); lower lateral line total 6(4), 7(13), 8(2), 9(1). Squ. caud. 3(1), 4(2), 5(4), 6(4), 7(1), 12(1), 15(1). Squ. prd. 8(1), 9(5), 10(8), 11(6). Squ. prv. 7(1), 8(1), 9(3), 10(11), 11(2), -(2). Squ. op. 8(1), 9(5), 10(5), 11(4), 12(1), -(4). Squ. sop. 2(1), 3(14), 4(4), -(1). Squ. iop. 2(10), 3(9), 4(1). From RMNH 31017(2), 31016(2), and 30996(11): Vertebrae 12+11(1), 12+12(14); supraneural 1(15). Cephalic lateralis pores distributed as shown in Fig. 40.

Colouration.

Description chiefly from RMNH 30998 specimens, 29.7 (male) and 31.9 (female) mm. Ground colour yellowish white, pure to dirty on abdominal sides, pure on abdomen medially and on chest; markings brown. Coun-

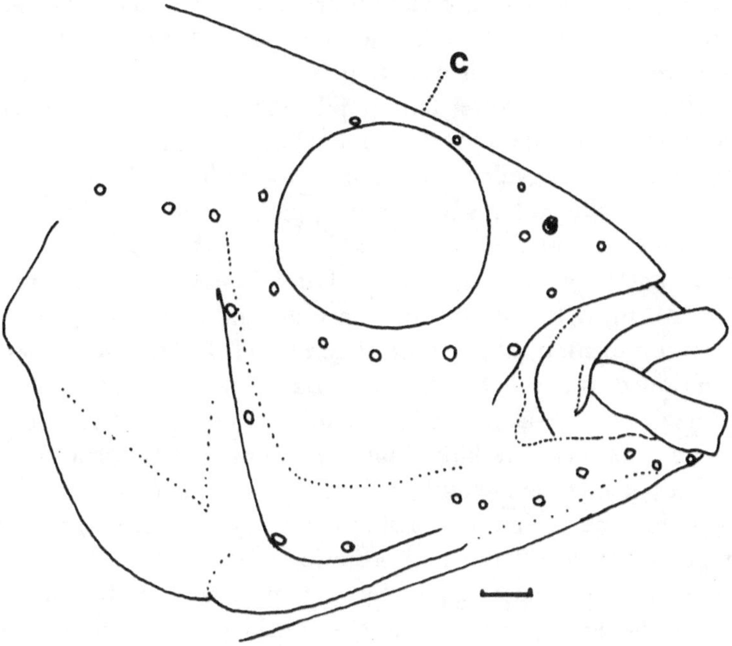

Fig. 40. *Apistogramma steindachneri*. Outline of head showing distribution of lateralis pores. *c*. coronalis pore out of view. Scale 1 mm. From NRM A82/1967122.3423, 35.9 mm SL.

tershaded, darkest on back and nape. Scales above lateral band narrowly dark rimmed. Below lateral band, especially above level of lower margin of caudal peduncle and anterior to anal fin base, a dark spot on base of each scale. In small specimens those scale spots may appear as a kind of abdominal striping, but these stripes are very thin and by far not as conspicuous as in, e.g., *A. ortmanni*; these spots are also often lacking. The dark scale-edging does not show in small specimens but is characteristic and conspicuous, forming a reticular pattern, in medium-sized and large specimens. Gill cover greyish, semitransparent; head anterior and ventral to suborbital stripe yellowish white. Lips greyish to yellowish white. Snout and nape greyish white.

Vertical bars not distinct but, as is characteristic of medium-sized specimens, with dispersed pigment. Narrowly along dorsal fin base, above lateral line, dark, continuously or as a series of indistinct spots making up dorsalmost portion of vertical bars. Bar 1 extending down to gill cover, Bars 2-3 to lateral band, Bar 4 to lateral band or anal fin origin, Bars 5-6 to lateral band or anal fin base, Bar 7 not apparent (across caudal peduncle in other specimens); broader than interspaces; pale brownish, above lateral band diffusely confluent, below, if apparent, with interspaces distinct. The bar pattern is very variable in distinctness, though never prominent. The lateral band may be slightly widened where crossed by bars, but usually it is not.

Preorbital stripe narrower than pupil. Suborbital stripe to junction of sub- and inter-opercula, of about pupil width, margins unsharp. Distinct supraorbital stripe, wider than pupil, to midline of nape; in large specimens usually not distinct, but apparent due to light interspace before Bar 1.

Postorbital stripe narrow, straight. Lateral band narrow (about a half scale deep), straight, with distinct margins; ending in Bar 6, with some pigmentation behind, above lower lateral line. In other specimens as a rule continued to Bar 7. Flank spot in Bar 3, from lateral band, tracing bar, dorsally to dorsal fin or lateral line. Flank spot dark brown and forming a prominent marking, but when continued to dorsal fin usually somewhat paler above lateral line; in young specimens spot restricted to lateral band and rather roundish. Variously developed light vertical zones immediately anterior and posterior to spot.

Although flank spot generally rather small and roundish or indistinguishable in juveniles, there is considerable variation in dorsalward extension of the spot in specimens over ca 20 mm. An extension to lateral line predominates, but varies within samples; no correlation to sex, size or locality found. In rare instances the spot reaches feebly below the lateral band. Three RMNH 31002 specimens have atypical, doubled and somewhat irregular Bars 4-6, and one specimen, a female, 26.2 mm, has a narrow 'extra lateral spot' in Bar 4 on the right side. Another female, RMNH 31015, 33.3 mm, also features an extra flank spot, but in Bar 2, on the right side. Abdominal stripes absent.

Pectoral spot simple, strong, faint or not appearing at all. Females with short black midventral stripe and genital papilla narrowly surrounded with black; males without abdominal markings. A few females have darkened chest. A large male, RMNH 30987, 51.2 mm, features darkened chest and midventral stripe.

Dorsal fin semitransparent, smoky, slightly darker towards base; anterior 2 membranes grey to black. Males usually with only first membrane distinctly darkened; both anterior membranes much more strongly pigmented in females. Light portions on soft part form 2-3 indistinct terminal spot stripes. Anal fin similar but not darkened anteriorly, and terminal spotting usually less distinct. Caudal fin like soft dorsal fin, with about 5-6 interspace-wide rather straight or slightly wavy cross stripes of darker dots except on marginal portions; these spots are nearly lost in large males, with greyish caudal fin; juveniles have distinct spots. Caudal spot variable in shape, round, ovate, or squarish, generally prominent and narrowly extended to fin margins; in juveniles connected with spots on upper and lower margins of fin base. Pelvic fin colourless to whitish in males; in females with dark outer margin of varying width.

Female, RMNH 31017, 33.2 mm, intensely pigmented and presumably showing female breeding pattern: very dark, but ground colour pale as usual, enhancing contrast in pigment intensity between dorsal and ventral regions; almost white along middle of chest and abdomen. Dark markings, and spot stripes in dorsal and anal fins prominent; lateral band and cross bars not intensified; dark scale margins and scale spots very faint. Lateral spot round, reaching little below lateral band, and up to lateral line. Midventral stripe from vent forwards to halfway to pelvic fins. Lower jaw dark grey. Dorsal fin anterior 2 membranes black; anal fin anterior 2 membranes and outer margin of fin black; pelvic fin 2 outer membranes black, but outer margin of spine white, inwards colourless. In another female, RMNH 20393, 33.2 mm, also suborbital and supraorbital stripes pronounced, and chest darkened. Other females, but no males, show similar intensification of the black markings.

Distribution (Fig. 37)

Apistogramma steindachneri is widely distributed in Surinam, collections ranging from the Suriname to the Corantijn systems, chiefly from the lowlands. Marowijne localities are imprecise, and the species is not expected to occur in the Marowijne except perhaps close to the mouth. Westward the species is well represented in collections from the lower Essequibo, Demerara and Mahaica (reported by Eigenmann 1912). The type locality is Georgetown, on the Demerara R. Collections from the Rupununi R., and adjacent R. Branco tributaries (Eigenmann 1912, Lowe-McConnell 1969) are tentatively identified as *A. steindachneri* (Kullander 1982a); however, a critical analysis of the *steindachneri* group material in the Rupununi should likely recognize more material of *A. rupununi*.

Local names

Krobia, Owroe wefi (Suriname R.; Richter & Nijssen 1980), Pikien Krobia (Surinam).

Notes

Apistogramma steindachneri is distinguished from all other *Apistogramma* species except *A. hippolytae* and *A. rupununi* by the dorsally extended lateral spot (flank spot) of adults, typically tracing Bar 3 to the lateral line or, fainter, to the dorsal fin. Males from ca 30 mm with marginal caudal fin rays produced or caudal fin long, truncate; males of *A. hippolytae* and

A. rupununi with subtruncate or rounded caudal fin. *Apistogramma steindachneri* is readily distinguished from *A. ortmanni*, the second Surinamese *Apistogramma* species, by, e.g., the possession of the flank spot, the caudal fin streamers of males, and the lack of abdominal stripes.

The flank spot, i.e. a dark spot tracing Bar 3 from the lateral band to or beyond the lateral line, is a synapomorphy of *A. steindachneri*, *A. hippolytae*, and *A. rupununi* (the *steindachneri* group; Kullander 1982a). In other *Apistogramma* species either there is no spot on the middle of the side or the spot is situated on the lateral band (termed lateral spot). As juvenile *A. steindachneri* display the latter condition, the two kinds of markings are probably homologous. The three *steindachneri* group species also have a very narrow lateral band compared to other *Apistogramma* species, probably the apomorphic condition.

Apistogramma steindachneri is the largest species in the genus, males reaching 65.3 mm SL in the wild. Under aquarium conditions it may attain 75.8 mm (ZMH 5682, holotype of *A. wickleri*). Females remain smaller; the largest female measured is 39.6 mm (RMNH 31015).

The syntypes of *A. steindachneri* are two adult males, 48.8 and 54.7 mm, from Georgetown, Guyana. The smaller specimen is in quite bad condition, the larger fine and agreeing with Regan's (1908) figure except for caudal fin shape. Both syntypes have the hind margin of the caudal fin damaged so that the shape is indeterminable, but it appears truncate rather than rounded, as at least in the smaller specimen marginal rays are slightly longer than the rest. Inappropriate restoration probably accounts for the rounded caudal fin of the fish figured by Regan (1908), which led Eigenmann (1912) to doubt the identification of his *A. steindachneri* material, and Ahl (1936) to describe Eigenmann's material and an aquarium specimen (ZMB 21147) with caudal fin streamers, as *A. ornatipinnis*, distinguished by the caudal fin shape.

Kullander (1980a) considered *Apistogramma ortmanni rupununi* Fowler (1914) a synonym of *A. steindachneri*. Reexamination of the type material, two adult males, ANSP 39347-39348, 37.6 (holotype) and 35.8 (paratype) mm, and also a large series from near Boa Vista on the R. Branco, suggests that it is a valid species distinguished from *A. steindachneri* by roundish-subtruncate caudal fin. *Apistogramma rupununi* differs from *A. hippolytae* especially in the smaller flank spot, reaching to the upper lateral line only, instead of to the dorsal fin base. The species was described from the Rupununi R. at 2-3° N, 50°20′W. These coordinates designate a line north of the Ilha de Maraca, on the coast of Brazil (Kanazawa 1966); 50 is probably a lapse for 59. This may be the species referred to as Zweifleck-*Apistogramma* by Kullander (1982a) and current aquarium literature (Schmettkamp 1982).

Apistogramma wickleri is based on some very large specimens held in aquarium and was distinguished from *A. steindachneri*, *A. ornatipinnis* and *A. cacatuoides* with reference to 'other head shape, other body measurements and other colouration', though the differences were not specified. The type series has the colour pattern and fin shape characterizing *A. steindachneri*. Because it consists of overgrown aquarium specimens, the material is not useful for comparison of morphometric data.

Hoedeman's (1951) Surinamese locality for the types of *A. cacatuoides* Hoedeman, an upper Amazonian species (Kullander 1986), probably resulted from confusion with a lot consisting of two male *A. steindachneri* with the same locality data and register number as are given for *A. cacatuoides* in the original description (ZMA 100.033). Two of three labels in the *A. steindachneri* bottle, both in ink, one giving the locality 'Dutch-Guiana (ditch near Paramaribo)', bear a manuscript name with 'Hoedeman 1950' as author. From one of those labels the part that had the second half of the register number written on it is torn off; the first half of the number is 100. The third label, in pencil, bears the name *Apistogramma cacatuoides* Hoedeman 1951, and on the reverse side 'Aant? Z.M.A. No. 100.033'. This suggests that there is some explanation for the particular but erroneous locality data of the *A. cacatuoides* types, although it remains unclear how the confusion may have come about. This *A. steindachneri* material is probably that actually reported by Hoedeman (1951) as *A. steindachneri*.

Apistogramma ortmanni (Eigenmann)

(Figs 37, 41-42)

Heterogramma ortmanni Eigenmann 1912, p. 506 (Erukin [holotype FMNH 53801]; Packeoo Fall, Kangaruma; Gluck Island; Konawaruk; Savannah Landing; Rockstone), pl. LXVIII, fig. 1 (monochrome photo of holotype).

Surinamese records

Apistogramma ortmanni; Kullander 1982a, p. 192 (distributional data from ms of present paper). — Kullander 1982c, p. 71 (distribution; compared with *A. gossei*).

Material

About 385 specimens, all from the Corantijn R. system.

Description

Unless otherwise noted, the qualitative data are from an adult male, RMNH 31023, 34.8 mm SL (Fig. 41). Morphometric data are from

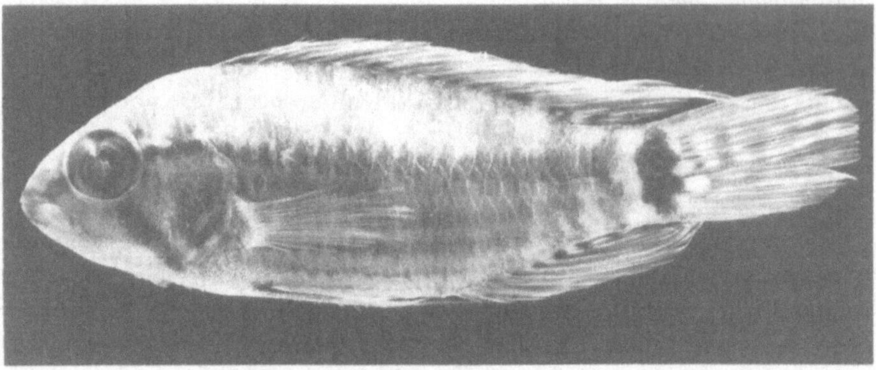

Fig. 41. *Apistogramma ortmanni*. Male, 34.8 mm SL, RMNH 31023, from Kabalebo R. tributary near Avanavero Vallen.

RMNH 31022-31023. Largest specimen examined a male 41.2 mm SL (USNM 225831); largest female examined 32.1 mm SL (RMNH 31023).

Shape: body moderately elongate, broadest and deepest just behind gill cover. Dorsal contour straight horizontal anteriorly, caudad straight sloping, with a strong terminating curvature; meeting about straight, horizontal caudal peduncle contour at an angle. Chest contour nearly straight, sloping, becoming horizontal just in advance of pelvic fin bases; abdominal contour almost straight, slightly deeper posteriorly; anal fin base contour about straight, upwards sloping, meeting straight, sloping caudal peduncle contour at an angle.

Head: moderately elongate, longer than deep, slightly shorter than body depth. Predorsal contour about straight ascending, with decurved snout and gradually levelled nape contour. Prepelvic contour straight sloping. Snout rounded, contours equal, but dorsal convex, ventral straight. Ascending processes of premaxilla reaching about 1/4 of orbit; angle about 45°. Angle of maxilla 55-60°; tip exposed, reaching to slightly beyond anterior margin of orbit. Lower jaw distinctly longer than upper, but equal anteriorly. Orbit diameter 37.6-42.6% ($x = 39.9 \pm 0.23$, $n = 23$) of head length; situated entirely in dorsal and chiefly in rostral half of head, tangential. Preoperculum entire in all specimens.

Scales: length of exposed part of body scales greatest on middle of flanks, but slightly lesser marginally, size decrease gradual (to 2/3 on caudal peduncle). Predorsal scales cycloid ventrally to line between dorsal fin origin and posterodorsal margin of orbit. Chest not naked anteriorly. Cheek completely scaled except for rostroventral corner;

scales cycloid. Most of opercular scales ctenoid. Subopercular scales in one series, cycloid. Interopercular scales cycloid. Circumpeduncular scale series 16. Upper lateral line to below penultimate dorsal fin spine, with 1-3 subserial pores in 17 of 22 specimens (RMNH 31022-31023); lower line to above third anal fin ray; a caudal canal or pore in 10 of 22 specimens (RMNH 31022-31023). Dorsal and anal fins naked. Basal 1/3 of caudal fin with scaled; proximally scales ctenoid, distally scales cycloid.

Fins: dorsal fin spines subequal in length from the 4th or 5th, the last longest. Lappets reaching slightly beyond spine tips, not produced; pointed in males, rather rounded in females. Soft part pointed, not produced, 2nd ray reaching to about middle of caudal fin. Soft anal fin similar, 3rd ray longest, to 1/3 of caudal fin. Soft dorsal fin in females rounded, reaching at most to slightly beyond caudal spot; in males usually pointed in the larger specimens, often produced, reaching middle or 2/3 of caudal fin. Soft anal fin always slightly shorter than soft dorsal fin, otherwise similar. Last dorsal fin and anal fin spines in males longer than in females (last dorsal fin spine length 16.5-18.3% of SL in females, 18.7-22.1% of SL in males); last anal fin spine length relatively shorter in large females (from 19.6 to 17.4% of SL), but always shorter than in males of corresponding SL (last anal fin spine length 18.8-20.1% of SL, no allometry apparent).

Pectoral fin rounded, 4th ray longest; in intact 30.8 mm specimen two dorsalmost and two ventralmost rays simple, remainder branched; reaching to about genital papilla (may reach to anal fin origin in other specimens).

Pelvic fin acuminate; slightly produced longest first ray, reaching to end of spinous anal fin base, as in other adult males. Females with pointed, but not produced pelvic fin, reaching to about anal fin origin, slightly shorter than in males of comparable size. Caudal fin rounded, dorsal lobe slightly broader than ventral; 2 marginal rays simple, inner rays with 2-4 branches. Clearly rounded caudal fin in females, but in adult males with characteristic tendency to subtruncate margin; longer in males than in females of comparable SL (36.9-40.9% of SL vs. 33.3-35.5% of SL).

Jaw teeth: slender, simple, conical with slightly recurved tips. Outer and anterior larger compared to inner and posterior. In 3 series (exceptionally a short inner 4th, in one specimen) in upper jaw; 24-31 teeth on one side in outer series counted in 5 specimens 30.8-37.5 mm; outer series continued laterally, next inner nearly as long, 3rd restricted to symphyseal region. In 3 series in lower jaw; 25-28 teeth on one side in outer series in 5 specimens 30.8-37.5 mm; outer series continued laterally, inner series short.

Tooth plates: lower pharyngeal bone examined in one specimen (RMNH 31023, 34.4 mm; Fig. 42). Anterior and lateral teeth simple, medially and posteriorly becoming bicuspid, with long rostrocaudally extended bases and a strong pointed, somewhat rostrad curved posterior cusp and a minor rounded anterior cusp. Gill rakers commonly absent from outer surface of first ceratobranchial. Always a short but distinct simple raker in angle between epi- and cerato-branchial; if a ceratobranchial raker present, it is nearly always minute and scarcely distinguishable from epithelial papillae. Pharyngeal tooth plate rakers also rather weak.

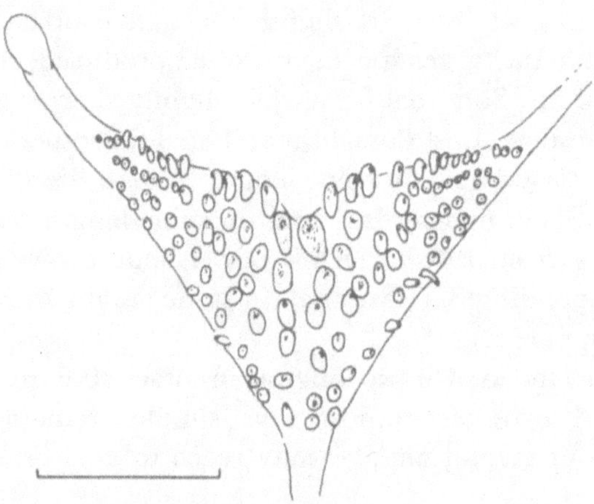

Fig. 42. *Apistogramma ortmanni*. Lower pharyngeal tooth plate in occlusal aspect. Scale 1 mm. From RMNH 31023, 34.4 mm SL.

Counts. From RMNH 31022-31024,31027: D. XV.6(3), XV.7(48), XVI.6(4), XVI.7(1). A. III.6(44), III.6.i(7), III.7(5). Squ. long. 21(9), 22(47). Cheek scale series 2(2), 3(54). Gill rakers 0(30), 1(26). From RMNH 31022-31023: P. 12(23). Upper lateral line canals 9(1), 10(5), 11(8), 12(4), 13(3), 14(1); upper lateral line total 12(1), 14(7), 15(13), 16(1); lower lateral line canals 2(1), 3(7), 4(9), 5(4), 6(1); lower lateral line total 6(5), 7(14), 8(3). Squ. caud. 4(7), 5(7), 6(3), 7(4), 8(1). Squ. prd. 8(3), 9(10), 10(7). Squ. prv. 8(2), 9(9), 10(8). Opercular scales 7(1), 8(1), 9(6), 10(5), 11(7), 13(1). Subopercular scales 3(17), 4(5). Interopercular scales 1(6), 2(10), 3(6). Lower pharyngeal tooth plate rakers 8(1), 9(12), 10(7), 11(1). From RMNH 31023: Vertebrae 12 + 11(1), 12 + 12(21); supraneural 1(22). Cephalic lateralis pore configuration as in *A. steindachneri*, see Fig. 40.

Colouration (description chiefly from the RMNH 31023)

Ground colour yellowish, markings brown. Males with silvery ventral gill cover and lateral chest region; pure ground colour on chest and ventrally on abdomen in females, but ventral extension of flank pigmentation varying. Countershading not pronounced. Snout and lips greyish to whitish. Ventral parts of head pale. Operculum greyish; otherwise gill cover and cheek pale brownish yellow.

Seven brownish vertical bars across side, distinct but faint; reaching ventrally to operculum (Bar 1), to line between upper margin of pectoral axilla and anal fin origin (Bars 2-4), anal fin base (Bars 5-6), and ventral margin of caudal peduncle (Bar 7). Bars somewhat broader than interspaces; not particularly intensified where crossing lateral band, but darkened close to dorsal fin base into which extended as dark spots, one on each interradial membrane.

Preorbital stripe dark brown. Dark brown spot on chin and anteriorly on lower lip. Suborbital stripe dark brown, of pupil width, straight, with even margins, ventrad to outer margin of gill cover at junction of sub- and inter-opercula. Supraorbital stripe to middle of nape, feebly darker than vertical flank bars, margins not well defined. Postorbital stripe narrow, narrower than diameter of pupil, continued posteriorwards by lateral band, which about scale deep and usually with even or only slightly uneven margins, reaching to Bar 7, dark brown or but little darker than vertical bars. No lateral spot.

Four abdominal stripes: (1) from pectoral spot to Bar 7 or slightly shorter, (2) from lower margin of pectoral axilla to about Bar 6, (3) from between pectoral axilla and pelvic fin base to about origin of anal fin base. In males usually a 4th abdominal stripe between pelvic fin base and about origin of anal fin; usually as a thin streak, whereas the other abdominal stripes made up of diffuse more or less confluent dots, one on each scale center. RMNH 31024-31026 specimens faded, abdominal stripes not or but vaguely apparent.

Pectoral spot simple, faint. Around base of pelvic fin a dark spot. A midventral stripe in both sexes, but in males short and faint; in females usually dark brown and reaching forwards from vent to pelvic fin bases or halfway, usually continued behind genital papilla by a spot of same colour, which occasionally expanded laterad. Those few males with a 'vent spot' show it only diffusely. No chest blotch.

Fins transparent, slightly smoky, except for pectoral fin, which clear. Anterior 2 dorsal fin membranes black. In males 2-3 terminal spot stripes across 3 last dorsal fin membranes, interspaces clear; these stripes at best indicated in females. Anal fin with 3-4 distinct terminal spot stripes in males, 2-3 faint in females. Caudal spot squarish to ovate, dark brown, usually extended dorsally and ventrally to margins of fin. Behind caudal spot in males up to 6 wavy broad vertical stripes of spots across middle portion, interspaces clear; females with immaculate caudal fin, or traces of dots appearing. Pelvic fin dark basally on outer part and along first membrane; darker in females than in males; inner portion of fin clear or faintly smoky.

Distribution (Fig. 37)

In Surinam *A. ortmanni* appears to be restricted to the Corantijn drainage, where it has been collected at only a few sites, in two regions (Avanavero Vallen; near mouth of Lucie R.). Numerous collections from other parts of Surinam, especially in the Suriname drainage, do not contain *A. ortmanni*, and the Corantijn may represent the eastern limit of the geographical range of the species.

Outside Surinam *A. ortmanni* has been encountered in Guyana and Venezuela. Guyana material examined is from Kuribrong, Amatuk, Erukin in the Potaro R. system; Rockstone, Gluck Island, Konawaruk, along the Essequibo R., and near Karanambo in the Rupununi system. Eigenmann (1912) also lists paratypes from Savannah Landing and Kangaruma in the Potaro system and "Packeoo Fall" in the Rupununi system. Four poorly preserved Venezuelan specimens (USNM 194106) from Rio la Clarita, Edo. Bolivar, are undoubtedly *A.ortmanni*; we have not been able to trace the locality on maps available, however, and there is no further information on the collection in USNM files (R. P. Vari, *in litt.*). Other Venezuelan material is from the R. Cuyuni drainage (MHNLS 4186, 4189).

Notes

Apistogramma ortmanni is is a species of the *A. regani* group, characterized by dark anterior dorsal fin membranes, absence of lateral spot, fully scaled prepelvic region, well-developed caudal spot restricted to caudal fin, and males without produced anterior dorsal fin lappets and caudal fin elongations. It is distinguished from all other species of that group except *A. regani* Kullander, especially by the prominent abdominal stripes. It is easily separated from *A. steindachneri* by having abdominal stripes but no flank spot.

Apistogramma ortmanni is most similar to *A. regani*, a species known only from the vicinity of Manaus, Central Amazonia. It differs especially in being more slender than *A.regani*, which has a body depth of 33.5-41.0% of SL, $\bar{x} = 36.7 \pm 0.30$ (n = 26, SL 23.9-34.5 mm; from material used in Kullander 1980a), and less boldly marked, with fainter vertical bars and caudal fin stripes not reaching dorsal or ventral fin margins at any size. Both species feature a dark spot at the base of the pelvic fin, but it is more intense in *A. ortmanni*. Differences from *A. gossei* listed by Kullander (1982c) include fewer gill rakers on ceratobranchial 1 (0-1 versus 2-3, but Guyanan and Venezuelan material has more gill rakers, 1-5, mode 3, n = 13) and lower pharyngeal tooth plate (8-11 versus 10-13). *Apistogramma ortmanni* also has more distinct abdominal stripes, larger caudal spot and spotted caudal fin.

GUIANACARA n. gen.

Type species

Guianacara owroewefi n. sp.

Diagnosis

Small (to 107 mm SL) South American cichlids with 6 preopercular and 5 dentary lateralis canal foramina, stochastic predorsal squamation, supraneurals 1-2, vertebrae 13 + 14 (occasionally 13 + 13). Similar to *Geophagus*, *Satanoperca* and *Biotodoma* but without lobe like expansion of first epibranchial; particularly similar to *Acarichthys* in having an anteroventral entopterygoid process, wide flattened lateral commissure, wide dorsal lamina on parasphenoid and basisphenoid, narrow fenestra between coracoid and cleithrum, and inner branch of first ray of pelvic fin longer than outer branch in adults. Uniquely distinguished by a dorsal shelf on urohyal, hyperdeveloped nasad process of lateral ethmoid, and mediad curved anterior margin of distal postcleithrum.

Description

The following list summarizes descriptive data common to all species. Body laterally compressed, in frontal aspect narrowly oval, deepest anteriorly, tapering chiefly due to slope of dorsal fin base contour. Greatest depth at region of pelvic fin bases, but ventral outline less arched than dorsal, highest at dorsal fin origin. Head short, deep, laterally compressed. Mouth terminal, low. Tip of maxilla well exposed. Jaws equal anteriorly. Lips fleshy, of about equal width, fold of upper interrupted symphysially, laterally curled inwards; lower lip fold discontinuous symphysially, laterally attaching to caudal-medial side of upper lip fold. Orbit round. Interorbital area convex in adults. Operculum deep; depth:length ratio about 2:1 in large specimens, longer (3:2) in smallest specimens.

Scales ctenoid except under gill cover, in median prepelvic and two parallel series bilaterally (may be ctenoid close to pelvic fins), around vent and genital papilla, anteriorly or on most of cheek; variably cycloid or ctenoid on gill cover. Scales largest anteriorly on flanks; slightly reduced in size in marginal areas (2/3 on caudal peduncle); squ. long. 25, rarely 24 or 26. Pectoral axilla naked except for thin, mostly cycloid basal scales. Predorsally typically 2 admedian series of ctenoid scales overlapping along midline, frequently one or a few minute cycloid scales intercalated medially near dorsal fin; irregularities occur; anterior scales of about flank scale size, caudad size decrease to about half that size. Cheek scales in 4-6 series, the dorsalmost longest, converging towards mouth angle; covering area between infraorbitals, lacrimal, preoperculum and fleshy mouth angle; one series behind orbit; posterior scales about half size of anterior flank scales, size decreases forwards. Preoperculum naked. Subopercular (largest nearly flank scale size) and interopercular scales in single series; opercular scales in two oblique series paralleling lower bone margin, largest scales about size of anterior flank scales. Upper lateral line in 3rd horizontal scale series above that containing the lower. Accessory lateral lines on caudal fin between rays D3 and D4, and rays V4 and V5, but both or either may be wanting. Fins naked except caudal fin, with distally deeply concave proximal scaled area.

Dorsal fin with 15 spines (rarely 14 or 16) and 10 rays (rarely 9, 11, or 12), spines increasing in length to 5th, shorter behind, but about last 3 spines again increasing in length. Anal fin with 3 spines and usually 8 rays (rarely 7 or 9). Pectoral fin hypaxial, asymmetrical, 4th ray longest. Pelvic fin with first ray longest, in adults inner branch slightly longer than outer, 2nd ray outer branch nearly as long. Caudal fin hind margin subtruncate or truncate.

Supraneurals 1-2; vertebrae 13 + 13-14; no ribs on caudal vertebrae; well-developed parhypural spine reaching to dorsal margin of ventralmost hypural. Four, occasionally 3 procurrent and 8 principal rays in each caudal fin lobe. Lacrimal with 4 lateralis canal foramina; succeeded by 4 infraorbitals to sphenotic, 2nd of which has a middle foramen. No microbranchiospines. Epibranchial 1 relatively short and broad, with medial arms of equal length; pharyngobranchial 1 widened ventrally. 1-4 epibranchial, 5-7 ceratobranchial gill rakers externally on first gill arch, median ceratobranchial rakers denticulate, all attaching to gill arch rather than basal skin over gill filaments. Ceratobranchial 4 with 3-4 tooth plates. No gill rakers on lower pharyngeal tooth plate (Fig. 46), which relatively wide, with stout medioposterior teeth. Preoperculum occasionally serrated, supracleithrum and posttemporal entire. Jaw teeth unicuspid, conical, not or slightly recurved.

Included species

Acara geayi Pellegrin, 1902
Guianacara owroewefi n. sp.
Guianacara sphenozona n. sp.
Guianacara oelemariensis n. sp.

Etymology

Guiana-, for the Guianas, principal area of distribution, and *-acara*, from Acara, the Tupi name for cichlids. The gender is feminine.

Distribution

The Guianas from the Cuyuni drainage eastward to the Oyapock drainage; in the Amazon basin in the upper Branco and Trombetas drainages; in the Orinoco basin in the Caroni drainage.

Guianacara n. subgen.

Type species

Guianacara owroewefi n. sp.

Diagnosis

Nominotypical subgenus of *Guianacara*, distinguished from *Oelemaria* by having 2 instead of 1 supraneural, a prominent dark stripe across the middle of the side (may be reduced to a dorsal spot in adults) instead of a dark midlateral spot on the side, and produced anterior dorsal fin lappets in adults. This group includes all known *Guianacara* species except *G. oelemariensis*.

Oelemaria n. subgen.

Type species

Guianacara oelemariensis n. sp.

Diagnosis

Distinguished from nominotypical *Guianacara* by having 1 instead of 2 supraneurals, a large dark spot posteriorly on the side instead of a vertical stripe, and in not having the dorsal fin lappets produced at any size.

The name *Oelemaria* is derived from the area of distribution of the single species and is to be treated as feminine in gender.

Notes

Only one *Guianacara* species has been recognized in the literature, viz. *G. geayi*, placed in the genus *Aequidens* by most authors, but provisionally referred to *Acarichthys* by Kullander (1980a). *Guianacara geayi* and its congenerics differ considerably from *Aequidens* species, and especially in the 5 dentary lateralis foramina, the caudal fin shape, relatively small scales, small ctenoid predorsal scales, dorsal fin shape, and basic colour pattern. The rationale for earlier workers grouping *G. geayi* with *Aequidens* appears to have been primarily the lack of an epibranchial lobe, in combination with 3-spined anal fin and absence of any notable morphological peculiarities.

Guianacara bears a strong superficial resemblance to *Geophagus*-like cichlids, and especially in the deep preorbital region, but also in the comparatively small scales, emarginate caudal fin and colour pattern. The genus differs from *Geophagus*-like cichlids in lacking an epibranchial lobe, and resemblances may be referable to plesiomorphies. *Guianacara* differs from *Cichlasoma*-like cichlids in having 5 instead of 4 dentary lateralis canal foramina. No character has been found that would indicate a close relationship to cichlids with an epibranchial lobe. Assignation of *Guianacara* species to *Aequidens* seems particularly ill-founded. To a major extent difficulties in discerning plausible relationships of *Guianacara* are due to a general lack of information on the anatomy of South American cichlids. Nearly all anatomical data on *Guianacara* discussed below are from *G. owroewefi*; very strong superficial similarity between species of the genus suggest that these data may be generally valid. *Guianacara* appears to be a very distinctive genus which presents a suite of plesiomorphic states. Derived characters suggest a close relationship to *Acarichthys*, the only other *Geophagus*-like cichlid without an epibranchial lobe.

Small scales, emarginate caudal fin, position of accessory caudal fin lateral lines, more than three procurrent caudal fin rays and, variably developed, preopercular serrations, are plesiomorphic features of *Guianacara* shared with geophagine and most other cichlids, but rare or unrecorded among cichlasomines.

(A) A well-developed calyx (cf. Greenwood 1985a, p. 146).

(B) A suture between the metapterygoid and anterior flange of the hyomandibula (cf. Greenwood 1985a,p. 145). In most South American cichlids the hyomandibula and metapterygoid develop interdigitating processes in this region.

(C) Toothed ceratobranchial 4. The usual condition among South American cichlids; the 3-4 tooth plates in *Guianacara* represent a reduction compared to long series in, e.g., *Cichla*.

(D) A long parhypural spine. The development of a parhypurapophysis is very variable among cichlids, though a long spine as in *Guianacara* seems to represent the usual percoid condition.

(E) A prominent premaxillary foramen. This character is insufficiently surveyed among cichlids; the foramen is often absent (e.g. in *Cichla* and *Satanoperca*).

(F) Suturally united vomer and mesethmoid. Dissociation of these elements is relatively rare among South American cichlids; see Greenwood (1985b) for discussion of this character.

(G) Double palatine-lateral ethmoid articulation. This character is discussed by Greenwood (1985a). In young *G. owroewefi* the posteromedial process of the palatine is not nearly in contact with the lateral ethmoid and there is no facet on the lateral ethmoid. In large *G. owroewefi* there is a small but distinct convex facet on the ventral edge of the lateral face of the lateral ethmoid articulating with the dorsomedial process of the palatine. Other cichlids which maintain a single, anterior, palatoethmoid articulation as adults, including all geophagines with epibranchial lobe, usually have much reduced posteromedial palatine wing.

(H) A small cartilage posterior to the cartilaginous basibranchial 4. This cartilage is relatively common among South American cichlids but is not reported from any other fishes. We have recorded it in *Satanoperca jurupari*, *Biotodoma cupido*, *Hypselecara temporalis* (Kullander 1986), *Apistogramma eunotus*, *Acarichthys heckelii*, *Teleocichla cinderella* (Kullander 1988). Its presence is intragenerically variable, being absent in *Apistogramma diplotaenia* (Kullander 1987) and *Teleocichla proselytus* (Kullander 1988). Although this cartilage may be taken to be a synapomorphy of a group of cichlids its mosaic distribution renders its polarity assessment problematic.

(I) Five dentary lateralis canal foramina. This is the common number in cichlids. In many the foramen below the middle of the rostral process of the anguloarticular is lost (fourth from anteriormost), as in *Cichlasoma*-like cichlids, *Acaronia* and some dwarf cichlids (*Apistogramma* species, *Taeniacara*).

Derived characters of *Guianacara* include:

1. The wide dorsal shelf on the urohyal. Modally in cichlids, the urohyal has the shape of an inverted Y in mesial cross-section, and most shape variation is found in the dorsal spine and in the relative width of the medial lamina and ventral wings (e.g., Kullander 1987, 1988). There is some variation in the width of the dorsal shelf, probably correlated with the development of the branchial musculature, though the *Guianacara* condition represents the extreme among cichlids so far studied.

2. The anteroventral projection of the entopterygoid. In other cichlids, except *Acarichthys*, the ventral margin of the entopterygoid is simply rounded.

3. The prominent nasad process of the lateral ethmoid. All cichlids with a deep snout, like *Guianacara*, also have a prominent nasad process, but in *Guianacara* the process is relatively longer than in any other cichlid.

4. The wide, flattened and mesially angled lateral commissure, forming a part of the posteroventral orbital wall. In most South American cichlids the lateral commissure is a simple straight rod and the trigeminofacialis chamber lies medial and anterior to the commissure whereas in *Guianacara* it lies essentially posterior to the commissure. The *Guianacara* condition otherwise corresponds with the tilapiine type of pars jugularis distinguished by Greenwood (1986), which occurs in other Neotropical cichlids.

5. The high dorsal crest of the parasphenoid and basisphenoid appear to be unique to *Guianacara* and *Acarichthys*.

6. Broad-tipped pelvic fin with inner branch of first ray longer than outer branch. In most cichlids the outer branch of the first ray is longer than the inner branch.

7. Wide coracoid and resulting reduced fenestra between coracoid and cleithrum. The shape of the pectoral girdle is little studied in cichlids. The very narrow fenestra between the coracoid and cleithrum is, however, a notable feature of both *Guianacara* and *Acarichthys*.

8. Mediad curved anterior edge of distal postcleithrum ventrad to the widening anterior to the overlapping part of the proximal postcleithrum. In other cichlids the distal postcleithrum is a lateromedially flattened element, thick axially and with variously shaped marginal lamina. The dorsal end of the postcleithrum is similar in shape in *Acarichthys* and *Guianacara*, but there is no anterior mediad curvature in *Acarichthys*.

Characters of doubtful polarity, or at most trivial, include the shape of the dorsal part of the distal postcleithrum, which features posterior and anterior projections somewhat as in *Cleithracara*, and which is similar in *Acarichthys*. Variation in postcleithral shape among cichlids has been given taxonomic emphasis (Stiassny 1982), and there is considerable variation between genera. The absence of microbranchiospines is a common loss feature. Naked dorsal and anal fins, reduced caudal fin lateral lines, relatively indistinct and often absent preopercular serrations, are derived conditions shared with many other cichlids.

The jaw morphology, specifically the long ascending processes of the premaxilla conform to the produced snout involving also a relatively long ethmovomerine region and deep lacrimal, and actually may be best characterized with reference to the most conspicuous feature of the produced snout, i.e. the nasad process of the lateral ethmoid (character 3). The teeth are subprocumbent and the gradual symphysiad size increase striking, suggesting an invertebrate picker; which is in contrast to the substrate sifting habit of *Geophagus*-like cichlids with similar snout shape but reduced (*Satanoperca*) or unspecialized jaw dentition. Trophic behaviour of *Guianacara* species is still unstudied; the relatively stout pharyngeal teeth suggest that relatively hard objects are prominent in the diet.

The provisional assignation of *Guianacara* species to the genus *Acarichthys* (Kullander 1980a) is supported by synapomorphies of these genera, but the groups are obviously quite distinct morphologically.

Acarichthys is monotypic with *A. heckelii* (Müller & Troschel), which occurs in the Amazon and Essequibo drainages (Kullander 1986). *Acarichthys* anatomy remains to be studied in detail. Preliminary data indicate overall similarity to *Guianacara* including plesiomorphic features C-I listed above. *Guianacara* and *Acarichthys* also agree in general shape and colour pattern features, e.g. in having a black spot anteriorly on the dorsal fin (otherwise only in *Papiliochromis*, *Apistogrammoides*, *Biotoecus* and some *Apistogramma* species).

Acarichthys has microbranchiospines on the external face of the three posterior gill arches, 14+15 vertebrae, abdominal hemal arches, serially sutured pectoral radials, few dorsal fin spines (12-14), and differs also in having a less developed calyx and in lacking a hyomandibular-metapterygoid suture (cf. characters A and B). The small interarcual cartilage in *Guianacara* is similar to that of *Cichlasoma*-like cichlids, whereas *Acarichthys* has a longer interarcual cartilage, in contact with epibranchial 1 and resembling the condition in *Geophagus*. *Acarichthys* has a single supraneural.

Characters 2, 4, 5, 6, and 7 are similar in *Acarichthys* and *Guianacara*, although in *Acarichthys* the anteroventral projection of the entopterygoid is relatively shorter and the lateral commissure is not as wide. The urohyal has a strengthened dorsal edge, without lateral projection (cf. character 1) and the nasad process of the lateral ethmoid although prominent, is closer to the modal cichlid appearance than to the produced condition in *Guianacara*.

The possibly derived feature of *Guianacara* shared with *Cichlasoma*-like cichlids that may be important is the much reduced interarcual cartilage, which is more or less elongate and articulated with the epibranchial in most *Geophagus*-like cichlids (Kullander 1987), with *Papiliochromis* exceptional by having a reduced *Cichlasoma*-like interarcual cartilage.

Guianacara oelemariensis and remaining *Guianacara* species obviously represent different phyletic lineages and are distinguished subgenerically.

Oelemaria is hypothesized as maintaining ancestral colour pattern, with a midlateral blotch which is a feature common to South American cichlids, and short dorsal fin lappets. The single supraneural, although not unique among cichlids, is an advanced feature compared to nominotypical *Guianacara*.

Nominotypical *Guianacara*, with two supraneurals, primitive to cichlids, have developed a unique flank colour pattern. Produced anterior dorsal fin lappets, characterizing nominotypical *Guianacara* species, occur in other geophagines, viz. *Apistogramma* and *Papiliochromis*, but as dorsal fin morphology of these is otherwise quite different the produced dorsal fin lappets of *Guianacara* are tentatively considered an autapomorphy of that subgenus.

In addition to the three Surinamese species, we recognize *G. geayi* in the Oyapock and Approuague river systems, as a *Guianacara* species. It still needs a more detailed study, but is conveniently distinguished from congenerics by colour pattern, especially the pointed flank stripe wider and uniformly pigmented, larger dorsal fin dots extending onto the spinous portion, and only young with anterior dorsal fin membranes blackish; also the dorsal profile of adults appears comparatively steeper. Material from the Trombetas, Caroni, Branco and Essequibo drainages consists to a large extent of juveniles or old, not well-preserved specimens of the subgenus *Guianacara*. The specific diagnoses below do not take that extralimital material in account. It appears likely that study of suitable material from those areas will reveal additional species in the genus. Eigenmann (1912) described the Guyana form, as *Aequidens geayi*; Steindachner (1915a) figured and described material from the R. Branco, identified as *Acara geayi*.

Guianacara owroewefi n. sp.

(Figs 43-65)

Surinamese records

Acara sp.? Klinckowström 1892, p. 104 (listed [NRM ZISP506, Surinam]).
Aequidens geayi; (?) Popta 1914-1917, p. 31 (colour decription; local names). — (*pt.*) Boeseman 1952, p. 198 (listed: Surinam, (Upper Saramacca?) [RMNH 18495]; Janbasi-gado (about 4°N) [RMNH 18494]; local names) — Boeseman 1956, p. 196 (listed: Gansee, Surinam River [RMNH 21421]). — Richter & Nijssen 1980, pp. 124, 125 (listed; Brokopondo reservoir; biomass; local name).
Acara geayi; (*pt.*) Puyo 1949, p. 239 (brief description, bibliography).

Holotype (Fig. 43)

ISNB 749. A male, 58.7 mm SL. Guyane française, R. Maroni system, R. Marouini below first rapids. 23 November 1969. Leg. Léopold III & J.-P. Gosse (Mission Guyane française, Sta 10).

Diagnosis

Distinguished from from *G. geayi* by comparatively small dots on dorsal fin, from both *G. geayi* and *G. sphenozona* by having the vertical flank

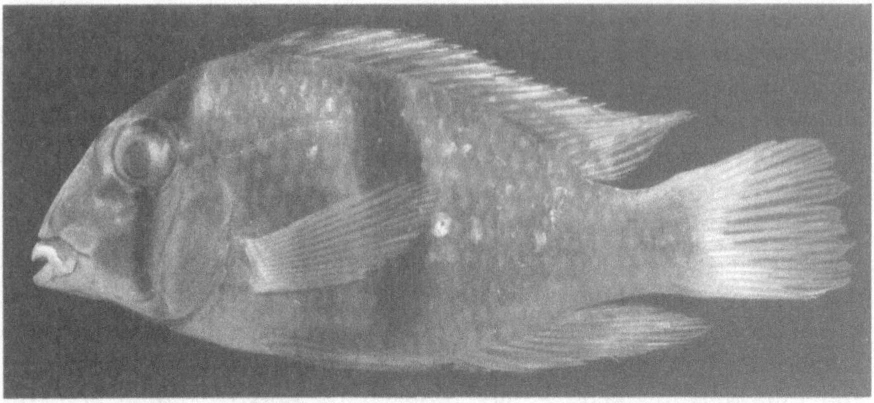

Fig. 43. *Guianacara owroewefi*, holotype, male, 58.7 mm SL, ISNB 749, from the R. Marouini.

stripe strongest on and just below the upper lateral line, and, in particular by a black blotch on anterior dorsal fin membranes also in adults.

Material

Holotype and over 1800 paratypes from the Marowijne, Suriname, Saramacca and Coppename river systems. Three specimens, 40-51 mm SL, out of NRM A82/1967052.3420 cleared and stained with alizarin red and alcian blue.

Description

Based on the holotype, and a 69.3 mm topoparatype, variation mainly from ISNB 748, RMNH 31041 (Fig. 44) + NRM A83/1964302.3014 (10, 62.3-75.9 mm, Gran Kreek) and RMNH 31029 (83, 20-47 mm, Feddiprati). Dorsal and anal fin, squ. long. and gill raker counts were obtained from 172 specimens; these were also checked for preopercular serrations; other counts are from 25 specimens measured.

The largest specimen examined is a male, 107.3 mm SL (ZMA 106.721); the largest female is 76.7 mm SL (NRM ZISP506).

Shape: moderately elongate, cross section outline in anterior trunk region with flattened chest and narrow, nearly keeled nape. Predorsal contour very steep, slightly curved in large specimens; ventral head contour straight, much less sloping.

Head: snout from above narrow, elongate, narrowly rounded; lateral outline acuminate, with steep dorsal profile and feebly sloping ventral profile. Small specimens more elongate, with less deep head and not so steep, rather equal head contours. Mouth width about equalling interorbital; maxilla not reaching to orbit in large specimens, to orbit in juveniles. Orbit in middle of head length, and entirely in upper half of head, most above operculum level, below forehead contour. Preopercular serrations in 11 of 172 specimens examined for this character, along free vertical margin, either as a series of distinct broad serrae or as a few hardly distinguishable rudiments.

Scales: Squ. long. 24(2), 25(168), -(2). Predorsally an admedial series of about 11-12 scale pairs crosswise back to dorsal fin origin, frequently one or a few very small cycloid scales intercalated medially near dorsal fin. Chest scales about half the size of anterior flank scales.

Cheek scales in 4(1), 5(20) or 6(4) series; those anterior to orbit cycloid. Opercular scales in holotype, aside from patch of 6 small cycloid or ctenoid scales over upper margin, in 3 oblique series paralleling lower margin; cycloid except for a few dorsal scales. Upper subopercular scales cycloid, lower ctenoid. One or 2 small cycloid and ctenoid scales on interoperculum next to preopercular corner. In other specimens usually 2 opercular scale series; interopercular scales usually all cycloid, but may also be mostly ctenoid.

Modal lateral line count 18/9; upper with 16(1), 17(5), 18(16), 19(3), lower with 2 pored + 7(1), pored + 7(1), 7(3), 8(7), 9(11), 10(2) scales, pores in small specimens only. Upper rising for about 5 scales, then descending slowly, in line with dorsal margin of caudal peduncle, at 4 1/2 (initially) to 1 1/2 (terminally) scales distance from dorsal fin; overlapping lower by 2-4 (usually 3) scales, extending to below 2nd to 6th dorsal fin ray. Lower line forward to above origin of soft anal fin, continued on caudal fin by one (2) or two (23) scales in slightly downwards inclined succession; marginal branches with up to 4 (upper) or 5 (lower lobe) canals, but both (in young) or either may be missing.

Caudal fin inner scales ctenoid, outermost cycloid, marginal squamation extending to middle, median to 1/3 of fin length.

Fins: dorsal fin with short first spine, length half that of last in small specimens, 1/3 to 2/5 in large specimens, relative spine length increasing to 5th (or 4th and 5th subequal), behind shorter to about 10th from which subequal, last 3, or only the last, again longer, last shorter than 5th (specimens ca 40-75 mm radiographed); first spine inserted above middle or end of upper margin of operculum. Anterior lappets pointed, long, posterior truncate, not much beyond spine tips; 2nd to 8th distinctly produced, 3rd or 4th longest, length beyond spine corresponding to about half spine length. Small specimens (ca 30 mm) also have distinctly pointed anterior lappets, but prolongation distinct only in large specimens (from about 60 mm SL). Soft fin rounded or pointed in small specimens, pointed in large specimens; 4th or 5th ray longest, reaching at most to end of caudal fin (large specimens). Soft anal fin similar, usually slightly shorter, 4th ray longest. D. XIV.10(5), XV.9(5), XV.10(157), XV.11 (4), XVI.9(1). A. III.7(4), III.8(160), III.9(8).

Fig. 44. *Guianacara owroewefi*, paratype, male, 69.8 mm SL, NRM A83/1964302.3014, from near mouth of Gran Kreek, Suriname R. drainage.

Caudal fin subtruncate with deep straight hind margin; truncate in young; one large specimen, 88 mm SL, with ray D7 slightly longer than the rest. Pectoral fin with tip broadly rounded; extending to above one of anal fin spine bases; with 13(2), 14(18), 15(5) rays. Pelvic fin pointed, but rather broad, 2nd ray not much shorter than longest first ray; few, large specimens have slightly produced first ray, reaching at most to base of 5th anal fin ray; small specimens with rounded or pointed fin, to about middle of spinous anal fin base.

Jaw teeth: in 15 specimens, 56.3-69.8 mm (RMNH 31041 + NRM A83/1964302.3014, holotype, and ISNB 748): upper jaw 9-18, median 14.5, lower jaw 19-23, median 20, in outer hemiseries; inner series 2 to 3 in upper, 3 to 4 in lower. From ISNB 748, 69.8 mm, with wide open mouth: Teeth of outer series much larger than those of inner series in the upper jaw, less in lower jaw, outer teeth of upper jaw distinctly longer than those of lower jaw. Outer series well separated from inner series. Lower jaw dental arcade rounded; anteriormost teeth in outer series subprocumbent; outer series extends nearly to end of lower jaw margin, one inner series long, but inner teeth chiefly restricted to a transverse band anteriorly in jaw. Upper jaw dental arcade more broadly rounded; slightly less than half of jaw margin without teeth posteriorly (large specimens otherwise with more restricted edentuolous portion), a narrow band of inner teeth anteriorly, one inner series about as long as the outer.

Gill rakers: an obtuse ceratohyal raker like papilla. Rakers on external surface of first gill arch short, reaching to base of raker next below or shorter, simple, conical. Two epibranchial, one in angle. In 172 specimens counted, 5(7), 6(135), 7(28), -(2) ceratobranchial rakers, lowermost minute. In 69.3 mm specimen (Fig. 45) external cerato-

Fig. 45. *Guianacara owroewefi*. First gill arch in lateral aspect, to show short gill rakers and prominent transverse folds. Outlines of epi- and cerato-branchials indicated by spaced dotted lines. Scale 1 mm. From ISNB 748, 69.3 mm SL.

branchial rakers extend by a mediad-caudad (upper) or mediad (lower) ridge, papilliferous on upper and medial surfaces, connecting with a similar ridge (though more knob like) from corresponding internal raker. Inner ceratobranchial rakers obtuse and short. Three tooth plates on 4th ceratobranchial, with 6, 8 and 11 pointed teeth respectively. Pharynx relatively smooth.

Tooth plates: lower pharyngeal tooth plate studied in one specimen only (NRM ZISP506, 66.6 mm SL; Fig. 46). Anteriormost teeth subconical, with anterior slightly recurved tip; along lateral margins simple, subconical; increasingly strong posteriorwards, bases rostrocaudally elongated; medioposterior teeth very strong, broad-tipped, with posterior obtuse cusp (submammiform); along posterior margin and in corners with strong forwards curved cusp and, variably, an anterior smaller cusp or merely a rounded slight expansion.

Vertebrae: 13 + 13(1), 13 + 14(11).

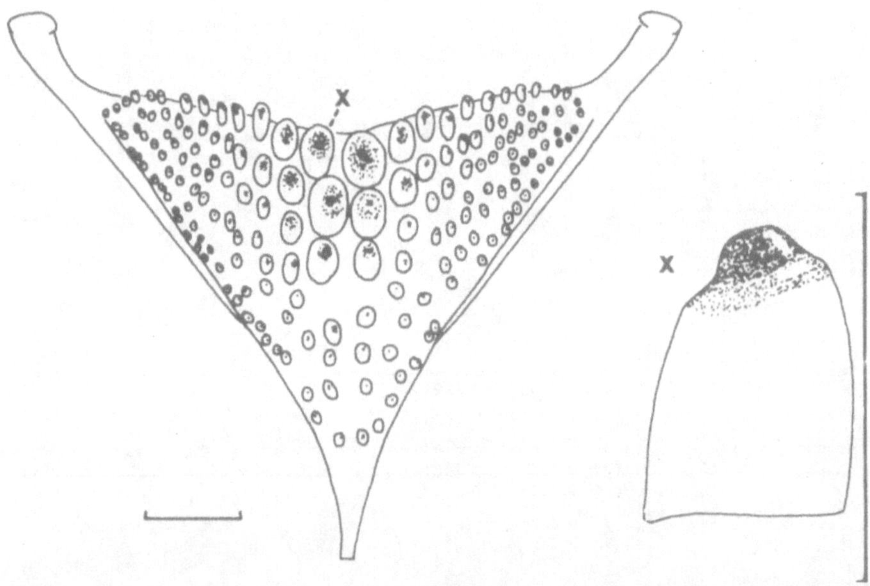

Fig. 46. *Guianacara owroewefi*. Lower pharyngeal tooth plate in occlusal view (scale 1 mm) and posterior series tooth (x) in medial aspect (scale 0.5 mm). From NRM ZISP506, 66.6 mm SL.

Osteology

Description based on three cleared and stained specimens, 40-51 mm SL; detailed data and figures all from the 51 mm specimen unless otherwise stated.

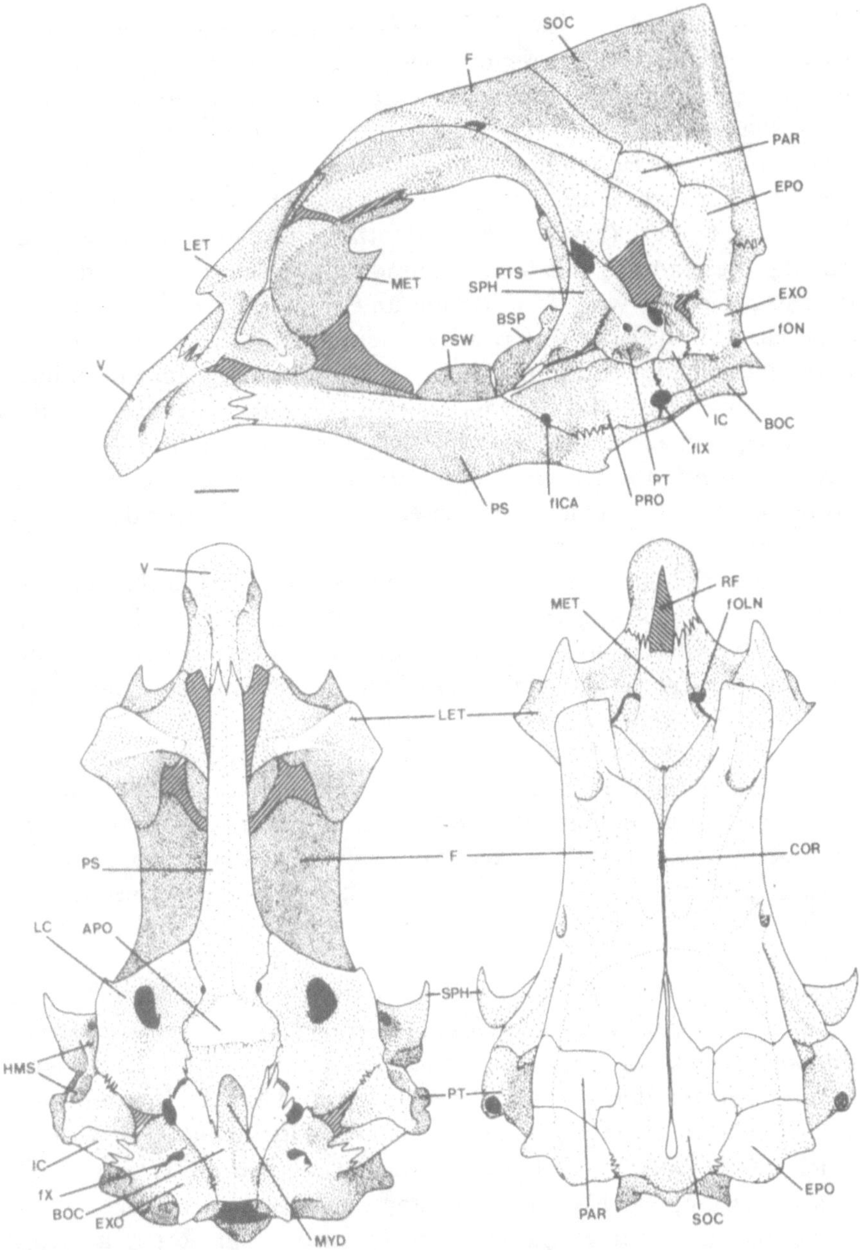

Fig. 47. *Guianacara owroewefi*. Neurocranium in approximately lateral (upper), ventral (lower left) and dorsal (lower right) aspect. From NRM A82/1967052.3420, 51 mm SL. Scale 1 mm. Cartilage striped; ethmoid cartilage passing over lower edge of lateral ethmoid not shown. *APO* pharyngobranchiad apophysis of parasphenoid, *BOC* basioccipital, *BSP* basisphenoid, *COR* coronalis foramen, *EPO* epioccipital, *EXO* exoccipital, *F* frontal, *HMS* hyomandibulad sockets, *fICA* foramen of internal carotid artery, *fIX* foramen of glossopharyngeal nerve, *fOLN* foramen of olfactory nerve, *fON* foramen of occipital nerve, *fX* foramen of vagus nerve, *IC* intercalar, *LC* lateral commissure, *LET* lateral ethmoid, *MET* mesethmoid, *MYD* posterior myodome opening, *PAR* parietal, *PRO* prootic, *PS* parasphenoid, *PSW* intraorbital parasphenoid wing, *PT* pterotic, *PTS* pterosphenoid, *RF* rostral fenestra (cartilage), *SOC* supraoccipital, *SPH* sphenotic, *V* vomer.

The neurocranium (Figs 47-48) is moderately deep and wide, with produced ethmovomerine region; neurocracial length (ncl) 14.7 mm, depth 66%, width 58%, interorbital width of frontals 29%, horizontal orbital diameter 45% of ncl.

The vomer, lateral ethmoid and mesethmoid form a prominent portion of the neurocranium. The vomer has an elongate, narrow cartilage filled rostral fenestra (Fig 47, RF), and forms strong sutures dorsally with both the mesethmoid and lateral ethmoid. The lateral ethmoid has a laterally projecting wing forming the anterior orbital wall which, ventrally, has a strong rostrad curvature, ending in a rostroventrad directed lacrimad facet; the ventral margin of the lateral wing bears a prominent ligament facet medially. The nasad process of the lateral ethmoid is hyperdeveloped and resembles a spinous process, and a thin wing ledge forms between it and the the lateral part of that anterior projection that sutures the lateral ethmoid to the vomer and which also bears the anterior palatoethmoid articulation facet.

The intraorbital median lamina of the mesethmoid has a characteristic shape with a dorsal projection that coossified with the anterior margins of the ventral intraorbital wings of the frontal; and a dorsocauadal notch into which enters the olfactory part of the connective tissue membrane separating the brain and orbital chambers.

Relatively high frontal crests rise from the level of the second frontal lateralis canal foramen and unite where the coronalis canal ends, the latter raised to an oblique rostrodorsad course, continuing to rise and succeeded caudad by the supraoccipital crest. The dorsal margin of the supraoccipital crest is shallowly grooved anteriorly and posteriorly, the groove appearing filled with thick connective tissue. The frontoparitetal crest is moderately wide.

The parasphenoid bears a moderately deep ventral median lamina, and a low anterior dorsal medial lamina projecting between the lateral ethmoids; the narrow lateral wings form a support for the lateral commissure. Intraorbitally, a high median crest (PSW in Fig. 47) with rounded margin rises just anterior to the basisphenoid. The anterior pedicle of the basisphenoid bears a medial laminar expansion on the anterior margin, which anteriorly contacts the posterior medial elevation of the parasphenoid. Posteriorly the parasphenoid forms a strong transverse apophysis for the pharyngobranchiad articular surface and ends in an incomplete roof for the very wide posterior myodome opening and laterally sutures with the basioccipital. The lateral awning and the hyomandibulad shell are both very shallow. In the junction of the exoccipital, prootic and basioccipital there is a very large opening for the glossopharyngeal nerve. The vagus nerve foramen is shielded by a

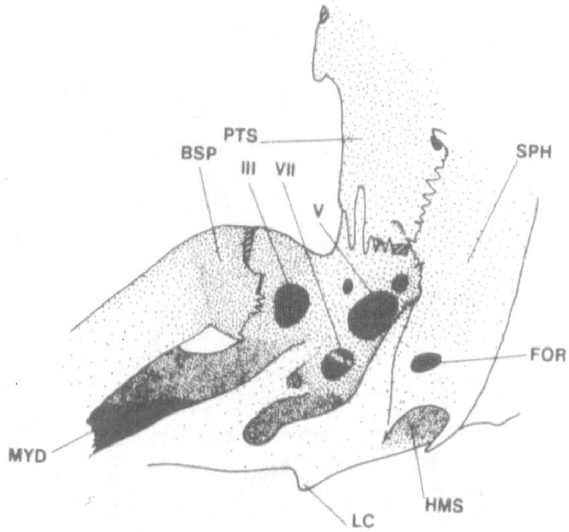

Fig. 48. *Guianacara owroewefi*. Approximately anterolateral aspect of pars jugularis. From NRM A82/1967052.3420, 51 mm SL. Cartilage striped. *BSP* basisphenoid, *FOR* foramen (artefact), *III* oculomotor nerve foramen, *LC* lateral commissure, *HMS* hyomandibulad socket, *MYD* myodome opening, *PTS* pterosphenoid, *SPH* sphenotic, *V* trigeminus nerve foramen, *VII* hyomandibular trunk nerve foramen.

laminar cover with blunt projections, contributing a calyx-like shape to the opening. The lateral wing of the sphenotic is wide and bears a long anterolateral spinous process.

The lateral commissure (Fig. 48) is wide and compressed, forming a continuation of the posteroventral orbital wall; it is slightly expanded medially on the lateral margin and there forming a blunt minor process. The trigeminofacialis chamber is hidden in lateral view by the lateral commissure and in anterolateral aspect appears as in in a narrow recess betweeen the anterolateral wall of the prootic and the dorsally stout basal portion of the lateral commissure.

The suspensorium (Fig. 49-51) is characterized by relatively deep ento- and meta-pterygoids, a well-formed calyx, a narrow but strong suture between the hyomandibula and metapterygoid. The preopercular lateralis canal has 5 foramina, the anguloarticular two and the dentary five.

The palatine (Fig. 50) has wide anteroventral and posteromedial wings and a low lateral margin. The posteromedial wing ends dorsally in a thickened tip and the lateral ethmoid-palatine ligament inserts dorsally

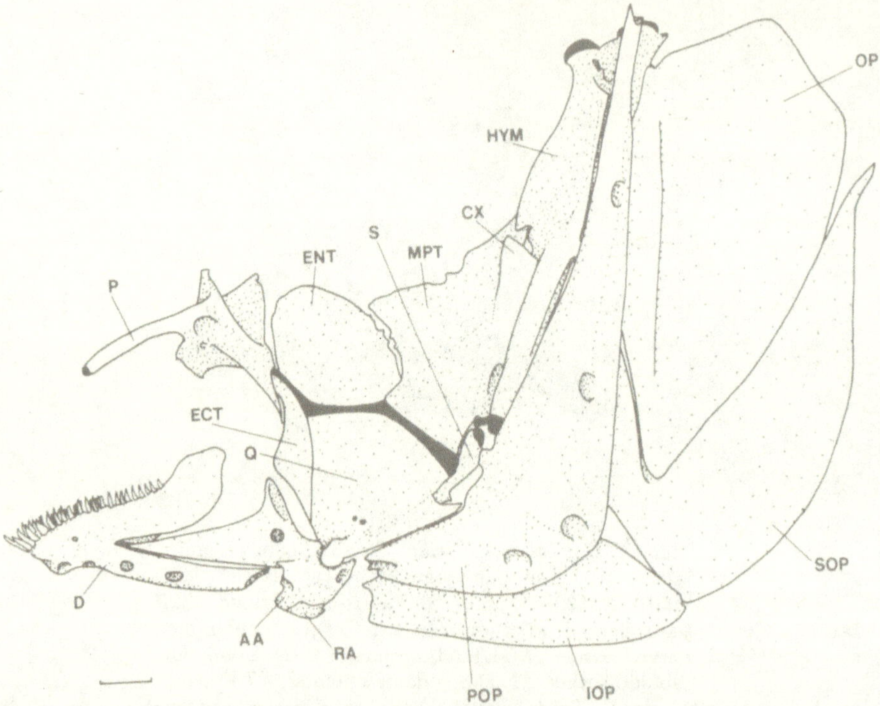

Fig. 49. *Guianacara owroewefi*. Left side suspensorium, operculars and lower jaw in lateral aspect. From NRM A82/1967052.3420, 51 mm SL. Scale 1 mm. Cartilage black. *AA* anguloarticular, *CX* calyx, *D* dentary, *ECT* ectopterygoid, *ENT* entopterygoid, *IOP* interoperculum, *HYM* hyomandibula, *MPT* metapterygoid, *OP* operculum, *P* palatine, *POP* preoperculum, *Q* quadrate, *RA* retroarticular, *S* symplectic, *SOP* suboperculum.

on the caudal margin. In the alizarin material there is no articulation between the palatine and the lateral ethmoid but in dissected larger specimens articulation facets are well-developed. The dorsal process bears a prominent articulation facet and anterolateral to it a facet for the palatopalatine ligament. The entopterygoid has an anterior ventrad extending tongue-shaped projection overlapping the quadrate (Fig. 51).

The ligament system was not surveyed in detail; the following orocranial ligaments were observed, however, numbers corresponding to those of Stiassny (1986). A short palato-maxillary ligament (X), a band-like palatopalatine ligament (XII), a very thin ethmomaxillary ligament (XI), a proximally wide band like ventral palatomaxillary ligament (IV?) extending from the palatine shaft to the ligament process on the lateral face of the maxilla (Fig. 50, LIPR), a medial palatovomerine ligament (V), an anterior palatovomerine ligament (VI; Fig. 50, LI PAL-V), and

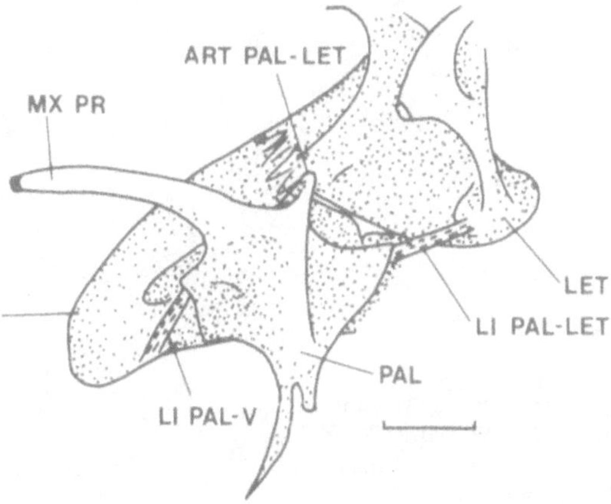

Fig. 50. *Guianacara owroewefi*. Left side palatine and associated ethmovomerine region in lateral aspect. From NRM A82/1967052.3420, 51 mm SL. Scale 1 mm. *ART PAL-LET* palatoethmoid articulation, *LET* lateral ethmoid, *LI PAL-LET* palatineateral ethmoid ligament, *LI PAL-V* palatovomerine ligament, *MX PR* maxillad process, *PAL* palatine, *V* vomer.

a posterior palatoethmoid ligament (Fig. 50, LI PAL-LET). There is no anteroventral palatomaxillary ligament; the palatolacrimal ligament runs between the anterodorsal corner of the lacrimal and a proximal point on the maxillad process of the palatine.

The premaxilla (Figs 52-53) has long slender ascending processes, which is nearly twice the length of the alveolar process. A canal penetrating the ascending process opens with a large foramen on the rostral face in the region between the medial and articular processs, and

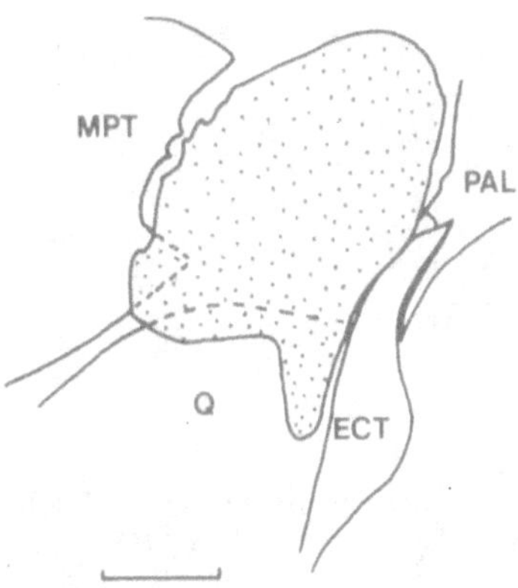

Fig. 51. *Guianacara owroewefi*. Medial aspect of part of suspensorium showing entopterygoid (shaded) and adjacent bones. From NRM A82/1967052.3420, 51 mm SL. Scale 1 mm. *ECT* ectopterygoid, *MPT* metapterygoid, *PAL* palatine, *Q* quadrate.

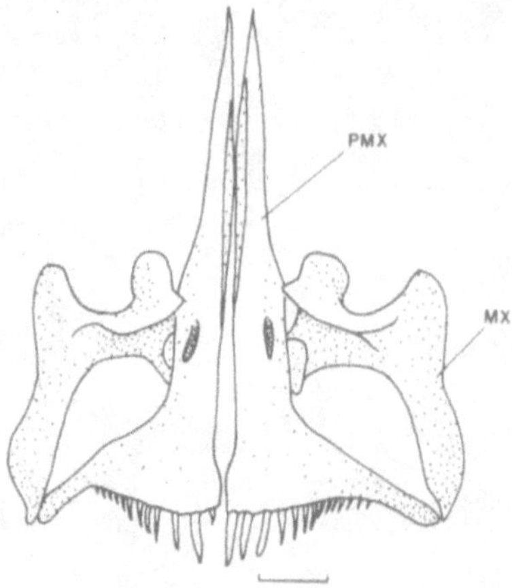

Fig. 52. *Guianacara owroewefi*. Upper jaw in anterior view. From NRM A82/1967052.3420, 51 mm SL. Scale 1 mm. *MX* maxilla, *PMX* premaxilla.

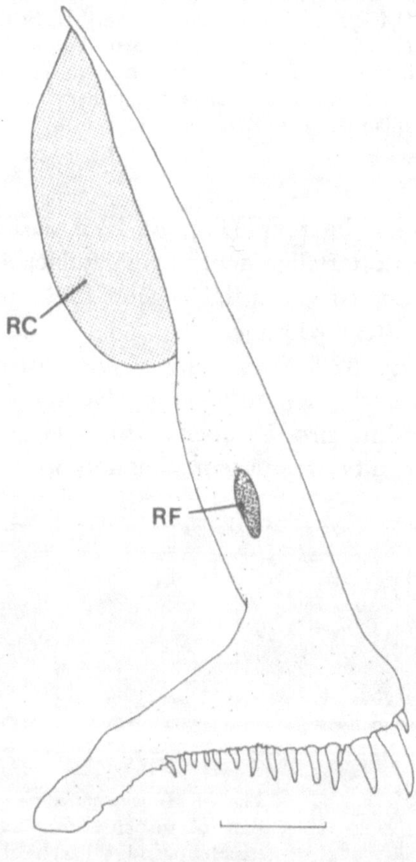

Fig. 53. *Guianacara owroewefi*. Right premaxilla in anterolateral aspect. From NRM A82/1967052.3420, 51 mm SL. Scale 1 mm. *RC* (shaded) rostral cartilage, *RF* rostral foramen.

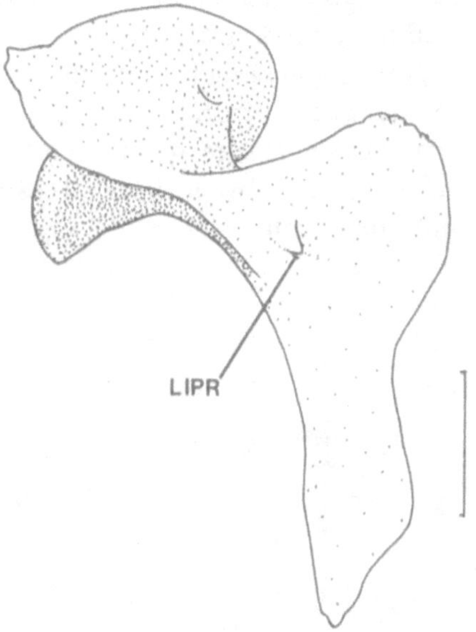

Fig. 54. *Guianacara owroewefi*. Left maxilla in lateral aspect. From NRM A82/1967052.3420, 51 mm SL. Scale 1 mm. *LIPR* process for primordial and ventral palato-maxillary ligament.

a smaller foramen on the posterior face opens near the base of the ascendning process.

The maxilla (Figs 53-54) is relatively short, with prominent caudal wing and short ventral process. A prominent spine on the lateral face near the head forms the attachment for both the primordial ligament and the ventral palatomaxillary ligament.

The gill arch skeleton (Figs 55-57) shows no particular features of interest. The first pharyngobranchial is spatulate in rostral aspect, with only a short dorsal portion cylindrical. The first epibranchial is short and stout, with the medial processes of about equal length and little divergent, the uncinate process much broader than the anterior process, however. Between the tip of the uncinate process of epibranchial 1 and the dorsal margin of the pharyngobranchial 2 lies a small interarcual cartilage of irregular shape, which is not in direct contact either with the epibranchial or the pharyngobranchial. The second epibranchial has two ventral processes, the medial of which articulates with pharyngobranchial 2 and the more lateral of which ends in an endochondral cartilaginous tip. Epibranchial 3 has a minute process on the rostral face slightly medial to the level of the ventral ligament process. A little posterior to the cartilaginous 4th basibranchial is a minute globular cartilage forming a 5th element in the copula series. Ceratobranchial 4 has a moderate medial process and bears 4 tooth plates on the dorsal face (Fig. 57).

The urohyal (Fig. 58) has a short dorsocaudad directed dorsal spine and the moderately developed medial lamina and ventral wings. Along the dorsal margin of the medial laminar portion the bone is strengthened and forms a wide shelf, nearly half the width of a ventral wing.

The hyoid arch (Fig. 59) shows no peculiarities. It is relatively slender with ceratohyal sutured on both sides. The 5th branchiostegal ray articulates on the medial side of the anterior ceratohyal.

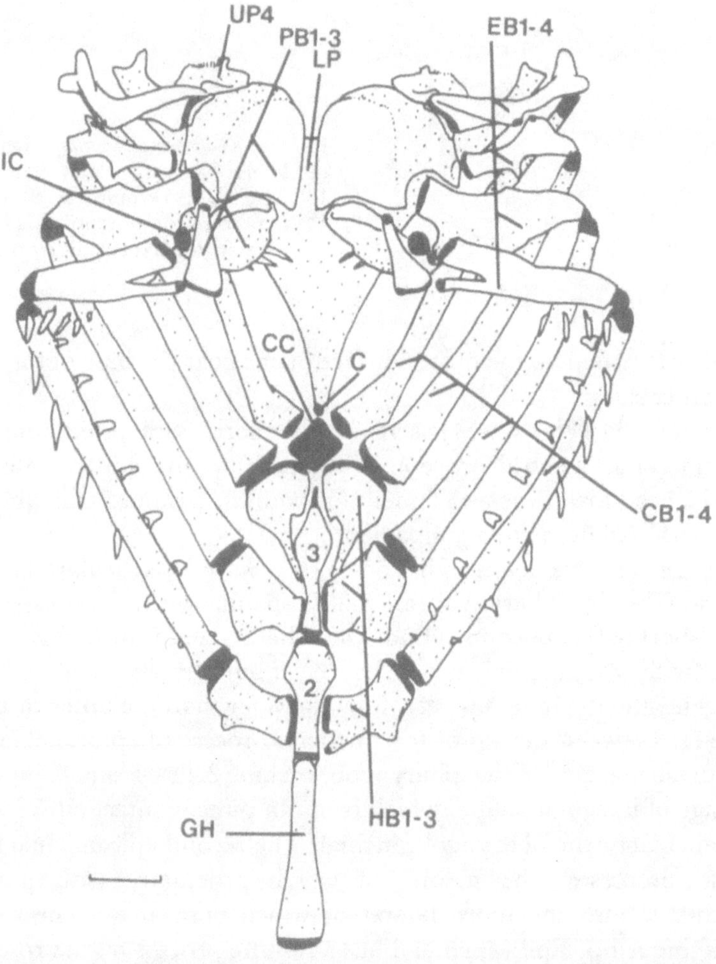

Fig. 55. *Guianacara owroewefi*. Gill arch skeleton in dorsal aspect. From NRM A82/1967052.3420, 51 mm SL. Scale 1 mm. Cartilage black. *CB1-4* ceratobranchials 1-4, *C* basibranchial 5?, *CC* central cartilage, *IC* interarcual cartilage, *EB1-4* epibranchials 1-4, *GH* glossohyal, *HB1-3* hypobranchials 1-3, *LP* lower pharyngeal tooth plate, *PB1-3* pharyngobranchials 1-3, *UP4* fourth upper tooth plate, *2-3* basibranchials 2-3 (basibranchial 1 beneath *GH*).

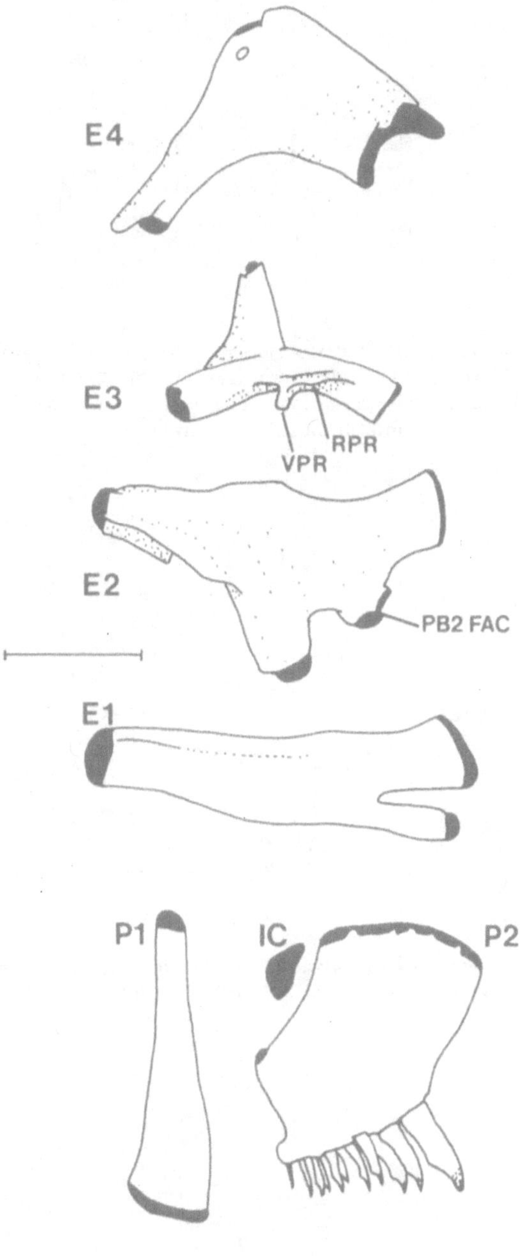

Fig. 56. *Guianacara owroewefi*. Dorsal gill arch elements in rostral aspect. From NRM A82/1967052.3420, 51 mm SL. Scale 1 mm. Interarcual cartilage and pharyngobranchial 2 shown in articulated position, remaining elements separate. Cartilage black. *E1-4* epibranchials 1-4, *IC* interarcual cartilage, *P1-P2* pharyngobranchials 1 and 2, *PB2FAC* pharyngobranchiad 2 articulation facet, *RPR* rostral process on epibranchial 3, *VPR* ventral ligament process of epibranchial 3.

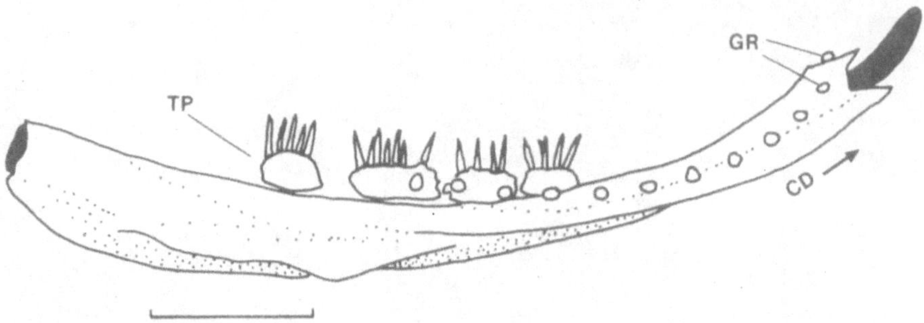

Fig. 57. *Guianacara owroewefi*. Ceratobranchial 4 in medial aspect. From NRM A82/1967052.3420, 51 mm SL. Scale 1 mm. Cartilage black. *CD* and arrow show caudad direction, *GR* gill rakers (only posterior two of series indicated), *TP* tooth plate (only anteriormost of four indicated).

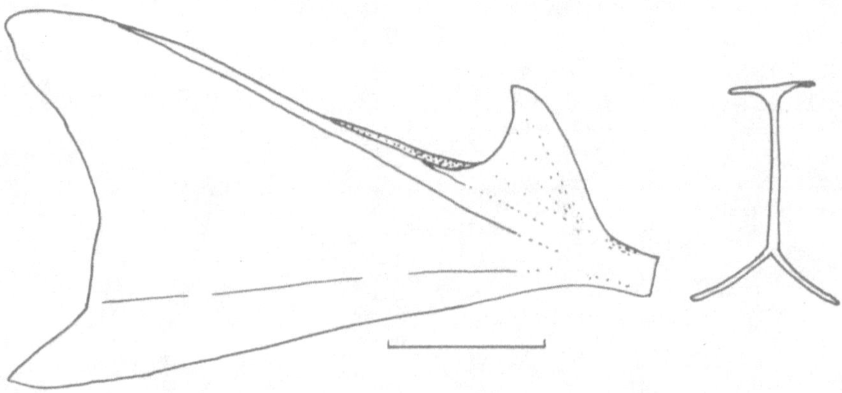

Fig. 58. *Guianacara owroewefi*. Urohyal in lateral and (to the right) cross section at approximately middle. From NRM A82/1967052.3420, 51 mm SL. Scale 1 mm.

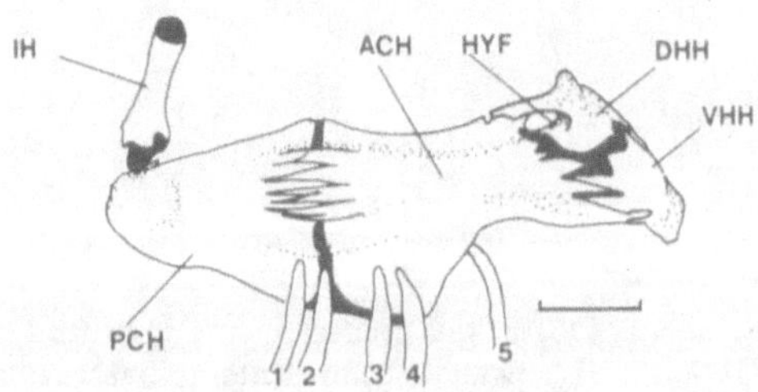

Fig. 59. *Guianacara owroewefi*. Right side hyoid in lateral aspect. From NRM A82/1967052.3420, 51 mm SL. Scale 1 mm. Cartilage black. *ACH* anterior ceratohyal, *DHH* dorsal hypohyal, *IH* interhyal, *HYF* hyoid foramen, *PCH* posterior ceratohyal, *VHH* ventral hypohyal, *1-5* branchiostegal rays.

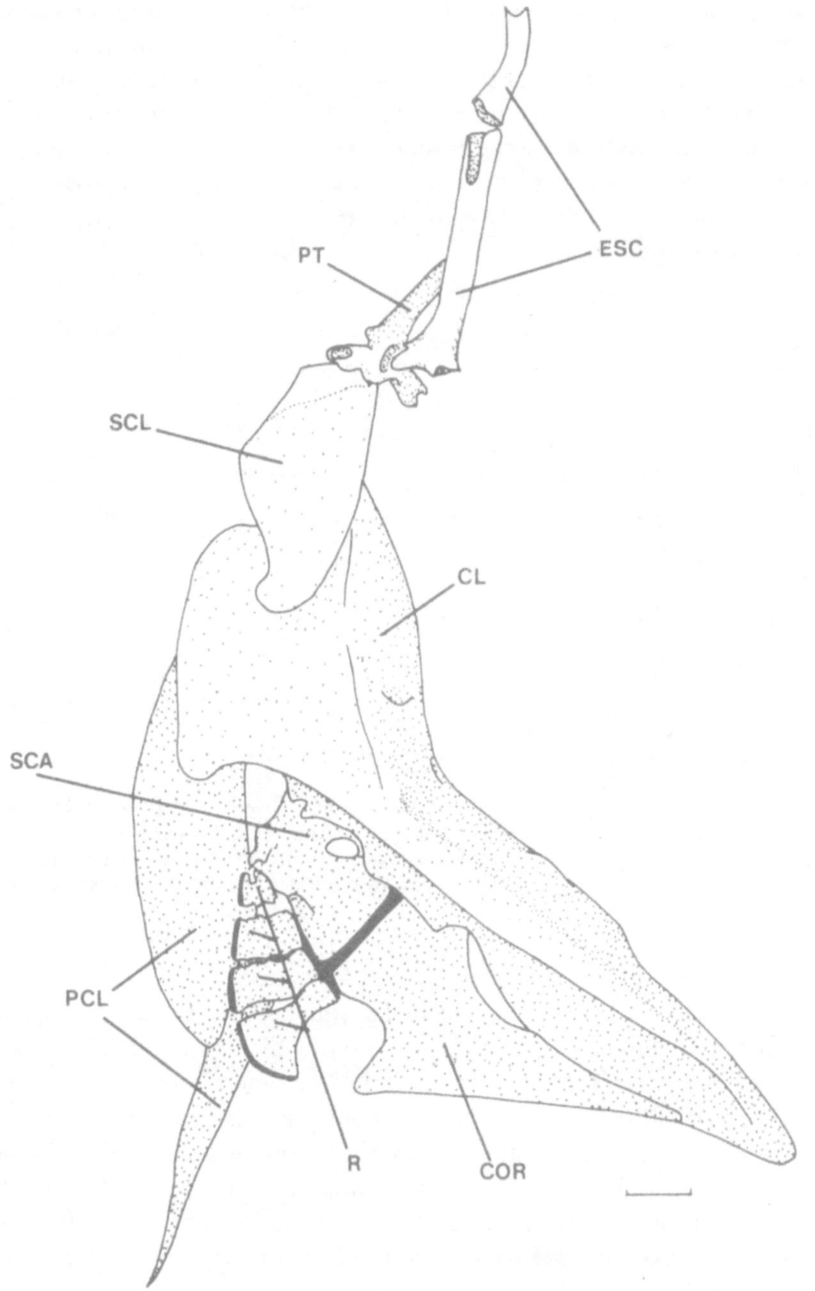

Fig. 60. *Guianacara owroewefi*. Pectoral girdle in lateral aspect. From NRM A82/1967052.3420, 51 mm SL. Scale 1 mm. Cartilage black. *CL* cleithrum, *COR* coracoid, *ESC* extrascapular, *PCL* postcleithra, *PT* posttemporal, *R* radials, *SCA* scapula, *SCL* supracleithrum.

Pectoral girdle (Figs 60-61): The anteroventral process of the coracoid is relatively wide and the fenestra between the cleithrum and coracoid rather reduced. The 3 ventral radials are in close contact proximally, though not sutured to each other. The distal postcleithrum has three projections proximally, one continuing the axis, one posteriorly and overlapped by the proximal postcleithrum, one anterior pointed rostrad, the proximal portion is relatively wide, tapering to a distal point, the anterior edge curved mediad.

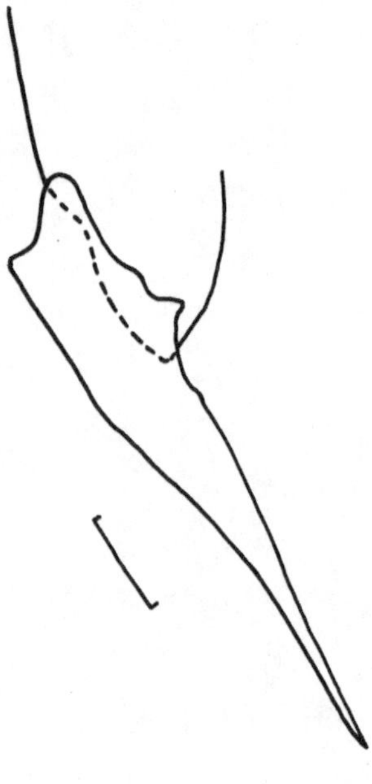

Fig. 61. *Guianacara owroewefi*. Medial aspect of distal postcleithrum. From NRM A82/1967052.3420, 51 mm SL. Scale 1 mm. Ventral tip of proximal postcleithrum with hidden outline indicated by broken line.

Suborbitals (Fig. 62): The lacrimal is squarish, with a deep incision in the anterior orbital margin at the bottom of which on the medial side lies a prominent process for the lateral ethmoid articulation facet. The canal-bearing part of the lacrimal occupies about the orad half of the bone, with 4 foramina. There are 4 infraorbitals, the second with a middle foramen. The first two infraorbitals are wide, tapering caudad, with a median canal; the posterior two infraorbitals are tubular. The first infraorbital overlaps the caudodorsal portion of the lacrimal and interdigitates with the second infraorbital.

There are 27 vertebrae of which the first two bear epineural ribs, the 3rd to 13th basapophyses and pleural ribs (pleural ribs only to 12th vertebra in one specimen); epipleural ribs are associated with the 3rd to 11th (12th in one) vertebrae. The third vertebra bears a pair of short

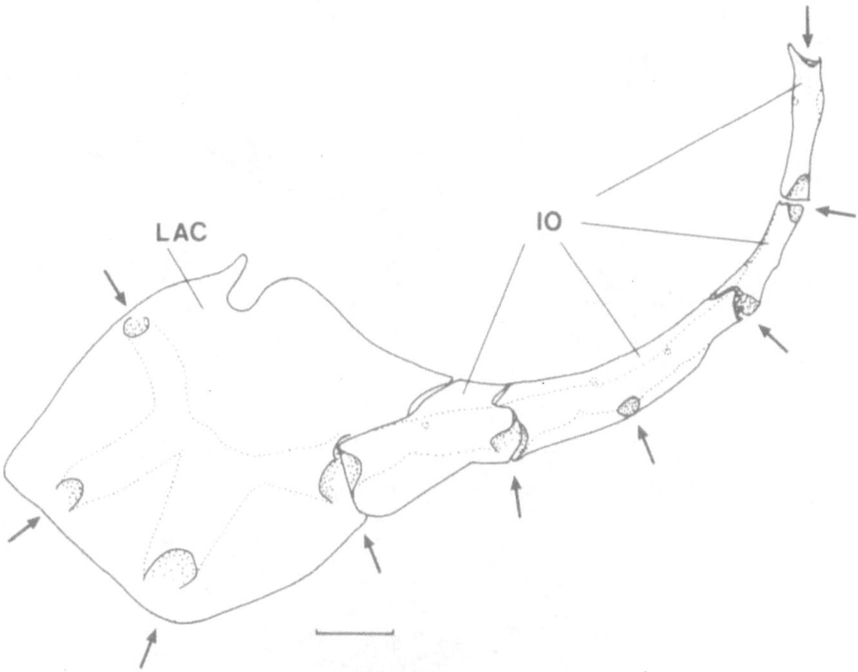

Fig. 62. *Guianacara owroewefi*. Suborbital series. From NRM A82/1967052.3420, 51 mm SL. Scale 1 mm. Arrows point to lateralis foramina ending in skin pores. *IO* infraorbitals, *LAC* lacrimal. Dotted circelets indicate nerve foramina actually on the medial or rostral side.

hypapophyses. The first hemal spine and arch is on the 14th vertebra, except that in one specimen there is a thin transverse bridge between the basapophyses of the 13th vertebra forming a hemal arch.

The caudal skeleton (Figs 63-64) includes the neural and hemal spines of the antepenultimate centrum, two posterior vertebrae, 5 hypurals, 2 epurals, uroneurals, parhypural and distal radials. The parhypural bears a long parhypurapophysis. There are 5 distal radials in the dorsal lobe, 3 in the ventral lobe, including the large plate anterior to the neural and hemal spine of the antepenultimate vertebra.

Colouration

Ground colour yellowish whitish; countershaded, pale brown on back and on upper margin of caudal peduncle, paler laterally, but flank rather uniform, abdomen whitish medially, chest ventrally of pure ground colour. Cheek brownish yellow. Opercularia, lower lip and lower jaw pale grey; preorbital, forehead, upper lip, grey. A light stripe from orbit to nostril.

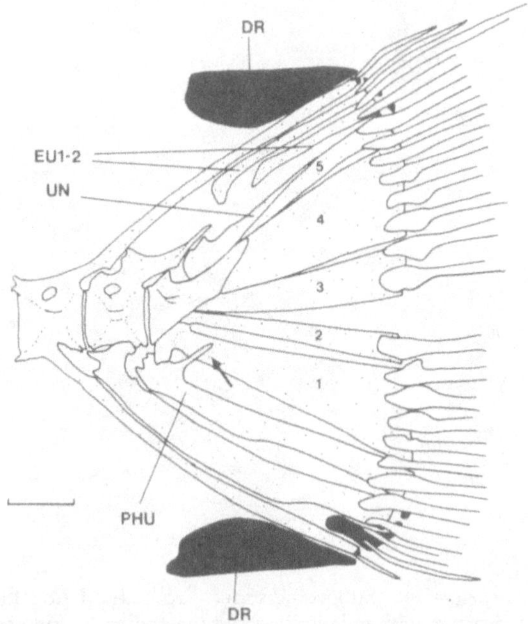

Fig. 63. *Guianacara owroewefi*. Caudal skeleton. From NRM A82/1967052.3420, 51 mm SL. Scale 1 mm. Cartilage black. *DR* anteriormost distal radial, *EU1-2* epurals 1-2, *PHU* parhypural, *UN* uroneural, *1-5* hypurals. Arrow points to parhypurapophysis.

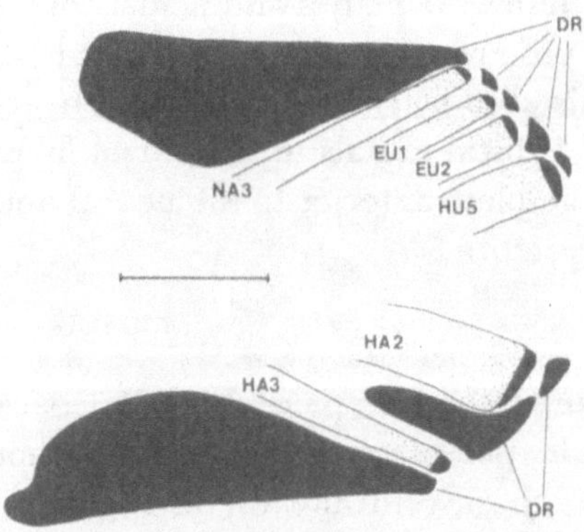

Fig. 64. *Guianacara owroewefi*. Distal caudal radials (cartilage, black, *DR*) and associated caudal skeleton element tips. From NRM A82/1967052.3420, 51 mm SL. Scale 1 mm. Cartilage black. Upper is dorsal, lower is ventral lobe. *EU1-2* epurals 1-2, *HA2* hemal spine of penultimate vertebra, *HA3* hemal spine of antepenultimate vertebra, *HU5* hypural 5, *NA3* neural spine of antepenultimate vertebra.

Nape grey brown, but slightly paler before supraorbital stripe and pale brownish yellow behind. Branchiostegal membrane brownish black except pale margin behind suboperculum. A dark brown stripe of about pupil width, running straight and vertically down from eye (slightly posterior to vertical diameter of orbit), onto infraorbital rim and cheek, through preopercular corner onto interoperculum and the lower preoperculad portion of suboperculum. Another dark brown stripe runs straight from upper margin of orbit (behind middle) to middle of nape, continuous with identical translateral stripe.

A strong dark brown vertical stripe passes through squ. long. scales 8-10, dorsally obliquely forward to 7th to 9th dorsal fin spines, slightly intensified on and just below lateral line, ventrally narrowing or becoming fainter to slightly below horizontal level of pectoral axilla or near abdominal margin. Pale zones anterior and posterior to cross stripe except close to dorsal fin.

Dorsal fin damaged, but black anteriorly, otherwise greyish, broad outer margin paler, tip of soft part hyaline; grey portion with small white dots. Anal fin with darkened lower margin, otherwise like dorsal fin though white dots sparse. Caudal fin brownish yellow basally, like body, otherwise greyish, except for smoky, almost hyaline, caudoventral portion and small white dots (about 8 to 10 along a median membrane). Pelvic fin densely brownish grey or greyish on outer margin, rapidly becoming fainter inwards, clear on innermost part.

A nearly uniform body with strong head stripe, oblique cross stripe on flank, black anterior dorsal fin (3 membranes) and dark branchiostegal membrane and pelvic fin, is characteristic of all specimens, except that young have a pale branchiostegal membrane. Sexes are isochromatic, but within samples females tend to have dark markings more strongly expressed.

Most specimens including holotype, have a paler or silvery centre on each flank scale, but not young (25-30 mm and smaller), and appearance in adults may be influenced by state of preservation. Suborbital stripe may be slightly curved (convex anteriorwards) and supraorbital stripe usually sloping slightly posteriorwards; in young, with more elongate head, these stripes clearly slanting backwards. Flank stripe appearing more oblique in young; always slightly stronger on and just below lateral line (crossing 8th to 10th or 9th to 11th scale) and often reaching to margin of belly.

Branchiostegal membrane darkened from about 40 mm SL, whitish in smaller specimens. Young have a dark spot (paler than head and flank stripes) around dorsal fin origin, extended by a point toward dorsal tip of preoperculum. Very small specimens also have extremely faint vertical bars behind flank stripe, 3 above anal fin base and one on caudal peduncle, and lack pale zones bordering flank stripe.

First 3 dorsal fin membranes black; a black spot of varying size and intensity, sometimes absent, continuing flank stripe through dorsal fin. In adults, dorsal fin otherwise dark on proximal 1/2 to 2/3, distally grey, margin black, always dotted (also variously anteriorwards on spinous part). Young with only faintly dotted fins. Very small specimens have a minute faint dark spot on middle of base of caudal fin. Commonly upper margin of caudal fin with a dark narrow margin and a broad whitish submarginal line. Pectoral fin base whitish or upper part darkened, inside of pectoral axilla dark.

In the Coppename and Saramacca R. material, medium and large specimens usually display four narrow, oblique, very faint dark bars on the side below the dorsal fin posterior to the flank stripe, and the dotting of the fins is not so extensive.

Distribution (Fig. 65)

Guianacara owroewefi appears to be fairly widespread in Surinam, having been collected in the upper Coppename, middle Saramacca, much of the Suriname, and in many Marowijne drainage localities. The species is sympatric with *G. oelemariensis* in the Oelemari R.

Etymology

Owroe wefi (alternative spellings ooroe wefi, owroe wiffi), meaning old wife, is one of the local names employed for this species (and other cichlids).

Local names

Ooroe wefi (Lijding 1959a, for *Aequidens geayi*); Crobia, Owroe wiffi (Popta 1914-1917, for *Aequidens geayi*); Ouru muje, Ouru wefee (Boeseman 1952); Krobia, owroe wefi (Richter & Nijssen 1980).

Notes

Colouration effectively separates *G. owroewefi* from the other nominotypical *Guianacara* species as in this species the flank stripe rather resembles a vertical stripe including a spot, than a narrow triangle; also the silvery or pale scale centers frequently seen in *G. owroewefi* have not been observed in any other species, and whereas the anterior dorsal fin membranes are black in young of all *Guianacara* species, only *G. owroewefi* and *G. oelemariensis* retain this marking as adults. *Guianacara owroewefi* and *G. sphenozona* differ in frequencies of gill raker numbers, but the difference is small.

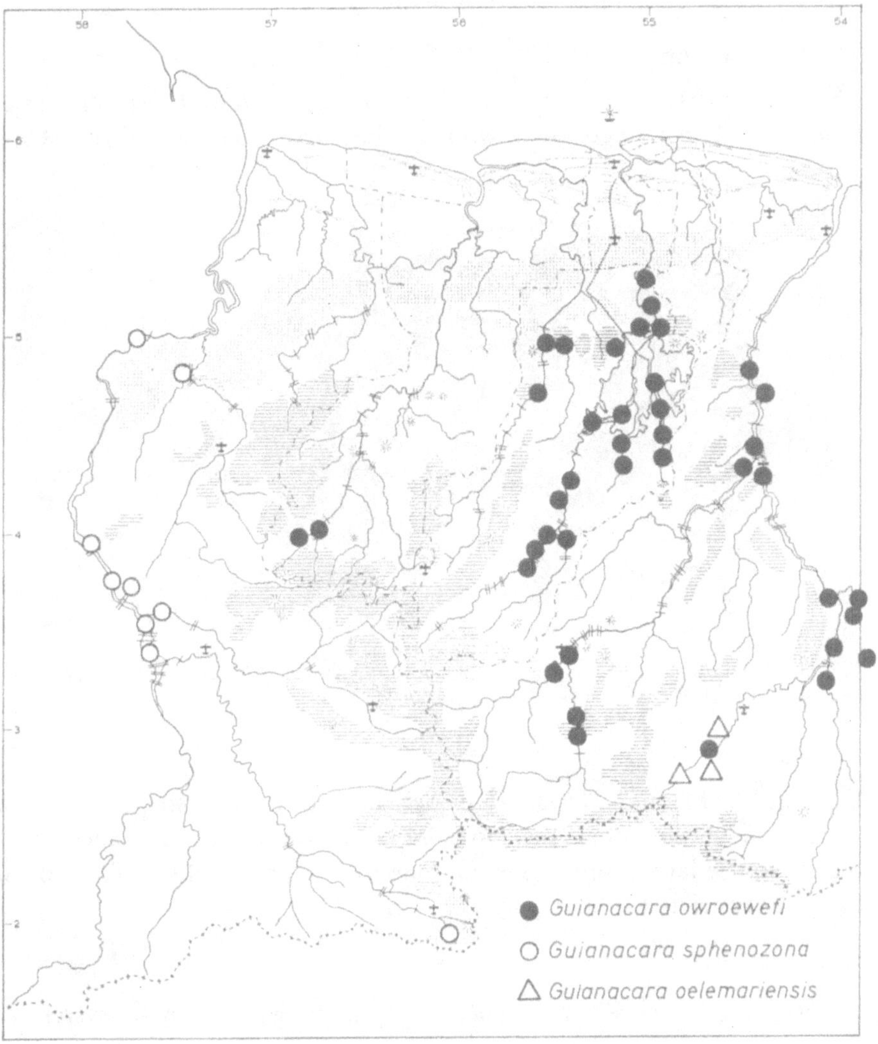

Fig. 65. Collecting localities of *Guianacara* species in Surinam.

Guianacara sphenozona n. sp.

(Figs 65-69)

Surinamese records

Aequidens geayi; (*pt.*) Boeseman 1952, p. 198 (listed: Lucie River [RMNH 18468]; Surinam, (Upper Corantyne River?) [RMNH 18469]; local names).

Holotype (Fig. 66)

RMNH 31047. A male, 73.4 mm SL. Suriname, Distr. Nickerie, Corantijn R. drainage, Sipaliwini. 23 January 1966. Leg. G.F. Mees.

Fig. 66. *Guianacara sphenozona*, holotype, male 73.4 mm SL, RMNH 31047, from Sipaliwini.

Diagnosis

Distinguished from *G. owroewefi* principally by the flank stripe, strongest on and above the lateral line, forming a narrow wedge or reduced to a blotch on the lateral line, from *G. geayi* by comparatively smaller dorsal fin dots, restricted to soft portion.

Material

Holotype and over 432 paratypes, all from the Corantijn R. system.

Description

Based on the holotype, variation recorded chiefly in the lots RMNH 31048 + NRM A83/1966037.3015 and RMNH 31050 (juveniles); see Figs 66-68 for general aspect of adults and young. Counts from specimens measured, except dorsal and anal fin, squ. long., and gill rakers from 136 specimens.

Fig. 67. *Guianacara sphenozona*, paratype, male, 84.8 mm SL, RMNH 31048, from Sipaliwini.

The largest specimen examined is a male, 84.8 mm SL (RMNH 31048, Fig. 67), the largest female is 81.1 mm SL (RMNH 31053).

Shape: moderately deep, cross section outline in anterior trunk region with flattened chest and narrow, keeled nape. Predorsal contour very steep, straight, little convex in advance of orbit, and sharply curved before dorsal fin; prepelvic head contour much less sloping, with projecting lower jaw articulation.

Head: snout from above narrow, rounded, in lateral aspect deep, rounded, with steep dorsal profile. Small specimens are more elongate, with longer head, and nape contour evenly curved, rounded rather than keeled nape. Largest specimen, 84.8 mm SL, has a very large head with

Fig. 68. *Guianacara sphenozona*, paratype, 34.2 mm SL, RMNH 31055, from the Avanavero Vallen.

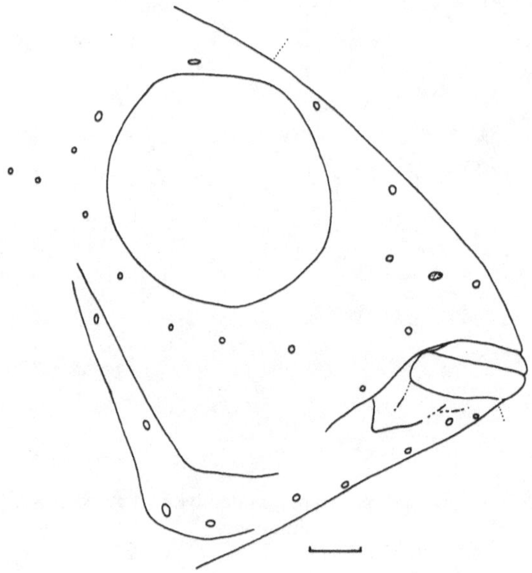

Fig. 69. *Guianacara sphenozona*. Distribution of cephalic lateralis pore in a 33.7 mm paratype (RMNH 31050); pattern representative for genus. Pores out of view (rostralmost dental, and coronalis pore) indicated by dotted lines.

straight predorsal slope and highest point just in advance of dorsal fin. Large specimens have rounded snout; small specimens have a pointed snout in lateral aspect.

Juveniles (NRM 31050) elongate, with nape outline nearly horizontal and sharply decurved snout. Mouth width little less than interorbital; maxilla not reaching to orbit (to orbit in juveniles). Orbit in middle of head length, but entirely in upper half of head, most above opercular level, little below forehead contour. Preopercular serrations only indicated in three of 104 specimens checked. Cephalic lateralis pores distributed as shown in Fig. 69.

Scales: squamation as in *G. owroewefi*. Squ. long. 24(1), 25(76), 26(3), -(2), in material from near Avanavero falls; 24(1), 25(52), -(1) in Sipaliwini material counted; total 24(2), 25(128), 26(3), -(3). Cheek scale series 4(20), 5(5), cheek scales cycloid except in superoposterior corner; subopercular and interopercular scales cycloid; interoperculum with 1-3 scales. Lateral lines modal count 18/9 (8 of 25 specimens); upper with 16(1), 17(6), 18(15), 19(3), lower with -(1), 8(4), 9(13), 10(7) scales; upper extending to below 3rd to 4th dorsal rays. Lower line continued on caudal fin by -(2), 2(22) or 3(1) canals on caudal fin; marginal caudal fin branches with up to 2 canals, but both or either usually missing. Caudal fin with ctenoid scales to 1/3 (at middle) and half (marginally) the length of fin.

Fins: dorsal fin with short first spine, 1/2-1/3 the length of last, relative spine length increasing to 5th, then decreasing to 11th from which subequal, last again a little longer, but shorter than 5th (specimens ca 50-60 mm radiographed); large specimen, 84.8 mm SL, has first spine 1/4 length of last, rising to 7th, subequal behind except 3 last which again increasing, last (15th) little longer than 5th. First lappet short and rounded; 3rd to 9th lappets slightly produced, 4th longest, 1/2-1/3 of spine length outside spine, behind truncate and short lappets. No sex difference in spine lengths or lappets; small specimens with short, rounded lappets, elongations beginning to appear at ca 35 mm and distinct from ca 45 mm. The 84.8 mm specimen is rather exceptional, with short, rather rounded lappets. Soft fin pointed, 4th ray longest, little produced and extending little past 3/4 of caudal fin in holotype; produced only in large males (like holotype) and reaching at most to end of caudal fin; pointed but not produced in large females (and young males), reaching to about middle of caudal fin; specimens less than ca 30 mm with rounded fin extending to little behind caudal fin base.

Soft anal fin similar to soft dorsal fin, but not produced, 5th ray longest, to middle of caudal fin; never produced, 4th ray usually longest, and always shorter than soft dorsal fin in backwards extension; juveniles with rounded fin to just behind caudal fin base. D. XIV.10(3), XIV.11(4), XIV.12(1), XV.-(1), XV.6(1, abnormal), XV.9(2), XV.10(118), XV.11(4), XVI.9(1), XVI.10(1). A. III.7(2), III.8(129), III.8.i(1), III.9(4).

Caudal fin very little emarginate in holotype, ray D6 longest, forming tip of otherwise square dorsal corner, ventral corner rounded; others with truncate hind margin, dorsal corner squared, but in large males with ray D7 feebly produced, ventral corner rounded. Pectoral and pelvic fin like *G. owroewefi*; but pelvic fin never conspicuously produced. P. 13(1), 14(15), 15(9).

Jaw teeth in 22 specimens, 28.0-84.8 mm: upper jaw 10 (young) - 19 (large), median 13, lower jaw 18-23, median 21.5 in outer hemiseries; inner series 2-3 in upper, 2 (young) to 5 (large) in lower. Otherwise as in *G. owroewefi*; young with only half of premaxilla toothed, large specimens with premaxilla toothed all along alveolar ramus.

Gill rakers: morphology and counts similar to that observed in *G. owroewefi*. First ceratobranchial rakers 6(45), 7(89), -(2); specimens from Avanavero with 6(28), 7(53), -(1), specimens from Sipaliwini with 6(17), 7(36), -(1).

Tooth plates: a 71.9 mm specimen (RMNH 31048) with lower pharyngeal tooth plate as in *G. owroewefi*, teeth only in more regular series; ceratobranchial 4 with 4 tooth plates with 2, 6, 8 and 9 pointed teeth respectively.

Vertebrae: 13 + 14 (7).

Colouration

Ground colour dirty whitish anteriorly on chest, anterior to pelvic fins and along abdomen; faintly yellowish on sides. Countershaded, greyish close to base of dorsal fin, on dorsal margin of caudal peduncle and on nape; sides obscurely brownish yellow, a little lighter on abdominal sides. Cheek brownish yellow. Preorbital and forehead grey; a whitish stripe from orbit toward, but not reaching to nostril. Upper lip, preoperculum and lower jaw pale grey, lower lip whitish. Operculum and suboperculum greyish white, but operculum brownish dorsally; interoperculum darker grey. Iris black except anteroventrally where silvery. Branchiostegal membrane dark grey, throat feebly paler. Pectoral axilla externally white ventrally, silvery dorsally, inside dark grey; pectoral fin with a faint greyish spot over upper 2/3 of basal part. Faint brown stripe from eye across cheek to preoperculum just above corner, margins indistinct, little curved. Blackish stripe, almost vertical, from upper margin of orbit (behind middle) to midline of nape, of about pupil width, not contrasted against dark grey nape.

A prominent black band from dorsal fin (8th to 11th spines) across lateral line scales 9-12, strongest across lateral line (one scale above lateral line series to one scale below), brownish and gradually narrower ventrad, to little below horizontal level of lower margin of pectoral axilla.

Dorsal fin greyish brown on basal 2/3 of spinous part, outer part smoky, lappet tips dark; soft part with wide greyish brown basal zone, outwardly smoky and with light spots along six membranes. Anal fin greyish brown on spinous and basal soft part, soft part outwardly smoky, two light spots on last membrane, outer margin dark. Caudal fin brownish yellow on muscular base, brownish over basal median part and along margin of lower lobe, dorsal lobe basally greyish; outwardly smoky, except along dorsal and ventral margins; light dots in regular series along membranes, about 10 along a median membrane. Pelvic fin grey brown, inner posterior margin broadly smoky.

Variation: Largest male, 84.8 mm SL, pale overall, greyish yellow, lacking cheek stripe and showing only a faint supraorbital stripe. Dorsal fin uniformly pale grey except for dark lappet tips and light spots on soft portion. Cross stripe on side reduced to a slightly ovate spot, extending from one scale above to one scale below lateral line scale series.

Vertical flank stripe always strongest on and above upper lateral line, expanded to dorsal fin, on which it is often continued, and forming a distinct narrow wedge, except that it may occasionally, and in both sexes, be reduced to a spot as described for the large male above; ventral extension varying somewhat and it may reach to but not onto the abdominal midline; usually crossing 9th to 11th lateral line scales. Supraorbital stripe sloping according to forehead steepness, more vertical in specimens with steep forehead, more oblique in specimens with curved nape. Intensity of suborbital stripe greatly variable; may be continued ventrad onto interoperculum. Females usually have dark markings more intense compared to males; especially the branchiostegal membrane, throat and lower jaw may be deep black, and also anterior to pelvic fin bases may appear distinct dark pigmentation. Light fin dotting varies in distinctness and extension, but is never continued onto spinous part of dorsal fin. Females always have fainter fin dotting. Anterior 3 dorsal fin membranes blackish in young only. Dorsal margin of caudal fin usually distinctly black, adjacent fin margin light.

Juveniles (RMNH 31050) pale, brownish yellow, with intense dark brown supraorbital, suborbital and flank stripes, anterior dorsal fin membranes and pelvic fin margin. Head stripes strongly caudad inclined from eye; flank stripe distinctly oblique, sloping caudad and ventrad, and often continued into dorsal fin. Branchiostegal membrane white in specimens up to about 30-35 mm SL, in which it becomes duskied. Fins light, faintly dotted or immaculate; basally on middle of caudal fin a diffuse dark spot, which may be seen also in some large fish. Indications of 2 vertical bars may appear behind flank stripe below dorsal fin. Upper margin of dorsal fin narrowly dark margined.

Distribution (Fig. 65)

Guianacara sphenozona is the only cichlid species besides *Krobia guianensis* and *Crenicichla sipaliwini* collected in the Sipaliwini; other localities are in the middle Corantijn drainage. The species probably occurs in Guyanan rivers.

Etymology

The species epithet, a noun in apposition, refers to the shape of the vertical flank stripe, and is derived from the Greek *sphen*, wedge, and *zone*, girdle.

Local names

Ouru muje, ouru wefee (Boeseman 1952).

Notes

See under *G. owroewefi*, p. 118.

Guianacara oelemariensis n. sp.

(Figs 65, 70)

Holotype (Fig. 70)

ISNB 750. A male, 81.0 mm SL. Suriname, Distr. Marowijne, Marowijne R. drainage, small right bank tributary to the upper Oelemari R. 14 November 1966. Leg. J.-P. Gosse (Mission Suriname 1966, Sta 163).

Diagnosis

Distinguished from other *Guianacara* species by its subgeneric characters, which include one instead of two supraneurals, midlateral spot instead of cross stripe over middle of side, and short dorsal fin lappets.

Fig. 70. *Guianacara oelemariensis*, holotype, male, 81.0 mm SL, ISNB 750, from Oelemari R. tributary.

Material

Holotype and 23 paratypes, 33.7-81.0 mm SL, all from the Oelemari R. drainage.

Description

Based on the holotype except where otherwise indicated. Measurements of 23 specimens. Counts below are from all specimens available. The largest specimen examined is the holotype, a male, 81.0 mm SL; the largest female is 71.3 mm SL (ISNB 751).

Shape: moderately elongate in facal aspect with flattened chest and narrow nape. Dorsal outline horizontal to about 6th spine. Predorsal contour gently arched from snout (where steepest) to dorsal fin origin; ventral head contour about straight, less steep.

Head: snout from above elongate, rounded off, lateral outline roundish, with steep dorsal profile and feebly sloping ventral profile. Small specimens similar in shape, also with rather steep dorsal head contour. Mouth width about equalling interorbital; maxilla reaching halfway between orbit and nostril (close to orbit in small specimens). Orbit mostly in upper 1/3 and posterior 1/2 of head, below forehead contour, far above pectoral fin and mouth levels. Small specimens with large eye, touching forehead contour, in upper half and middle of head length. Interorbital area convex; nearly flat in small specimens. Preoperculum entire in all specimens.

Scales: Squ. long. 24(1), 25(23). Predorsal scales in 2 bilateral series from between orbits, along midline a series of ca 9-10 scale pairs crosswise back to origin of dorsal fin, 2 (in others 1-3) minute scales intercalated medially near dorsal fin; often irregularities which may include a scale over the midline anteriorly. Lateral chest scales about 2/3 size of anterior flank scales. In small specimens scales of median prepelvic series and 2 bilateral parallel series cycloid except near pelvic fin bases, in large specimens more or less ctenoid on posterior half of ventral chest.

Cheek scales in 3(1), 4(12) or 5(11) series; only most superoposterior scales ctenoid. Opercular scales, aside from two small scales over the upper margin, in 2 oblique series paralleling the lower margin, cycloid and ctenoid. Six subopercular scales, 2 upper cycloid. Interoperculum with 2 small cycloid scales near preopercular corner. Other specimens with similar lateral head squamation; with 2 opercular scale series, occasionally a 2nd subopercular series, 1-2 interopercular scales.

Lateral lines, 16/10(1), 17/7(2), 17/8(4), 17/9(5), 18/7(1), 18/8(5), 18/pore + 8(1), 18/9(4), 19/8(1); upper line to below 1st to 3rd soft dorsal fin ray, overlapping lower by 2 scales (right side) or not at all (left side), at 3 1/2 (initially) to 1 1/2 (terminally) scales distance from dorsal fin; lower line reaching forward to above 3rd anal fin soft ray, continued on caudal fin by one (7) or 2 (17) canals immediately behind. Two dorsal lobe canals on caudal fin in holotype, but no ventral lobe canals; one or 2 ventral canals in 5 paratypes, 1-3 dorsal in 5 paratypes. Caudal fin scaled for 1/4 medianly, nearly half marginally, all scales ctenoid.

Fins: dorsal fin spines increasing in length to 5th, 6th and succeeding little shorter, subequal, last 2 spines injured, but were probably a little longer than preceding, incompletely regenerated last spine nearly as long as 5th. Paratypes, 61.7 and 74.5 mm SL, with first spine about half as long as last, increasing to 5th to 6th or 5th to 7th (subequal), decreasing behind, 8th or 11th to 14th subequal, 15th (last) a little longer and little shorter or little longer than 5th, respectively. Dorsal fin lappets free, reaching beyond spine tips, truncated; in other specimens lappets similar, never produced. Soft dorsal fin pointed, 4th ray longest, but point largely formed by 3rd to 5th ray tips, reaching to above middle of caudal fin; in young specimens soft dorsal fin rounded, reaching hardly to 1/3 of caudal fin; in large specimens pointed, to at most 3/4 of caudal fin (74.9 mm male), little shorter in females compared to males. Soft anal fin pointed, 4th ray longest, reaching to about 2/5 of caudal fin; variation otherwise similar to that of soft dorsal fin. D. XV.8(1), XV.9(4), XV.10(18), XVI.9(1); A.III.8(24).

Caudal fin subtruncate, lower lobe a little shorter than upper and with more rounded corner; similar or more truncate in others. Pectoral fin subacuminate, reaching to above 1st anal fin spine (to 1st or 2nd anal fin spine in large specimens; 2nd to 3rd in young). P. 14(22), 15(2). Pelvic fin pointed, first ray not produced, reaching to 3rd anal fin spine (never produced, may reach to 2nd anal fin ray).

Jaw teeth: upper jaw with 8 (young) to 20 (large), median 15, teeth in outer hemiseries; lower jaw with 15-21, median 19; 2-3 inner series in upper jaw, 2 (young) to 4 (large) in lower. Holotype with anteriormost outer teeth in both jaws subprocumbent; lower jaw dental arcade rounded, with a wide interspace between outer series and band of inner teeth, all teeth nearly straight, inner decreasing in size posteriorly and found chiefly anteriorly in jaw, one inner series slightly caudad extending; premaxilla completely toothed, outer teeth increasing in size toward symphysis, largest very little recurved, dental arcade rounded, inner teeth much smaller than outer, one inner series continued little caudad, but inner teeth chiefly in a transverse band anteriorly in jaw, inner and outer teeth widely separated anteriorly. Smaller specimens differing chiefly in having upper jaw edentulous near distal end.

Gill rakers: first ceratobranchial rakers 6(20), 7(4). A dissected specimen, 56.5 mm, has gill rakers as described for *G. owroewefi*.

Tooth plates: lower pharyngeal teeth as in *G. owroewefi*, but posterior enlarged teeth not so blunt. Three small tooth plates on ceratobranchial 4.

Vertebrae: 13 + 14(7).

Colouration

Ground colour yellowish white; countershaded, dark brown on back, lighter laterally, chest and belly ventrally of dirty ground colour. Flanks with light scale margins, giving a mild patchy appearance to sides; below level of lower margin of pectoral axilla also scale centers light. No vertical or horizontal bands. Operculum dorsally brownish, ventrally like other operculars and lower jaw, greyish. Cheek brownish yellow. Preorbital and forehead brownish grey; nape above eyes dark brown. Lips greyish white. Faint light stripe from orbit to nostril. Iris silvery except above pupil where black.

A dark brown (but dorsally lighter brown) stripe of pupil width from eye down across preopercular corner to junction of sub- and inter-opercula, slightly curved. Indistinct dark brown stripe from posterior half of upper margin of orbit obliquely posteriad to nape midline before dorsal fin origin, little wider than pupil. Black oblong spot in squ. long. scales 9-13 and half of those above and below, margins indistinct. Branchiostegal membrane broadly margined with dark brown. Only faintly brownish between pelvic and pectoral fin bases.

Dorsal fin uniformly brown except for three anterior blackish membranes, and posteriormost soft part membranes with colourless dots. Anal fin brownish, posteriormost membranes with colourless dots. Caudal fin base like adjacent body, beyond scaled area clear, ventral margin smoky, with about 12 not quite regular cross series of brownish dots, as wide as interspaces and rather close-set, to distal margin. Pelvic fin with brown outer margin, inwardly of fainter pigmentation, inner margin hyaline.

Variation (ISNB 751): Young specimens (ca 50 mm) show traces of cross bars, all straight vertical and wider than interspaces: (1) on caudal peduncle, (2) between ends of dorsal and anal fin bases, (3) and (4) above anal fin, (5) including flank spot, above vent, (6) between flank spot and pectoral fin base, (7) above pectoral fin base up to dorsal fin origin. On the smaller, and some of the larger specimens, may also be traced a faint horizontal band from gill cleft straight to caudal fin base, about 1 1/2 scale wide, on and above lower lateral line. A dark spot below light pre-orbital stripe close to orbit. Small specimens have a dark spot around dorsal fin origin, extended narrower toward tip of preoperculum.

Suborbital stripe always slightly curved, usually stronger on ventral-most part. Supraorbital stripe of varying intensity, always caudad inclined. Flank spot always strongly expressed, roundish, more or less oblong, more rounded in smaller specimens, of orbital diameter (small specimens; in squ. long. scales 9-12) or 1 1/2 orbital diameter long (large specimens; in squ.long. scales 9-13). A dark brown spot anterior to and occasionally between pelvic fin spines, extended dorsally to inside of pectoral axilla in all. Branchiostegal membrane more or less dark margined, more strongly pigmented in large specimens.

Dorsal fin anterior 3 membranes black, fin otherwise uniformly grey, in some young with indication of 2 horizontal series of lighter dots.

Dotting of soft part variously developed; but large specimens have more vivid dorsal , anal and caudal fin dotting than small specimens, and females less of dots than males, unpaired fins may be nearly immaculate in females. Caudal fin base pigment concentrated to an unsharp spot on middle in smaller specimens and some large; caudal fin richly dotted except along lower margin in large males, in young and females only along middle and/or upper margin. Pelvic fin always brown margined.

The small ISNB 754 specimens show horizontal and vertical bands on side distinct, and have the branchiostegal membrane light.

Distribution (Fig. 65)

Known exclusively from a few localities in the Oelemari R., a Surinamese Lawa (Marowijne) tributary.

Etymology

Named for the river to which this species appears particular.

CICHLASOMA Swainson

Plesiops (Cichlasoma) Swainson 1839, p. 230 (type species by monotypy *Labrus bimaculatus* Linnaeus, *ex L. punctatus pt.* Linnaeus).

Synonymy

Plesiops (Cichlaurus) Swainson 1839, p. 173 (variant equivalent to *Cichlasoma*, suppressed by first reviser, Swain (1883)).
Cichlosoma Regan 1905c, p. 61 (unjustified emendation of *Cichlasoma*).

Cichlasoma includes small to moderately large (to 136 mm SL) South American cichlids, with 4 dentary lateralis pores; Baudelot's ligament attachment lateral on basioccipital; caudal peduncle short, not including any complete vertebral centrum; short or absent parhypurapophysis; adults with dark spot posterosuperiorly on cheek; asymmetrical caudal fin dot pattern and 5 instead of 4 vertical bars posterior to midlateral spot except in some species with secondarily immaculate caudal fin and two bars more or less confluent; short accessory lateral lines between caudal fin rays D2 and D3, and V4 and V5, may be wanting; predorsal scale pattern triserial or, secondarily, uniserial.

Recognized most conveniently on the scaled soft dorsal and anal fin bases in combination with a light-ringed spot superiorly on the caudal fin base.

The genus includes 12 allopatric species and ranges over most of tropical South America, including the Amazon, Plata, Orinoco, Tocantins and São Francisco basins, northeastern Brazil, Trinidad and the Guianas.

The genus *Cichlasoma* was revised quite recently (Kullander 1983a) and we refer to that paper for further details on the genus as well as the only species encountered in Surinam, *C. bimaculatum*.

Cichlasoma bimaculatum (Linnaeus)

(Figs 71-72)

Labrus bimaculatus Linnaeus 1758, p. 285 (M. Mediterraneo).

Synonymy

Acara Gronovii Heckel 1840, p. 361 (Flüsse in Surinam).
Sparus filamentosus Gray 1854, p. 60 (Surinami).

Fig. 71. *Cichlasoma bimaculatum*, male 123.3 mm SL, ZMA 106.961, from swamp along Brokopondo lake near Afobaka.

Surinamese records

Cichlasoma bimaculatum; Kullander 1983a, p. 65 (revison; see for earlier references). — Fernholm & Wheeler 1983, p. 253 (holotype listed), p. 252 (*Labrus punctatus* Linnaeus composite, including *C. bimaculatum*).

Diagnosis

Cichlasoma bimaculatum is a relatively common fish in Surinam, and reaches at least 123 mm SL. It is readily distinguished from other Surinamese cichlids by usually having four instead of three spines in the anal fin; rare indvividuals with 3 anal fin spines can be identified by the scaled dorsal and anal fins, triserial predorsal squamation and caudal fin base spot. For further data, refer to description in Kullander (1983a).

Distribution (Fig. 72)

Along the Guianas coast from the Moruka R. in Guyana eastward at least to Sinnamary in French Guiana, also in the Rupununi R. system and upper R. Cuyuni and R. Branco systems. Collected in Surinam at Albina, in the Cottica system, Suriname R. system from Paramaribo upstream to Gran Kreek, at Totness (Coronie), and in the lower Nickerie and Corantijn R., one record from the Lucie R. (Kullander 1983a)

Local names

Crobia, Owroe wiffi (Popta 1914-1917), Krobia, Manja koko (Lijding 1959a), Krobia, Owroe wefi (Richter & Nijssen 1980: Brokopondo reservoir).

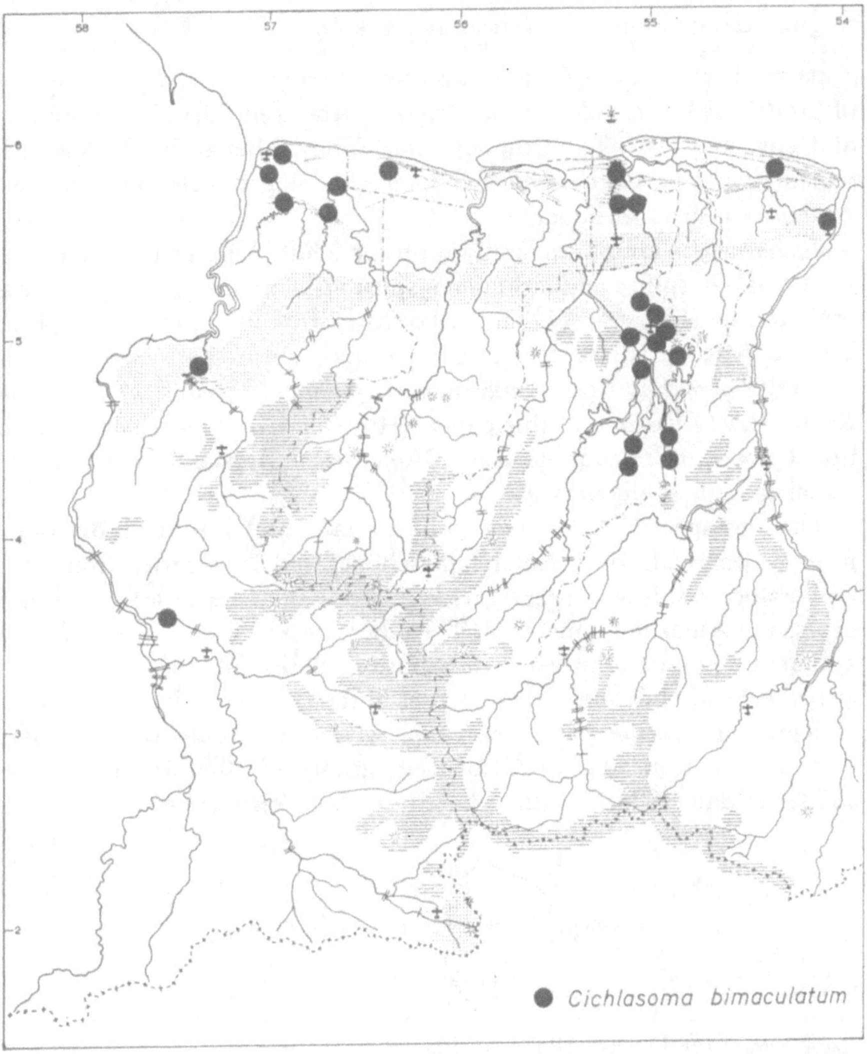

Fig. 72. Collecting localities of *Cichlasoma bimaculatum* in Surinam.

AEQUIDENS Eigenmann & Bray

Astronotus (Aequidens) Eigenmann & Bray 1894, p. 616 (type species by original designation *Acara tetramerus* Heckel).

Aequidens is recognized for a group of a little more than a dozen species of *Cichlasoma*-like cichlids in the Orinoco, Amazon, Parnaiba, Tocantins and Paraguay basins and the Guianas. They differ from *Cichlasoma* in naked dorsal and anal fins and long caudal peduncle including 2-3 vertebrae instead of none.

Aequidens was long a catch-all group for South American cichlids with three anal fin spines and lacking conspicuous characters. The genus was reviewed by Kullander (1983a), who distinguished a number of species groups. Most of these species groups have since been recognized as genera: *Bujurquina*, *Tahuantinsuyoa* and *Laetacara* in Kullander (1986), and *Krobia* and *Cleithracara* in this paper. Also *Guianacara* species have traditionally been included in *Aequidens*. Nonetheless, *Aequidens* 'sensu stricto', remains vaguely diagnosed.

The type species, *A. tetramerus*, which attains 160 mm SL, is one of the most widespread South American cichlids, and is recorded from most Guianan rivers. It was recently redescribed (Kullander 1986) on the basis of western Amazonian material. The other species reach 100-120 mm SL and are much more limited in distribution. Besides *A. paloemeuensis*, described herein, only one of them occurs in the Guianas, viz. the Guyanan endemic *A. potaroensis* Eigenmann. *Aequidens paloemeuensis* departs significantly from the provisional diagnosis of *Aequidens* (Kullander 1983a, update 1986), which is based on *Aequidens tetramerus*.

Aequidens tetramerus (Heckel)

(Figs 73-76)

Acara tetramerus Heckel 1840, p. 341 (Rio-branco).

Synonymy

Chromys uniocellata Castelnau 1855, p. 15 (rio Ucayale, mission de Sarayacu).
Aequidens stollei Ribeiro 1918, p. 13 (Rio Jamary).
Acaronia trimaculata Allen, in Eigenmann & Allen 1942, p. 380 (Iquitos).

Surinamese records

Aequidens tetramerus; (?) Boeseman 1952, p. 198 (listed: Nanni creek, Doksen savanna [not found in RMNH]; local names). — Kullander 1984, p. 159 (compared with *A. plagiozonatus*, data from ms of present study). — Kullander 1986, p. 346 (comparison with W Amazonian material; data from ms of present study).

Material

Over 86 specimens from the Corantijn, Nickerie, Coppename, Suriname and Marowijne river systems.

Fig. 73. *Aequidens tetramerus*, 93.0 mm SL, ZMA 107.292, from kreek on right bank of R. Maroni, E Mooisantitabbetje.

Description

Chiefly from a 93.0 mm specimen (ZMA 107.292, Fig. 73), variation chiefly from others in the same lot. Counts below are from 22 specimens measured. Largest specimen examined 125.0 mm SL, a female (ZMA 107.292).

Shape: body moderately elongate, moderately compressed laterally; dorsal and ventral contours about equally arched. In frontal aspect narrowly elliptic with narrowly rounded nape and chest. Caudal peduncle moderately long, ventral margin longer than dorsal.

Head: short, moderately deep, with slightly extended snout. Predorsal contour moderately steep, straight ascending to little in advance of dorsal fin; prepelvic contour less steep, straight. Orbit in upper and most in anterior half of head, below forehead contour. Mouth terminal, narrower than slightly convex interorbital. Maxillary tip exposed, below anterior margin of orbit (or not to orbital margin); ascending premaxillary processes reaching to anterior margin of orbit; lower jaw articulation below anterior half of orbit.

Scales: Squ. long. 24(20), 25(2). Circumpeduncular scale series 16. Predorsally 4 median anterior and 4 posterior scale pairs overlapping on midline (rarely irregular); anteriormost scales ctenoid, 4/5 size of anterior flank scales, smaller and cycloid posteriorly (all cycloid in smaller specimens). Chest scales ctenoid except under gill cover and smaller scales on base of pectoral axilla; largest 4/5 size of anterior flank scales. From between pelvic fin bases forwards about 14 scales, cycloid like those in 2 parallel series on each side; posterior 2/3 size of anterior flank scales, smaller anteriorly.

Cheek completely scaled; scale series 3(19), 4(3); cycloid save for a few posterosuperiorly (all cycloid in smaller fish); largest, posterosuperior, scales, 2/3 the size of anterior flank scales. Operculum with 2 series of large (like flank scales) ctenoid scales; one series on suboperculum (usually 3 scales), cycloid, lowermost largest, almost flank-scale size; frequently a scale on suspensoriad process. Interoperculum with 1 or 2 scales, cycloid, posterior 2/3 size of anterior flank scales. Preoperculum invariably naked. Upper lateral line at 3 (initially) to 1 1/2 (plus rudiment; terminally) scales distance from dorsal fin, in 3rd scale series above that containing lower lateral line; overlapping lower by 2 scales, to below 2nd dorsal fin ray. Lateral lines 16/2 pores + 7(1), 17/2 pores + 6(1), 16/7(1), 16/8(2), 16/9(3), 17/7(1), 17/8(7), 17/9(5), 18/8(1); 1-2 canals continuing lower line on the base of the caudal fin, but no dorsal or ventral lobe lateral line sequences.

Fins naked, except caudal fin; last 4 dorsal fin radial muscle apexes exposed; along dorsal fin base several minute scales and larger scales with indented margins. About 1/3 of caudal fin scaled, hind margin of scaled area concave; simple series of interradial scales, outer scales cycloid.

Fins: first dorsal fin spine above first lateral line scale, length not 1/3 that of last; spines equal in length from 5th, but last 3 again slightly increasing. Soft fin pointed, 4th and 5th rays slightly produced,reaching past middle of caudal fin (frequently beyond caudal fin in large specimens). D. XIV.11(1), XIV.12(1), XV.10(3), XV.11(14), XVI.10(2), XVI.11(1). Soft anal fin pointed, not produced, 4th ray longest, to beyond middle of caudal fin (shorter in small specimens). A. III.8(6), III.8.i(1), III.9(15).

Caudal fin hind margin with slightly concave middle, but well-rounded corners; 16 principal rays (in others similar, rounded, or sub-truncate). Pectoral fin moderately acuminate, 4th ray longest, to above first anal fin spine (usually to above 3rd anal spine). P. 12(1), 13(10), 14(11). Pelvic fin pointed, first ray slightly produced, reaching first anal fin spine base (in other specimens to at most middle of soft anal fin.

Jaw teeth: simple, pointed, slightly recurved apically. An outer series of larger teeth, to near jaw ends, and a wide band of smaller inner teeth, one series of which extends to near posterior end of outer series. Slight, but distinct size increase forwards of outer series teeth. Small specimens with upper jaw teeth series reduced to about anterior half of each jaw half. In largest specimen, one inner series extending as far back (to jaw ends) as outer series. 13-21/15-23 (mean 15.8/19.4) outer hemiseries teeth in 19/19 upper/lower jaws counted (SL range 47.7-125.0 mm); 2 or 3 inner series in upper jaw (3 in larger fish), 2 (small) to 3 in lower jaw.

Gill rakers: 5(5), 6(16), 7(1).

Gill arches and tooth plates in 97.5 mm specimen: first gill arch (Fig. 74) externally with 2 papilliform short rakers, one in angle and 4 upper short, broad, compressed, subcutaneously apically distinctly denticulated ceratobranchial rakers, and 2 rudimentary lower ceratobranchial rakers; upper 4 ceratobranchial rakers connected to rudimentrary medial swellings. Internally 2 epibranchial and 8 ceratobranchial rakers, 7 lower conical denticulate, connected basally to weak medial swellings, uppermost and epibranchial rakers papilliform. Right ceratobranchial 4 eden-

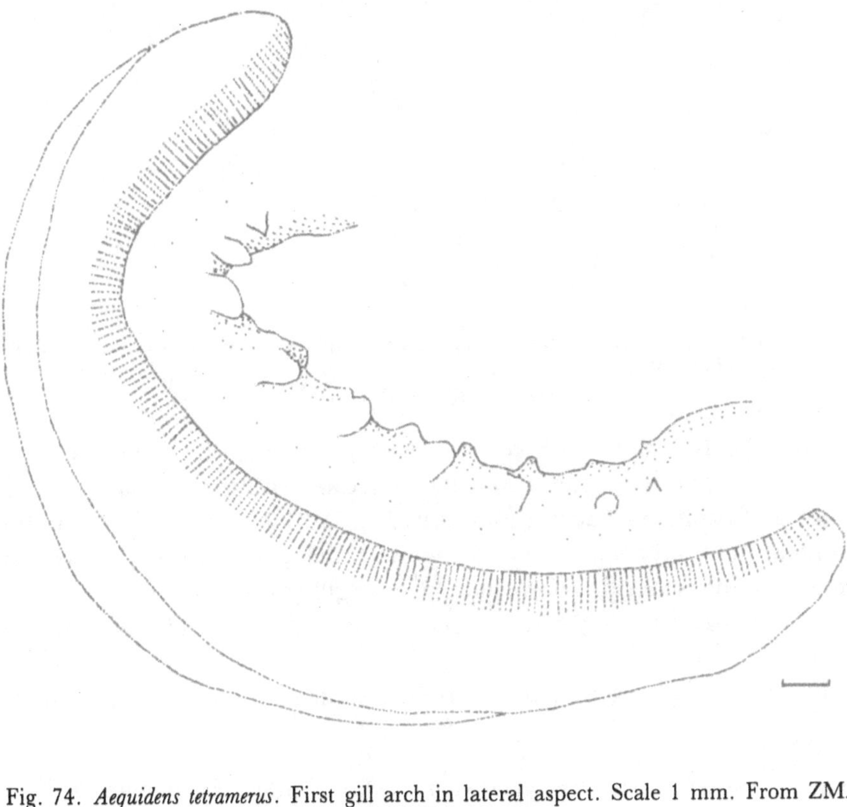

Fig. 74. *Aequidens tetramerus*. First gill arch in lateral aspect. Scale 1 mm. From ZMA 107.292, 97.5 mm SL.

tulous, left ceratobranchial 4 with a single tooth plate with 12 pointed teeth. Epibranchial 1 slender, with little separated medial processes; pharyngobranchial 1 moderately stout, but not expanded ventrally.

Lower pharyngeal tooth plate (Fig. 75) with 5 or 6 irregular rows of teeth on each half. Anterior teeth straight, with posterior point and more or less developed swelling subapically on rostral margin; teeth along margin slender, posteromedially stouter, strongly compressed, with posterior antrorse cusp and prominent anterior shelf.

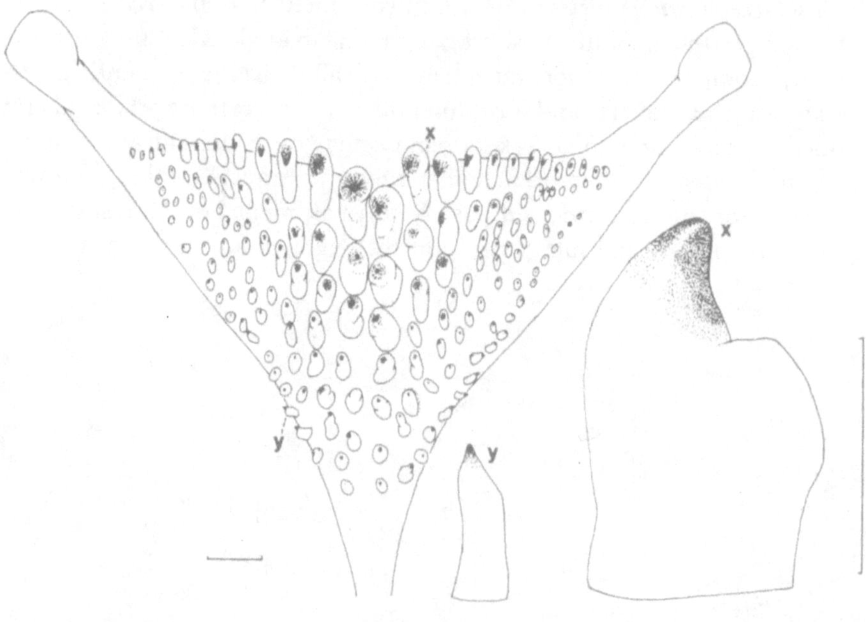

Fig. 75. *Aequidens tetramerus.* Lower pharyngeal tooth plate in occlusal view and teeth in medial (*x*) and lateral (*y*) aspect. Scales 1 mm; short for tooth plate, long for teeth. From ZMA 107.292, 97.5 mm SL.

Juvenile, 18.7 mm, with very large eye, tangented by predorsal contour, snout pointed, interorbital flat, maxilla reaching behind anterior margin of orbit. Soft dorsal fin pointed, to 1/3 of caudal fin. Pelvic fin pointed, not produced, to second anal fin spine; pectoral fin to above 3rd anal fin spine, other fins damaged. Cheek, gill cover, predorsal, anterior back, chest and prepelvic scales cycloid; lateral line chiefly with pored scales.

Vertebrae: 13 + 13(4); 2 entire vertebral centra within caudal peduncle.

Colouration

Warmly brown, turning to yellowish on chest, anterior to pelvic fins and along abdominal midline, to very dark brown or blackish along back, on nape and forehead. Gill cover and lower jaw light grey, darker dorsally. Preorbital and lips darker grey. Cheek brownish yellowish. Distinct white line from eye to nostril. Scales on dorsal sides finely black margined, fainter ventrally, light margined on chest. No distinct vertical bars.

Blackish brown spot posterosuperiorly on cheek, spot varying in shape from roundish to triangular (expanded ventrally). Intense large blackish brown spot on middle of side in 9th to 11th (8th to 10th or 9th to 12th in other specimens) squ. long. scales up to include lateral line scales above. Lateral lines light. Indistinct brownish horizontal band about one scale wide, chiefly in squ. long. scale series, from below lateral line origin to just short of end of caudal peduncle (band varying in intensity, but never prominent, and may not show in overall unformly colored specimens).

Spinous dorsal fin blackish; a white dot on each membrane between spines of varying clarity, distally variously evident one, posteriorly two series of whitish elongate dots along middle, and dark lappets with clear tips; soft fin lighter posteriorly, to clear, with roundish to rhombic dark dots on membranes, up to about 6 along a membrane. Caudal fin with basal blackish brown spot on dorsal lobe from fin margin to just above lower lateral line level, light margined anteriorly and posteriorly; fin base below brownish, mottled with yellowish; rest of fin clear with strongly oblique dark stripes across membranes, forming dorsad-caudad running stripes across fin up to ray D4, above stripes running ventrad-caudad, distal margin of fin smoky. Anal fin brownish, last membranes lighter, with brownish dots. Pelvic fin brownish, lighter inwardly.

Juvenile, 18.7 mm, overall dark. Back brownish, abdomen, chest, pre-pelvic area dirty yellowish, like bar interspaces. Nape, forehead, snout dark brown; gill cover, lower jaw greyish, cheek lighter with brown spot between eye and lower limb of preoperculum. Narrow faint brownish band along side from gill cover to Bar 1, interrupted in bar interspaces. 5 indistinct vertical bars, wider than interspaces: 1) on caudal peduncle; 2) between dorsal and anal fin base ends; 3-4) above anal fin base; 5) little in advance of vertical from anal fin origin, containing deep blackish brown midlateral spot from band up to lateral line; sides anteriorly darker than bars. End of caudal peduncle light. Dorsal fin greyish, soft part clear with dark base and dark stripe across middle. Caudal fin smoky; basally, except for ventral margin, a dark brown bar, light margined posteriorly on dorsal lobe. Anal fin smoky, with dark stripe across lighter soft part. Pelvic fin slightly smoky.

Medium-sized specimens, 50-60 mm, with faint cheek spot, may show indicated vertical bars, as in juvenile, occasionally also a bar anterior to that containing midlateral spot distinguishable.

Distribution (Fig. 76)

Surinamese collections are from the Corantijn, Nickerie, Coppename, Suriname and Marowijne drainages. There is a notable gap in the distribution comprising the Saramacca drainage, and it is curious that there are only two localities in the Suriname drainage which is otherwise well collected. As a rule few, mostly large specimens are taken at each locality, suggesting an universal bias in the collecting procedure (cf. also *Nannacara anomala*, *Cichlasoma bimaculatum* and *Cleithracara maronii*). The species is widespread in Amazonia, and recorded also from the Orinoco,

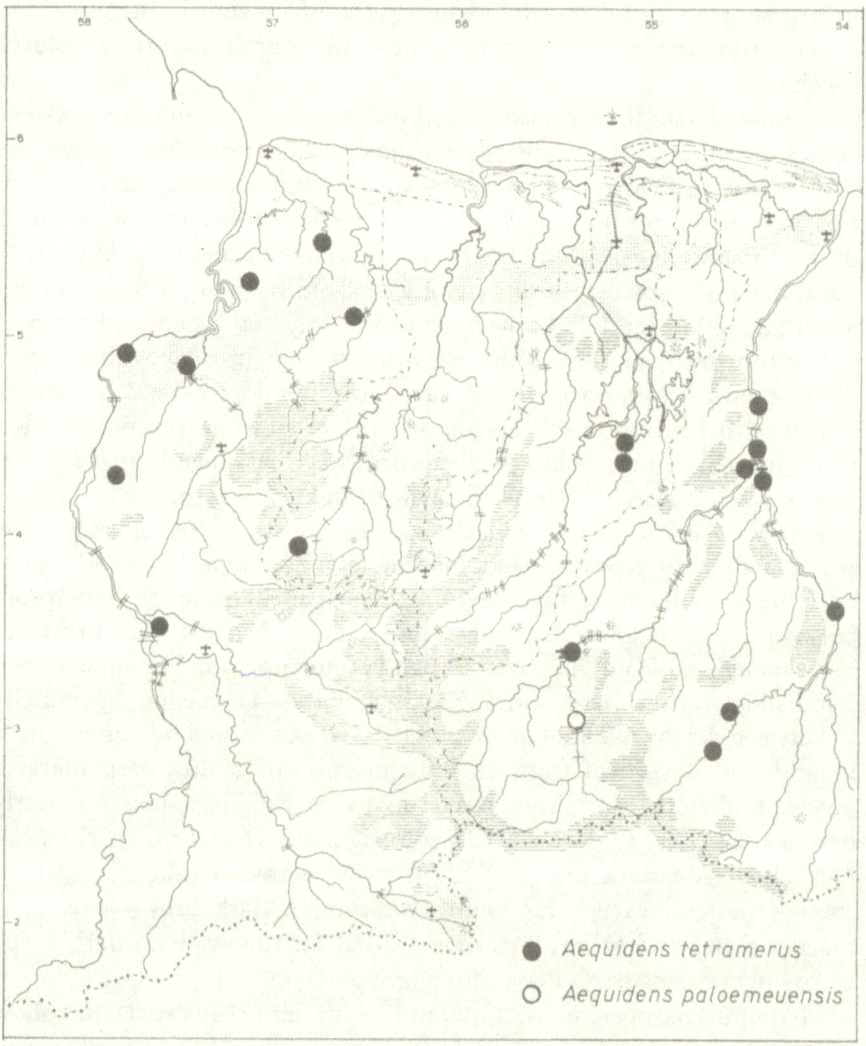

Fig. 76. Collecting localities of *Aequidens* species in Surinam.

Parnaiba and Tocantins drainages. The Guianan distribution includes the Oyapock, Approuague, Cuyuni and Essequibo drainages in addition to Surinamese localities.

Local names

Koenapari, Koemaparoe (Lijding 1959a); Kunapari, Kamaraparu (Boeseman 1952).

Notes

Surinamese *Aequidens tetramerus* are slenderer and overall darker, and have more contrasted caudal fin spot pattern and average shorter pectoral fin than western Amazonian material. Dorsal and anal fin count modes are higher (D. XV.11 vs. XV.10; A. III.9 vs. III.8) than in Ucayali-Amazonas *A. tetramerus*, though the same as in western Bolivian *A. tetramerus*. See Kullander (1986) for further discussion of *A. tetramerus*.

Aequidens tetramerus differs from all other Surinamese cichlids in the combination of naked vertical fins, large scales (squ. long. 24) and triserial predorsal squamation. One of the radiographed specimens (ZMA 107.292) has a small mammal molar tooth in the intestine.

Aequidens paloemeuensis n. sp.

(Figs 76-79)

Holotype (Fig. 77)

ISNB 755. An adult male, 95.3 mm SL. Suriname, Distr. Marowijne, R. Marowijne system, small brook tributary to right bank of R. Paloemeu on road between Trombaka Noord and Trombaka Zuid. 8 November 1966. Leg. J.-P. Gosse & Léopold III (Sta. 156).

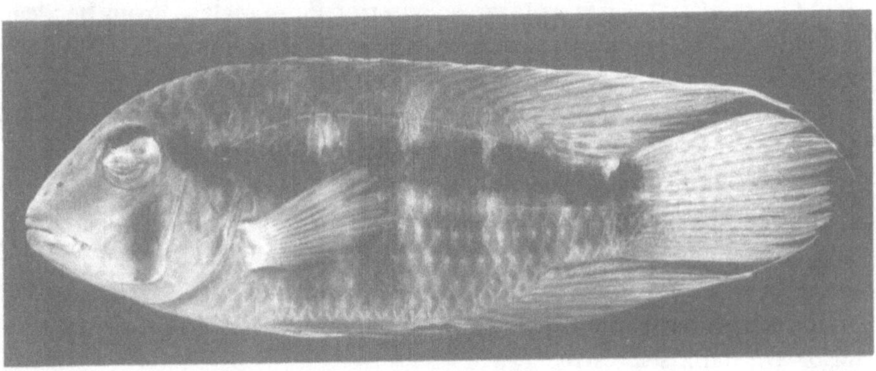

Fig. 77. *Aequidens paloemeuensis*, holotype, 95.3 mm SL, ISNB 755, from Paloemeu R. tributary.

Diagnosis

Distinguished from other *Aequidens* species by uniserial predorsal squamation. Similar to *A. potaroensis* in retaining the suborbital stripe as adult, differing from that species in having uniserial predorsal midline scale-arrangement, and lateral band intense throughout its length instead of fainter anterior to midlateral spot than posterior to it.

Material

Only one sample known, comprising holotype and 14 paratypes, 15.6-90.8 mm SL, from a small tributary to the Paloemeu R.

Description

Based on the holotype, largest known specimen, 95.3 mm SL; counts from all specimens available.

Shape: elongate, moderately compressed laterally; dorsal and ventral outlines about equally and little arched. In frontal aspect elliptic, with rounded nape and chest.

Head: moderately long. Predorsal contour rising straight to above middle of orbit, caudad gradually more level; ventral head contour little curved, less steep. Snout broad, rounded, little extended, with dorsal contour much steeper than ventral. Mouth narrower than interorbital, which curved. Maxillary tip exposed, not reaching to orbit; lower jaw articulation below anterior half of orbit; ascending processes of premaxilla not reaching to orbit. Young (ca 40 mm) deeper than adults, with narrower, more flattened interorbital, and maxilla reaching behind anterior margin of orbit.

Scales: Squ. long. 23(15). Predorsal scales ctenoid (also in young specimens), a median series of usually 8 (occasionally 7 or irregular posteriorly) posteriorly gradually smaller scales, anterior of flank scale size. Chest scales nearly as large as anterior flank scales. From between pelvic fin bases forwards 12 scales in a median series; posterior 2/3 flank scale size, smaller rostrally.

Cheek with 3(15) scale series; all scales cycloid, posterosuperior 4/5 flank scale size, smaller ventrally. Opercular scales ctenoid, size varying, little smaller or little larger than anterior flank scales. Scales cycloid in young, ca 40 mm. Suboperculum with two large cycloid scales, lower larger, of about flank scale size, and a smaller proximally on suspensoriad process (2-3 scales in paratypes, suspensoriad process scale present or absent). Interoperculum with 2 (2-3 in paratypes) cycloid scales, posterior 4/5 flank scale size.

Upper lateral line reaching to below 2nd or 3rd dorsal fin rays, overlapping lower by 2-3 scales. Lower lateral line to above 3rd anal fin spine, continued by 1-2 canals on caudal fin, but no dorsal or ventral caudal fin lateral line branches in any of the 15 specimens available. Lateral line counts: 2+14 pores/7 pores(1, 14.6 mm), 15/9(1), 16/2 pores + 7(1), 16/8(7), 16/9(2), 16/10(1), 17/8(2).

Fins, except caudal fin, naked. One or a few vestigial scales close to soft dorsal fin base occasionally lie mostly on the fin itself, however. Caudal fin with deeply concave basal squamation, marginally scales extending to 1/3 of fin; interradial scales in single series and distally cycloid.

Fins: first dorsal fin spine above first lateral line scale, length 2/5 that of last spine; spines equal in length from 5th but last 2 again increasing. Lappets short, free, little beyond spine tips, anterior with rounded tips, posteriorly truncate with posterior point. Soft fin pointed, 5th and 6th rays longest, 5th with long filamentous tip reaching beyond caudal fin base. D. XIII.11.i(1), XIV.9(1), XIV.10(6), XIV.11(1), XV.9(4), XV.10(2). Soft fin rounded in 37.2 mm specimen, in others (larger) pointed, to 1/3 of caudal fin, from 72 mm produced, reaching 3/4 to past caudal fin.

Soft anal fin pointed, 3rd and 4th rays longest, latter produced, reaching to end of caudal fin. A. III.7(1), III.8(11), III.9(3). Soft anal fin similar to soft dorsal fin but not reaching quite as far, and never produced in paratypes, to half to 3/4 of caudal fin in large specimens.

Caudal fin hind margin very little concave, corners well-rounded. In paratypes, hind margin feebly rounded, subtruncate or as in holotype. Pectoral fin tip rounded, reaching to above 2nd anal fin spine in holotype, to above 2nd or 3rd anal fin spine in paratypes. P. 13(14), 14(1). Pelvic fin pointed, first ray produced, reaching to base of 5th anal fin ray; in young, ca 40 mm, pointed, not produced, to 2nd anal fin spine; in large paratypes first ray slightly produced, reaching to 3rd anal fin spine to 4th anal fin ray.

Jaw teeth: in 14 specimens 37.2-95.3 mm, 10-20/13-21 teeth in upper/lower jaw outer hemiseries, irregular increase in upper, decrease in lower, with increasing size; 2 (young) to 4 (large) inner series anteriorly in upper, 3 (young) to 4 (large) inner series anteriorly in lower jaw. In holotype an outer series of stronger teeth to near jaw ends and an inner band of small teeth, wide only anteriorly, in lower jaw along anterior half of each jaw half, in upper jaw one inner series nearly as long as outer. In small specimens only anterior half of upper jaw toothed and inner series to jaw end only in specimens 85 mm and larger.

Gill arches and tooth plates examined in 78.1 mm paratype. First gill arch (Fig. 78) externally with 2 papilliform epibranchial, a broader raker in angle and 5 ceratobranchial rakers, upper 4 of which broad, blunt, well-denticulated (a few teeth projecting through skin), connected medially to transverse narrow soft swelling, lowermost minute, hardly visible. Inner 2 + 1 + 9 rakers short, denticulated, with medial accessory swellings. Epibranchial 1 slender, without ventral expansion, medial arms close together. Pharyngobranchial 1 slender, rod-like. Microbranchiospines externally on 2nd 4th arches. In entire material, 1-2 epibranchial and 4(2), 5(11) or 6(2) ceratobranchial rakers.

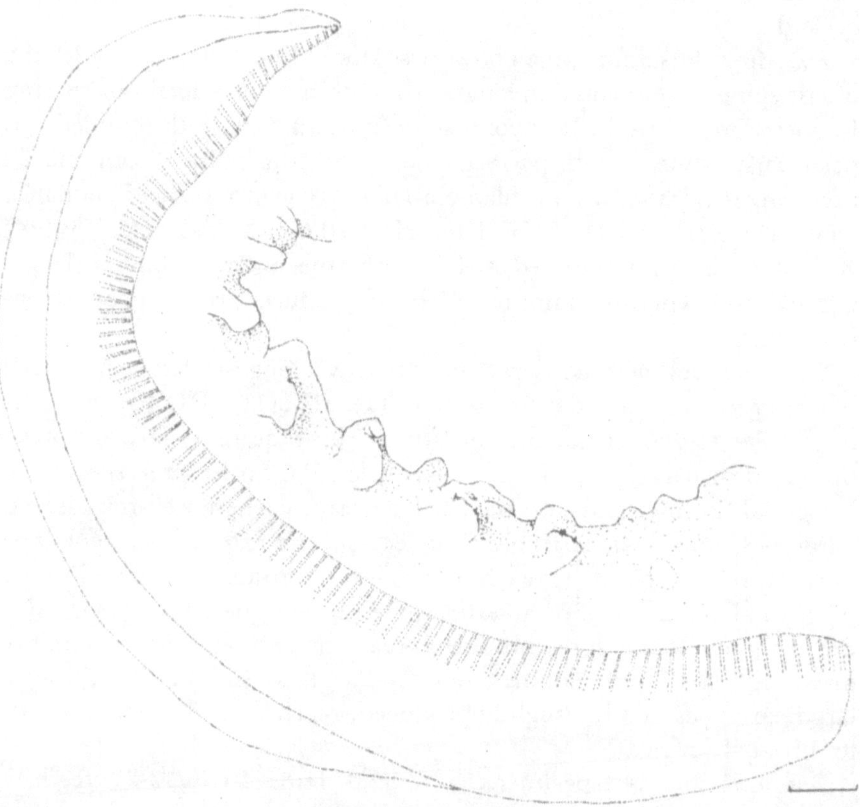

Fig. 78. *Aequidens paloemeuensis*. First gill arch in lateral aspect. Scale 1 mm. From ISNB A87/1966452.3943, 78.1 mm SL.

Lower pharyngeal tooth plate (Fig. 79) broad, with about 6 series of moderately strong teeth; anteriormost simple, conical, marginal small, slender, with posterior pointed cusp and subapical rostral curvature; median and posterior stronger, compressed, with posterior long antrorse cusp and shelf-like rostral expansion, few teeth noticeably worn. Ceratobranchial 4 with 3 tooth plates with 8, 8, and 6 teeth respectively.

Vertebrae: 12 + 13(7), 3 vertebral centra within caudal peduncle.

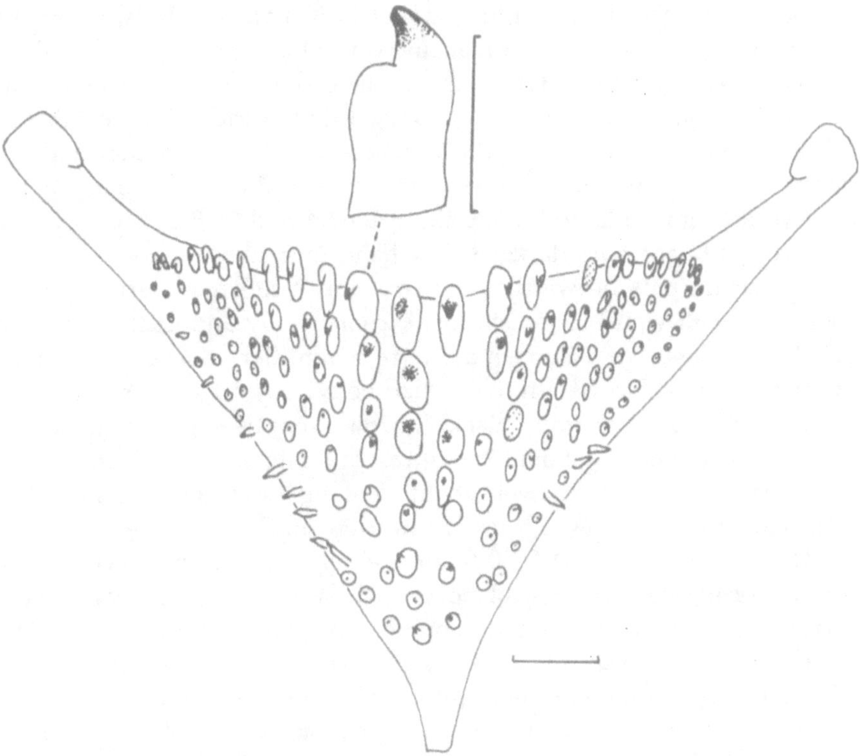

Fig. 79. *Aequidens paloemeuensis*. Lower pharyngeal tooth plate in occlusal view, and posterior tooth in medial aspect. Scales 1 mm. From NRM A87/1966452.3943, 78.1 mm SL.

Colouration

Ground colour light greyish white, on chest, preventrally, along midline of abdomen, and on underside of head; cheek and flanks more yellowish. Snout and preorbital light grey; dark stripe between orbits over dorsal (nasad) half of preorbital and around snout tip, another between orbits forwards curved, above nostrils, a third directly between orbits; 2 much fainter and narrower stripes below eye, one along buccad preorbital margin, the other paralleling ventrally. Operculum greyish brown.

Prominent brown stripe from dusky area next below orbit down to lower margin of preoperculum, somewhat narrower ventrally. Nape scales basally very light brownish grey, but major portion brown, margin narrowly black-lined. Scales above horizontal band finely black margined, below light margined. Light, yellowish, dot on scale bases below horizontal band forming regular series of dots.

Blackish brown prominent horizontal band from just behind orbit caudad to caudal fin base spot; margins uneven, especially as narrowing in bar interspaces behind Bar 5. Light brown vertical bars, wider than interspaces, across sides, not reaching ventral margin of body (numbered from posteriormost rostrad): 1) on caudal peduncle; 2) above end of anal fin base; 3) above spinous anal fin base; 4) in advance of vertical through vent; 5) posterior to vertical through pelvic fin base. On back and nape around dorsal fin origin a brown area, not extending down to lateral line; brown area above lateral line uniting Bars 4 and 5; Bar 3 brown above horizontal band; bars darker below band than above.

Dorsal fin pale brownish; outer margin of soft part narrowly white; soft part with white roundish dots except basally and distally. Anal fin slightly darker, lower margin dark grey; faint light dots on last 3 membranes. Caudal fin basally light, like body, but major portion of each scale dark brown; dorsally blackish brown spot of irregular shape, continuous with horizontal flank band; rest of fin lighter than dorsal fin, and with clear round dots all over, about 12 along middle membranes. Pelvic fin outer margin brownish grey, inwardly smoky.

Larger paratypes (71.9-90.8 mm) similar; but a spot behind orbit usually separate from rest of horizontal band which then commencing with blotch posterosuperiorly on operculum. Bars 4 and 5 tend to be best developed (on lower sides); Bar 2 divided vertically in one specimen. Back above lateral line rather uniformly brown back to Bar 4. Dorsal fin with more distinct light lappet tips; anal fin with clear hind margin. Caudal fin base spot well-delimited, separate or almost separate from horizontal band, more or less ovate, restricted to dorsal lobe, not reaching dorsal margin of fin, fin base below brown or of same colour except for lower margin.

Young (37.2-58.7 mm) lighter; whitish to yellowish ground colour, markings fainter. Narrow line on cheek missing, but preorbital dark stripe more intense than in larger specimens. Stripe below eye more backwards inclined. Somewhat oblong, blackish brown blotch in Bar 4 where crossed by horizontal band dominant body marking; horizontal band brown, more intense than vertical bars. Bar 4 more intense than other bars. One specimen with Bar 3, another with Bar 2 divided vertically. Light spots on scale bases missing from specimens smaller than 48.6 mm. Caudal fin base spot as in adults, but separated from horizontal band by ligh stripe marking end of caudal peduncle. Fins lighter than in adults; immaculate caudal fin in those 37.2-48.6 mm, lightly dotted in 58.7 mm specimen. Pelvic fin light, slightly darkened along outer margin.

Juvenile 14.6 mm somewhat faded. Yellowish whitish with dark scale margins producing reticulated effect. Head markings faint. Horizontal band faint, spot in Bar 4 distinct; Bars 4 and 5 more distinct than the other bars (on lower sides). Fins immaculate, almost clear, pelvic fin clear; caudal fin base spot small, behind horizontal band.

Distribution (Fig. 76)

Known only from the type locality, where syntopic with *Krobia itanyi*.

Etymology

Named for the Paloemeu River.

Notes

Aequidens paloemeuensis is assigned to *Aequidens* pending revision of the group. In some characters it rather resembles *Krobia*.

Aequidens paloemeuensis has an unmodified hyoid similar to that of *Aequidens*, distinguishing the species from *Krobia*. The species differs from *Krobia* also in (1) naked vertical fins, (2) reduced number of vertical bars (3 instead of 4 posterior to midlateral spot), and (3) the lateral band running to the caudal fin base instead of to or toward the dorsal fin base.

Characters shared with *Krobia* but not with most other *Aequidens* include: (1) uniserial instead of triserial predorsal squamation, (2) vertebral count 12+13 rather than 13+13, (3) three facial stripes instead of none, (4) wide suborbital stripe retained in adult. The low squ. long. count, 23 (24 or more in other *Aequidens*), may be correlated with the low vertebral count.

The lateral band of *A. paloemeuensis* has a course intermediate between that in *Aequidens tetramerus* in which it posteriorly runs close to the lower lateral line, and that in *Krobia* in which it ends at the end of the base of the dorsal fin. The *A. paloemeuensis* band minus caudal peduncle section is nearly as the band of *Krobia guianensis*. In some *Aequidens* species (e.g,. *A. pallidus*), however, the lateral band has a posterior course very little removed from the lower lateral line, representing intermediacy between *A. tetramerus* and *A. paloemeuensis*.

Aequidens paloemeuensis strongly resembles *A. potaroensis* in colour pattern, shape and meristics. *Aequidens potaroensis* has been collected so far only in the Potaro R. and nearby Essequibo R. in Guyana. These species differ chiefly in predorsal scale pattern and lateral band appearance. *Aequidens potaroensis* has triserial predorsal scale pattern, and the lateral band is much fainter anterior to the midlateral spot than posterior to it.

KROBIA n. gen.

Type species

Acara guianensis Regan, 1905.

Diagnosis

Moderate-sized South American cichlids (to 128 mm SL) with four dentary lateralis foramina, distinguished from all other cichlids by characters in combination: uniserial predorsal squamation; narrowly scaled or naked vertical fins; blotchy lateral band from head toward end of base of dorsal fin; lateral line sequence dorsally on caudal fin situated between rays D3 and D4; similar to *Bujurquina*, *Tahuantinsuyoa* and *'Aequidens' pulcher* in having ledges forming a furrow for the hyoid artery dorsally on the ceratohyal, though lower than in the listed genera.

Other characters of *Krobia* include lip folds of American type, both discontinuous; jaw teeth caniniform, little recurved; premaxilla without rostral foramen; 12+13 vertebrae; 2 supraneurals; 3-4 vertebrae within caudal peduncle; parhypural spine present; paired short hypapophyses on 4th vertebra; microbranchiospines externally on 2nd to 4th arches, variably internally; externally on first gill arch 1-2 epibranchial and 3-5 ceratobranchial gill rakers; 2-3 toothplates on 4th ceratobranchial; wide epibranchial 1; rod-like pharyngobranchial 1; minute interarcual cartilage; moderately wide cephalic lateralis canals and foramina; flank and chest scales ctenoid, large, squ. long. 23-24, modally 23; upper lateral line in 3rd horizontal scale series above the lower, at 3 (initially) to 1 1/2 (terminally) scales distance from dorsal fin; prepelvic scales cycloid in median and 2 parallel series on each side; naked preoperculum; 16 circumpeduncular scale series; operculum with 2 vertical series of scales; dorsal fin scaled; pectoral and pelvic fins naked; accessory caudal fin lateral lines between rays D3-D4 and V4-V5 or absent; pectoral fin asymmetrical, 4th ray longest; pelvic fin with first ray longest; dorsal fin with 13-15 spines, 9-11 rays (mode D. XIV.10); anal fin with 3 spines, 7-9 (modally 8) rays; caudal fin subtruncate or slightly emarginate, with 3 procurrent and 8 principal rays in each lobe; 4 vertical bars posterior to the one containing midlateral spot; three facial stripes between eyes; suborbital stripe present at all sizes; no blue stripes or dots on head sides; caudal fin base spot superior; lateral band running to or toward end of base of dorsal fin; caudal fin immaculate or dusky with clear dots, which are smaller in ventral than in dorsal lobe.

Included species

Acara guianensis Regan, 1905
Aequidens itanyi Puyo, 1943
Two undescribed species are recognized in material from French Guiana.

Etymology

Krobia is a Surinamese vernacular name applied on several cichlid species, including *K. guianensis*. It is to be considered as of feminine gender.

Distribution

The generic range includes the Guianas and the Xingu river drainage.

Notes

Kullander's (1983a, p. 280) " '*Aequidens*' *guianensis* group" corresponds to *Krobia* except for removal here of *A. potaroensis*, '*A.*' *hoehnei* (Ribeiro), and one undescribed species listed by Kullander.

Krobia species are similar to *Aequidens* s. str. in a general fashion, including long caudal peduncle, cycloid prepelvic scales, and three anal fin spines. The genera differ drastically in colour pattern, hyoid shape, fin squamation and predorsal scale pattern. *Aequidens paloemeuensis* is superficially similar to *Krobia*, but has the hyoid similar to *Aequidens* and *Cichlasoma*, i.e. without notch or prominent groove.

Bujurquina, *Tahuantinsuyoa* and the '*A.*' *pulcher* group are similar to *Krobia* in having a high-walled groove dorsally on the anterior ceratohyal, leading the hyoid artery to a dorsomedial course and shielding it from lateral view. Those genera and *Krobia* are similar also in having an oblique lateral band running to or toward the end of the dorsal fin base, uniserial predorsal squamation, lack of a large foramen in the premaxilla and retention (in some form) of the suborbital stripe in adults. These character states distinguish *Krobia* from *Aequidens* and *Cichlasoma* but are not unique.

In *Bujurquina*, *Tahuantinsuyoa* and '*A.*' *pulcher*, the lateral wall of the hyoid artery groove is high and shields the hyoid artery which caudally runs on the medial side of the hyoid. In *Krobia*, as verified in both *K. itanyi* and *K. guianensis*, the lateral groove wall is lower and the groove is best described as dorsal in course. In regard to ceratohyal shape, *Krobia* thus seems to be less derived than the other genera with ceratohyal groove walls and thus may represent their sister group.

Scaled instead of naked soft vertical fins, dorsal accessory caudal fin lateral line between rays D3 and D4 instead of D1-D2; vertebral hypapophyses on 4th instead of 3rd vertebra, long instead of absent or rudimentary parhypural spine, anterior and posterior lamina on the first epibranchial, substrate instead of mouth brooding, and none of the lateralis foramina or canals of the head expanded, are characters states distinguishing *Krobia* from *Bujurquina* and partly also from *Tahuantinsuyoa* and the '*A.*' *pulcher* group, but which are likely plesiomorphic for *Aequidens*-like cichlids. The fins are narrowly scaled in comparison with most *Cichlasoma* species.

Blue iridescent stripes and spots on the head sides are found in most South American cichlids, including *Bujurquina*, *Tahuantinsuyoa*, '*A.*' *pulcher* group, *Laetacara*, and some *Aequidens* species. Such stripes are absent from *Krobia*, *Cichlasoma*, *Cleithracara*, *Nannacara* and some *Aequidens* species (e.g. *A. tetramerus*). Plain sides represent a derived character state, but being a loss state it is doubtfully homologous in the groups listed. In the genera other than *Krobia* there is commonly a dull metallic, greenish to yellowish sheen on the head sides, not on record for *Krobia*.

Another head marking, the three frontal stripes, is shared only with *Aequidens paloemeuensis* and *A. potaroensis*, though variably expressed among *Krobia* species. In *Laetacara*, the only other group with distinct facial stripes, the ventralmost stripe is relatively distinct, though narrower; posterior to it there is, like in remaining *Aequidens*-like cichlids, only one, wide stripe between the orbits.

Krobia has an epibranchial 1 shape similar to that of *Cichlasoma*, *Aequidens*, and '*A.*' *pulcher*, i.e. with appreciably wide lamina along the anterior and posterior aspect; in *Bujurquina* and *Tahuantinsuyoa* the epibranchial 1 is flattened dorsally and the lamina are absent or (the posterior) reduced in width giving the bone a slenderer aspect than in the other genera listed.

Krobia species described here have 12+13 vertebrae. Among other cichlasomines that number is recorded also in '*Aequidens*' *biseriatus* (Regan) (syntypes, n = 7), '*A.*' *pulcher* (Gill) (n = 2), *A. chimantanus* Inger (n = 5) and *A. paloemeuensis* (n = 7).

Most *Bujurquina* species and *Tahuantinsuyoa macantzatza* do have 12 preanal vertebrae though the total count is usually 26; in these, however, the first two hemal spines project between the first two interhemals, and it appears that the first hemal-spine bearing vertebra is a modified abdominal vertebra.

With his interpretation it follows that *Bujurquina* and *Tahuantinsuyoa* species have a basic count of 13 + 13 vertebrae, like nearly all other cichlasomines, and that variation in the total count is due to loss or addition of caudal vertebrae. The only *Bujurquina* specimens noted to have 12 abdominal vertebrae are the holotype of *B. zamorensis*(12 + 14), the *B. oenolaemus* type series (in 5 of 8, 11 instead of 12 preanal vertebrae) and one *B. vittata*. The latter two species (with 23-25 vertebrae) represent a further reduction of the vertebral column.

Aequidens and *Cichlasoma* species have modally 26 (13 + 13) vertebrae. Lower counts (24-25) represent reduction of the caudal vertebrae (13 + 12, 13 + 11), with significant exceptions only in *C. pusillum* Kullander (12 + 12, 12 + 13, 13 + 12) (Kullander 1983a, table 14).

The only other cichlasomines with 12 abdominal vertebrae so far studied are '*A.*' *sapayensis* (Regan) (12 + 14, n = 1), *Cleithracara maronii* (12 + 12), and *Laetacara* species (12 + 12).

The vertebral count in *Krobia* may be explained as derived from the 13 + 13 condition by loss of one abdominal vertebra, or from the pseudo-12 + 14 condition by loss of the 13th vertebra. Further study of osteological material might reveal characters by which individual vertebrae can be identified, what seems not possible from radiograph studies; then the transformations of vertebral counts could be traced more precisely. On the basis of lower frequency among cichlids, the 12 + 13 count in *Krobia* is a derived character.

Krobia guianensis (Regan)

(Figs 80-91)

Acara guianensis Regan 1905b, p. 340 (Guiana [BMNH 1851.5.2:8]).

Surinamese records

Acara tetramerus; Klinckowström 1892, p. 104 (listed [NRM ZISP505, Surinam]).
Aequidens vittatus; Boeseman 1952, p. 197 (listed: Coropina creek, Republiek [RMNH 19371, 2 specimens, one according to Boeseman]; Zanderij I, upper part of Coropina creek [not found in RMNH]; Zanderij I, savanna streamlet [not found in RMNH]; Upper Saramacca River [RMNH 18322] Surinam, (Upper Saramacca?) [not found in RMNH]; local names). — Boeseman 1954, p. 23 (listed: probably neighbourhood of Paramaribo [probably RMNH 23825]; note on abnormality). — Boeseman 1956, p. 196 (listed: Surinam /RMNH 21380/). — Keenleyside & Bietz 1981, p. 87 *et seq.* (field studies on reproductive behaviour; near Onverdacht).
Aequidens potaroensis; Boeseman 1954, p. 23 (descr.; neighbourhood of Paramaribo [RMNH 23825]).

Material

1345 specimens from the Corantijn, Nickerie, Saramacca, Coppename, Suriname, and Cottica drainages.

Fig. 80. *Krobia guianensis*, 98.6 mm SL, RMNH 31118, from Saramacca Kreek.

Description

Based on a 98.6 mm female (RMNH 31118, Saramacca Kreek; Fig. 80). Dorsal and anal fin, squ. long., and gill raker counts from 106 specimens from the Suriname, Coppename, Nickerie and Corantijn R. systems; other counts from 26 measured specimens. Largest specimen examined 127.9 mm (ZMA 106.991).

Shape: moderately elongate; moderately compressed laterally. In rostral aspect narrowly ovate with narrow rounded nape and well-rounded chest. Dorsal and ventral contours similarly arched, dorsal slightly more curved. Predorsal contour distinctly downcurved on snout,

Fig. 81. *Krobia guianensis*, 111.2 mm SL, RMNH 31090, from Bigi Birri Kreek near Cupido.

continued ascend slightly arched; dorsal fin base contour horizontal from 3rd dorsal fin spine, an initially distinct but weak then increasing slope from 7th spine, abruptly bent terminally. Ventral head contour straight, not as steep as predorsal; chest contour slightly curved.

Head: moderately long and, save for snout, uniformly wide. Snout from above narrowing rostrad, tip narrowly rounded; in lateral aspect broadly rounded with narrow lower jaw and much steeper dorsal than ventral contour; slightly extended. Mouth appearing almost subterminal, very ventral on snout; narrower than interorbital width. Tip of maxilla exposed, not reaching backwards to a vertical from anterior margin of orbit. Ascending processes of premaxilla short, not reaching orbit. Interorbital area indented medially, otherwise slightly curved. Eye large, situated in middle of head length, in upper half of head, just inferior to predorsal contour.

Scales: Squ. long. 23(101), 24(4), -(1). Predorsally 3 series posteriorwards from between orbits, median of 7 scales plus a scale pair just anterior to dorsal fin origin, all cycloid; 8 scales, interorbital cycloid, in lateral series; along midline of about uniform flank scale size. A median series of 2 scales between and 10 anterior to pelvic fins; larger, posteriorly, of about 2/3 size of flank scales, but not much smaller anteriorly. Chest scales nearly size of flank scales; pectoral axilla naked save for small cycloid basal scales.

Cheek scales in 2(1), 3(21), 4(4) series, and one series up behind orbit, posterior feebly granulate; posterosuperior of about flank scale size, anteroventrally feebly smaller. Operculum with 3 ctenoid scales in anterior series, 3 cycloid in posterior. Two large (about flank scale size) cycloid subopercular scales. On posterior 2/3 of of interoperculum 2 cycloid scales, smaller than flank scales.

Upper lateral line overlapping lower by 2 scales, extending to below base of first dorsal fin ray. Lower lateral line extending to above base of 3rd anal spine, immediately continued on caudal fin by 1 (left side) or 2 (right side) tubed scales; on dorsal caudal fin lobe by a tubed scale between rays D3 and D4, on ventral caudal fin lobe by a tubed scale between rays V4 and V5. Lateral lines comprising 13/10(1), 14/9(1), 15/-(1), 15/8(6), 15/9(4), 16/-(1), 16/7(1), 16/8(9), 16/9(2) scales, lower continued by 2, rarely one canal on caudal fin, commonly 1-2 tubed scales on dorsal and/or ventral lobe.

Fin scales (Fig. 82): marginal minute body scales extend over soft dorsal fin base; otherwise soft dorsal fin only feebly scaled, a few minute ctenoid scales along base and on right side a few interradial scales. Anal fin naked save for a small ctenoid scale between bases of 1st and 2nd rays on right side. Caudal fin scaled basally; proximally a layer of larger scales, about

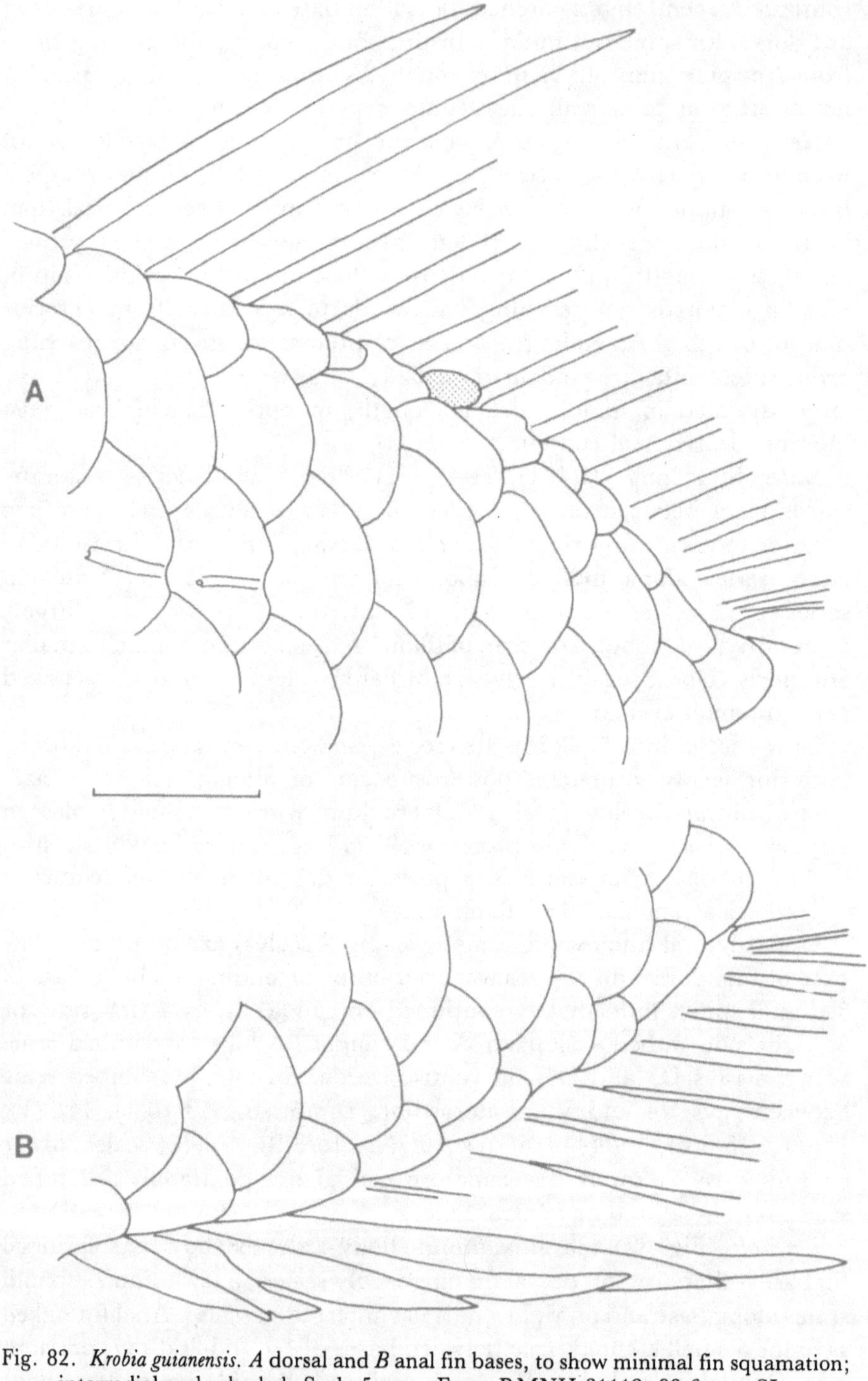

Fig. 82. *Krobia guianensis*. *A* dorsal and *B* anal fin bases, to show minimal fin squamation; interradial scale shaded. Scale 5 mm. From RMNH 31118, 98.6 mm SL.

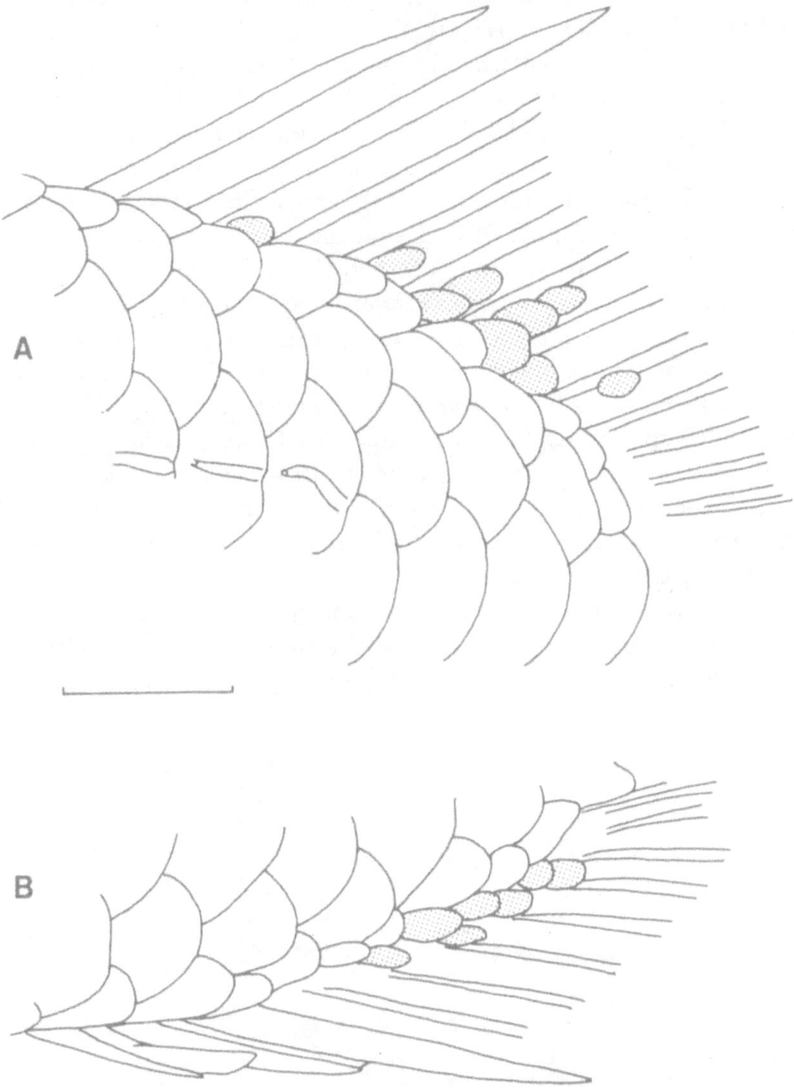

Fig. 83. *Krobia guianensis. A* dorsal and *B* anal fin bases, to show well-developed fin squamation; interradial scales shaded. Scale 5 mm. From RMNH 31070, 84.9 mm SL.

3 scales long, distally small ctenoid interradial scales; distal margin of scaled area concave, medially 1/4, marginally 1/3 of fin scaled,

Fins: first dorsal fin spine, in advance of a vertical from hind margin of operculum, little longer than 1/3 length of last; spines equal in length from 5th, last 3 again slightly increasing in length. Lappets free, pointed, posteriorly longer, but little beyond spine tips. Soft dorsal fin pointed,

4th ray longest, with a filament extending to end of caudal fin. D. XIII.9(1), XIII.10(1), XIII.11(3), XIV.9(5), XIV.10(90), XIV.11(3), XV.9(2), XV.10(1). Soft anal fin pointed, 4th ray longest, with filament extending past 3/4 of caudal fin. A. III.7(1), III.8(92), III.8.i(1), III.9(11), IV.7(1). Caudal fin with emarginate hind margin with well rounded corners. Pectoral fin subacuminate, reaching to above base of 3rd anal fin spine. P. 13(12), 14(14). Pelvic fin pointed, first ray produced, extending to base of 2nd anal fin ray.

Jaw teeth: with tips rounded off or blunt. An outer series of stronger teeth along most of jaw rim, forwards size increases; inner band of smaller teeth anteriorly, one inner series continued posteriorwards almost as far as outer. In lower jaw anterior 2/3 with outer series of stronger teeth, inner band of smaller teeth along anterior half of each jaw half. In 19 specimens 43.1-111.2 mm, 10-17 teeth in upper jaw outer hemiseries, variation irregular, inner teeth in about 3 series anteriorly; in 25 specimens 33.8-111.2 mm, 14-23 teeth in lower jaw outer hemiseries, distinctly decrease with increasing size, inner teeth in about 3, occasionally 4 series anteriorly.

Gill rakers: externally on first ceratobranchial 3(5), 4(88), 5(12), -(1) Gill arches as in *K. itanyi*. Microbranchiospines externally on 2nd to 4th arches, absent or present ventrally internally on 4th arch.

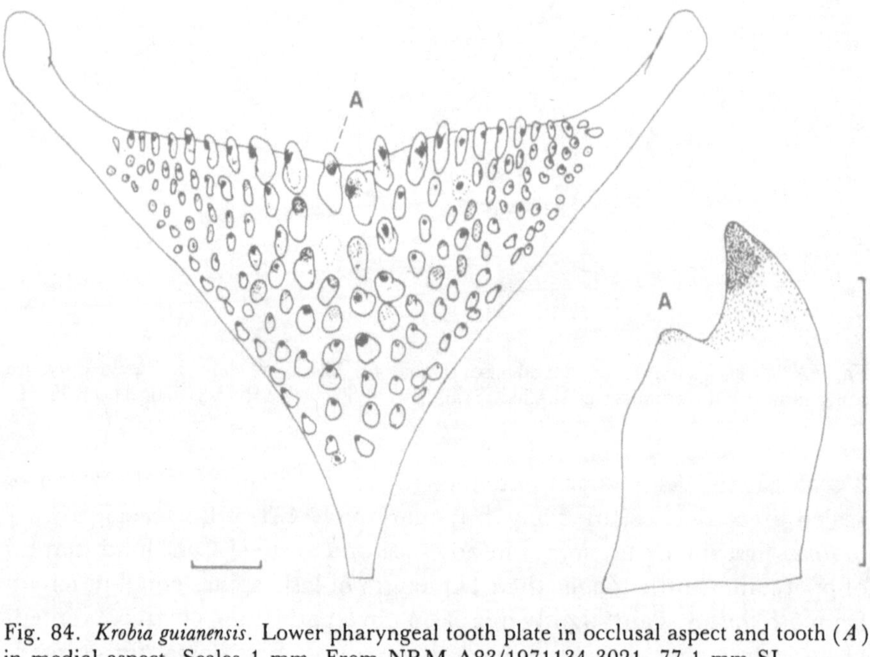

Fig. 84. *Krobia guianensis*. Lower pharyngeal tooth plate in occlusal aspect and tooth (*A*) in medial aspect. Scales 1 mm. From NRM A83/1971134.3021, 77.1 mm SL.

Lower pharyngeal tooth plate (Fig. 84) with many apically abraded medial teeth, anterior teeth subconical with erect or slightly caudad bent tip, remainder compressed, gradually larger toward posteromedian, with posterior antrorse cusp and 1-2 minute cusps on anterior margin.

Vertebrae: 12+13(12).

Variation: shape very variable, from slender like *K. itanyi* to very deep and compressed (Fig. 81), a 111.2 mm specimen with depth 50.8% of SL extreme. Squamation, dentition and fins varying as in *K. itanyi*. Extremes in dorsal and anal fin squamation of large specimens shown in Figs 82 and 83. No sexual dimorphism in fin shape or otherwise; caudal fin, however, rarely with slightly produced rays D4 and D5.

Juveniles of 15-20 mm comparatively elongate with rather short snout and mouth wider than flat interorbital; eye relatively large and most in anterior half of head. Predorsal, back, anterior sides and chest scales cycloid. Fins naked. Soft dorsal and anal fins rounded, reaching little beyond caudal fin base; caudal fin hind margin slightly convex.

Osteology

Two cleared and stained specimens, 29.2 and 37.7 mm SL are similar to *Bujurquina*, *Tahuantinsuyoa* (Kullander 1986) and *Cichlasoma* (Kullander 1983a) in general osteology. Both have 25 vertebrae, with pleural ribs to the 12th, epipleural ribs to the 10th, a hemal fenestra and hemal spine on the 13th vertebra, and paired short hypapophyses on the 4th vertebra. Vertebrae contained in the caudal peduncle 3-4. The parhypurapophysis is long and spine-like.

There are no distinctive characters in the neurocranium (not figured), suspensorium, jaws or pectoral girdle. There is no large foramen in the rostral aspect of the premaxilla. The posteriomedial wing of the palatine bears a dorsal weakly developed ethmoidad articulation facet (Fig. 85), and the corresponding facet on the lateral ethmoid is equally indistinct. In the smaller specimen there is no contact between vomer and mesethmoid but in the larger specimen an anterolateral projection from the mesethmoid makes contact on the right side and reaches nearly to the vomer on the left side. The lateralis canal system resembles that of *Cichlasoma*; the infraorbitals are relatively slender and no foramina or canal branches are enlarged (Fig. 86).

The branchial skeleton (Figs 87-88) includes no unusual features. The length of epibranchial 1 is 43-44% that of ceratobranchial 1; anteriorly and posteriorly run laminar wings contributing considerably to the width as seen in dorsal and ventral aspect. The arms are short, subequal in length, the posterior medial arm slightly thicker than the anterior, and little divergent. The interarcual cartilage is small, suspended in a con-

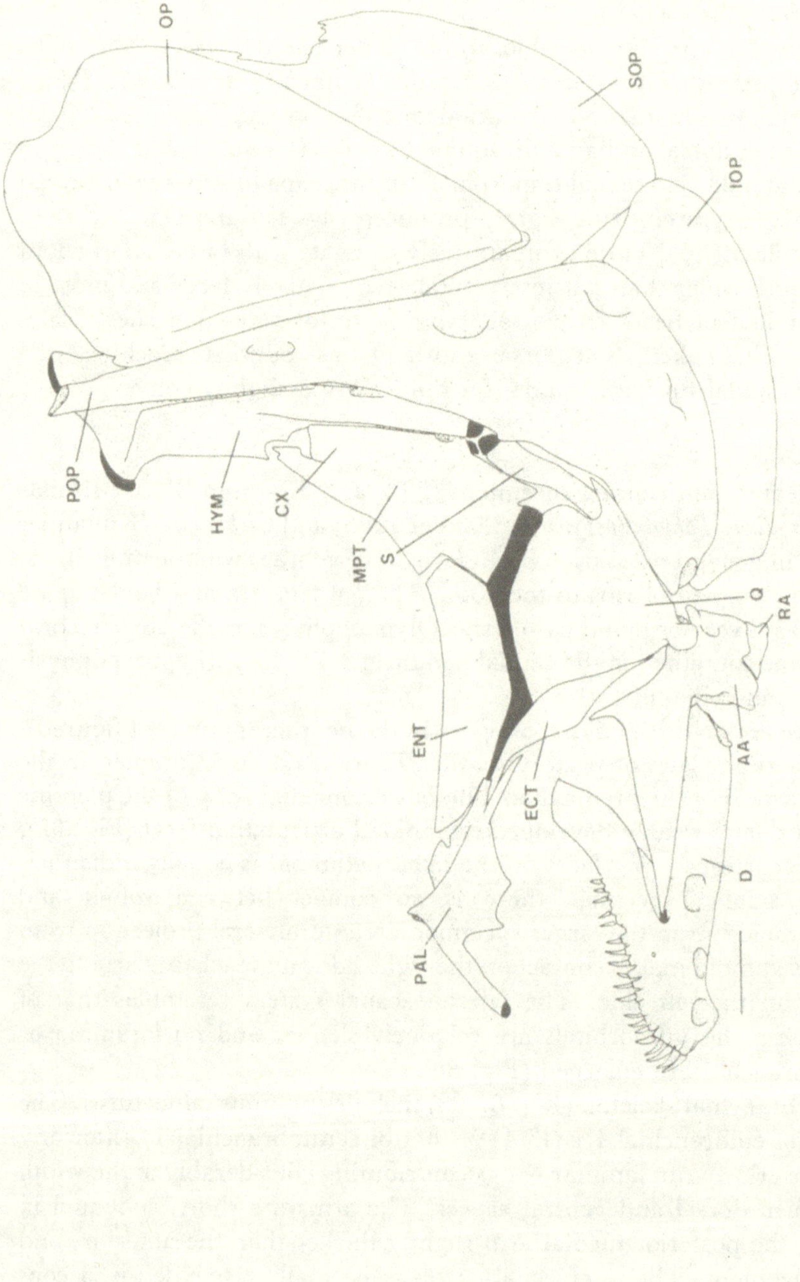

Fig. 85. *Krobia guianensis*. Suspensorium, lower jaw and operculars. *AA* anguloarticular, *CX* calyx, *D* dentary, *ECT* ectopterygoid, *ENT* entopterygoid, *HYM* hyomandibula, *IOP* interoperculum, *MPT* metapterygoid, *OP* operculum, *PAL* palatine, *POP* preoperculum, *Q* quadrate, *RA* retroarticular, *S* symplectic, *SOP* suboperculum. Cartilage black. Scale 1 mm. From NRM A83/1975297.3023, 37.7 mm SL.

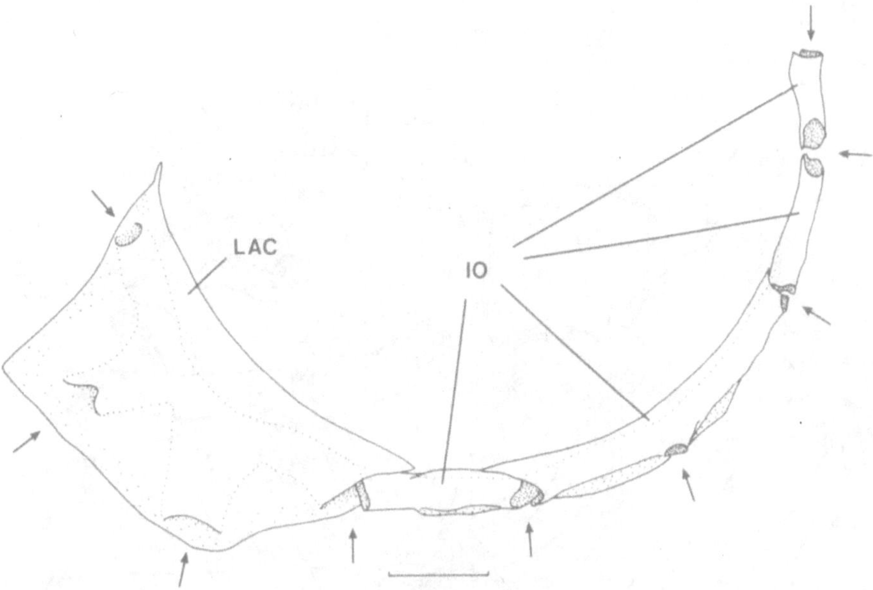

Fig. 86. *Krobia guianensis*. Suborbital series. *IO* infraorbitals, *LAC* lacrimal. Arrows point to lateralis canal openings. Scale 1 mm. From NRM A83/1975297.3023, 37.7 mm SL.

nective tissue strap. Pharyngobranchial 1 is rod-like. There is no distinct medial process of ceratobranchial 4, which has 2 or (on one side in one specimen) 3 small dorsal tooth plates.

The hyoid (Fig. 89) shows some variation in the development of the lateral wall of the dorsally positioned furrow for the hyoid artery, being most indictinct in the figured specimen. The urohyal has a long, dorsocaudad directed anterior process which flattened along the rostral margin, and a relatively wide shelf topping the laminar portion.

Colouration

Ground colour pale dirty yellowish, on chest, preventrally, along middle of abdomen, interoperculum, lower jaw, suboperculum. Cheek light brownish. Operculum greyish brown, lighter ventrally. Nape and back brown, sides lighter with light brown vertical bars and feebly lighter interspaces. Snout and forehead dark grey; preorbital dark grey, shading to brownish towards cheek; lips grey; light stripe between orbit and nostril only indicated, and no evidence of snout stripes as in *K. itanyi*. Scales of dorsal and middle sides with inconspicuous narrow black border. Brown wide stripe from orbit down to lower margin of preoperculum.

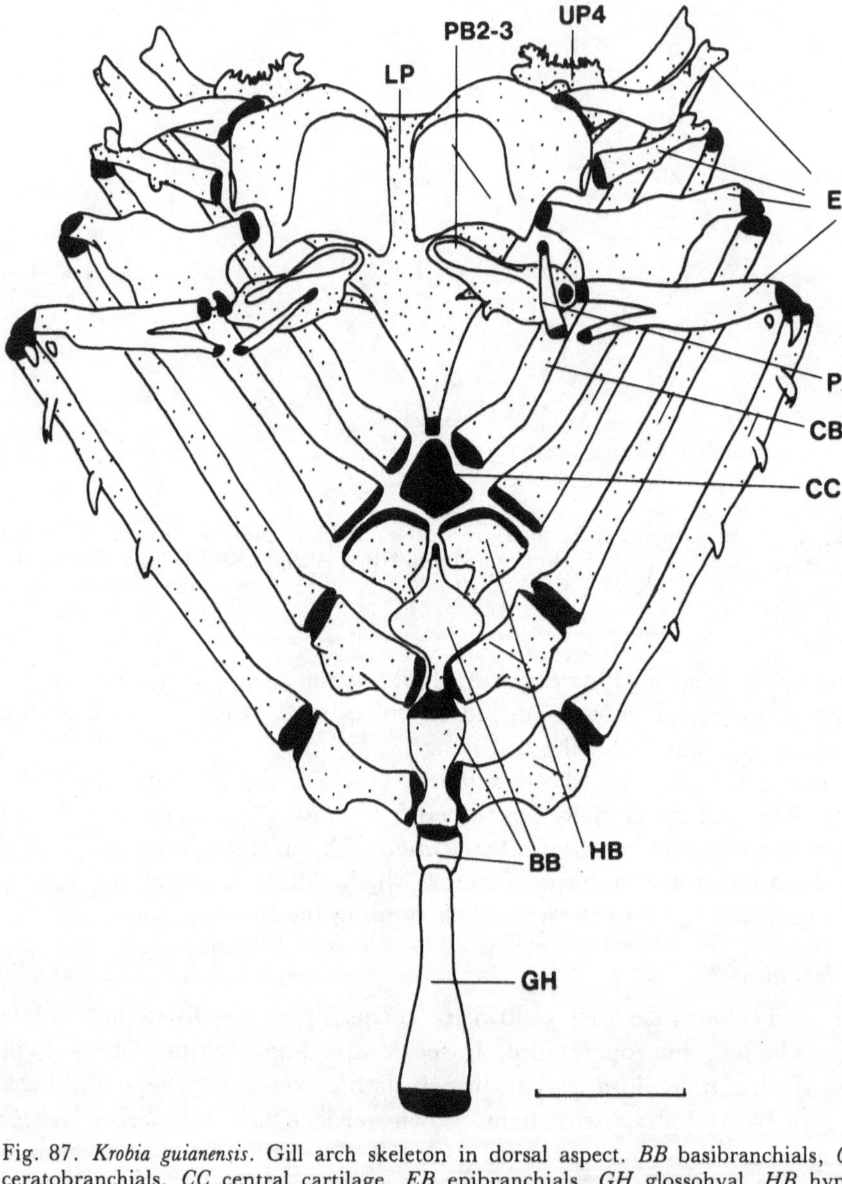

Fig. 87. *Krobia guianensis*. Gill arch skeleton in dorsal aspect. *BB* basibranchials, *CB* ceratobranchials, *CC* central cartilage, *EB* epibranchials, *GH* glossohyal, *HB* hypobranchials, *LP* lower pharyngeal tooth plate, *PB1* pharyngobranchial 1, *PB2-3* pharyngobranchials 2-3, *UP4* fourth upper tooth plate. Cartilage black. Scale 1 mm. From NRM A83/1975297.3023, 29.2 mm SL.

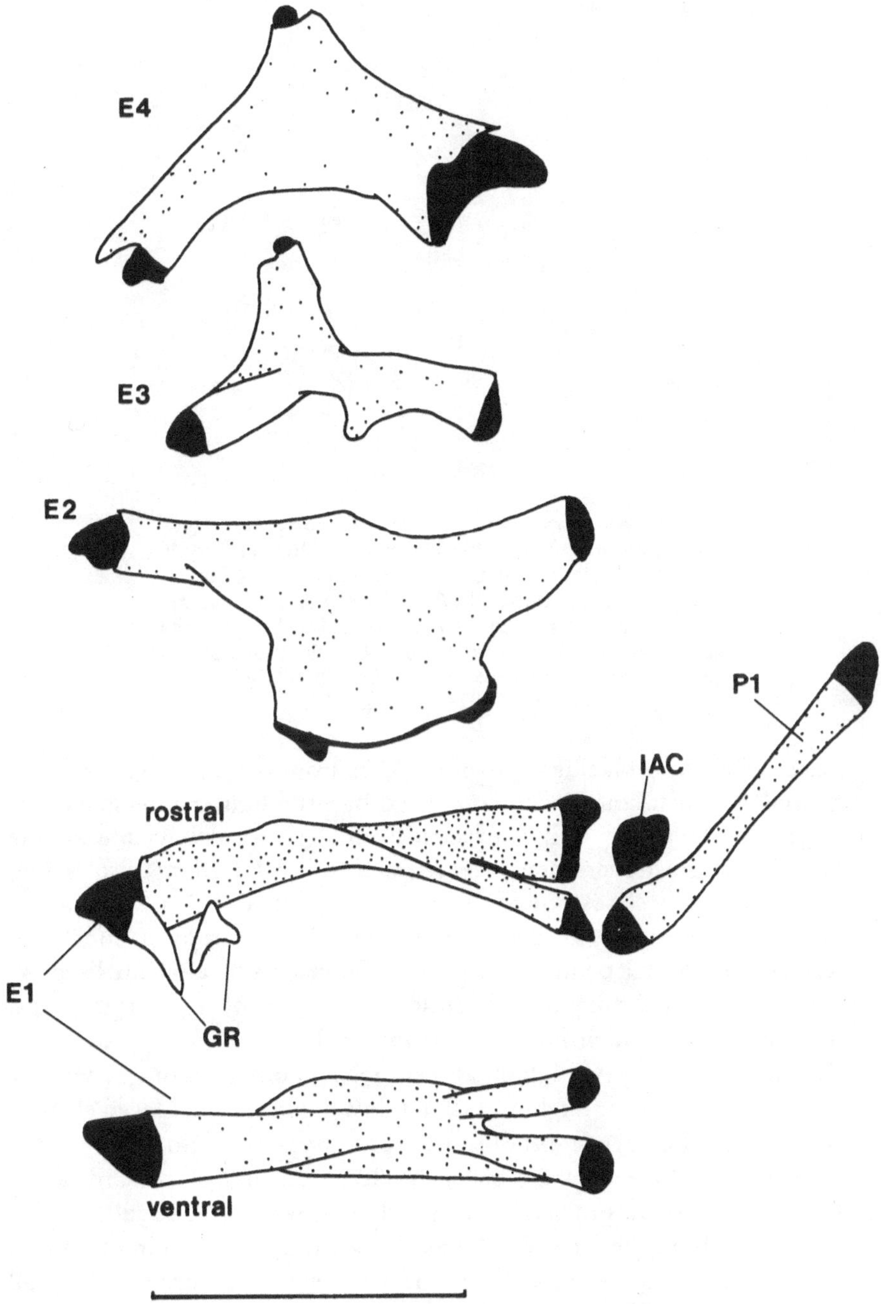

Fig. 88. *Krobia guianensis*. Right side upper gill arch elements in rostral aspect (*E2-E4* epibranchials 2-4), or both rostral and ventral aspect (*E1* epibranchial 1). *IAC* interarcual cartilage, and *P1* pharyngobranchial 1 shown in associated position with rostral aspect of epibranchial 1. *GR* gill rakers associated with epibranchial 1. Cartilage black. Scale 1 mm. From NRM A83/1975297.3023, 29.2 mm SL.

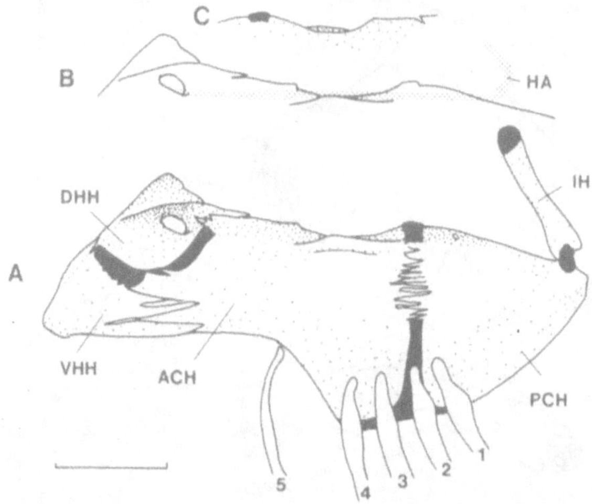

Fig. 89. *Krobia guianensis*. Left side hyoid. *A* full lateral aspect, *B* dorsal margin to show course of hyoid artery, *C* medial aspect of dorsal margin at region of artery furrow, *ACH* anterior ceratohyal, *DHH* dorsal hypophyal, *HA* hyoid artery (shaded), *IH* interhyal, *PCH* posterior ceratohyal, *VHH* ventral hypohyal. *1-5* branchiostegal rays. Cartilage black. Scale 1 mm. From NRM A83/1975297.3023, 37.7 mm SL.

Series of dark to blackish brown blotches from orbit to upper margin of caudal peduncle anteriorly, connected by little lighter interspaces: 1) on head between eye end supracleithrum; 2) just below lateral line anteriorly; 3) most prominent, squarish, in Bar 5 (squ.long. scales 7-9) and four scales in series above; 4) in Bar 4; 5) in Bar 3; 6) in Bar 2.

Faintly expressed vertical bars: 1) posteriorly on caudal peduncle; 2) anteriorly on caudal peduncle; 3) above anterior soft anal fin base; 4) above spinous anal fin base; 5) little in advance of 4, merging with abdominal colour which is similar to that of bars.

Dorsal fin light grey, lighter anteriorly, spinous portion posteriorly with a series of clear or slightly duskied dots along middle; large clear to slightly smoky dots over most of soft portion, upper margin lighter. Anal fin grey brown, lower margin blackish; clear dots on last 4 membranes. Pelvic fin outwardly greyish brown, lighter inwardly. Caudal fin with large black, slightly ovate spot on base between upper margin of fin and lower lateral line level, base below brown, rest of fin grey with small (lower lobe) to large (dorsal lobe) round dots, but lower lobe immaculate ventrally and distally.

Young without light spots in spinous dorsal fin, instead margin dark, lappet tips light; large specimens commonly with 2-3 series of light dots along spinous dorsal fin and lappet tips light. Caudal fin less vividly spotted in young, and commonly dotting restricted to basal half also in adults, distally pigment only along membrane margins.

Considerable variation in the flank blotch series with regard to degree of confluence, but always extending back to anterosuperior margin of caudal peduncle, and Bar 5 spot (midlateral spot) always dominant. Usually abdominal side scale bases with inconspicuous dark vertical stripe.

Facial stripes, as in *K. itanyi*, evident only in faded or overall light specimens, and even then the middle stripe hardly evident, otherwise forehead and snout uniformly dark.

Juveniles of 15-20 mm are much like adults, but much lighter; unpaired fins immaculate or with few dots, caudal fin base spot only a narrow dark stripe on dorsal lobe base. Dark dorsocaudad extended spot above orbit; cheek stripe more backwards inclined.

Distribution (Fig. 90)

Krobia guianensis is widely distributed in Surinam, and ranges west into the Essequibo drainage.

Local names

Krobia, ooroe wefi (Lijding 1959a; for *Aequidens vittatus*), Krobia, Ouwe wefi (Boeseman 1952; for *Aequidens vittatus*).

Notes

Krobia guianensis is easily distinguished from *K. itanyi* by having the lateral blotch series extended to the end of the dorsal fin base rather than terminating at the end of the lateral line or shorter, the usually obsolete facial stripes, and more extensively light dotted fins, adults also with light dots on the spinous dorsal fin. It is also deeper, especially at adult sizes (Fig. 91; depth usually over 45% of SL in specimens over 50 mm SL).

The original description of *Krobia guianensis* is based on a single specimen, 80 mm SL, from 'Guiana' (BMNH 1851.5.2:8), presented by the Berlin Museum. With a depth of 50.6% of SL, it is near the extreme for the species. Reexamination of the specimen confirms Regan's (1905b) data save that the lower lip fold is interrupted rather than continuous, and that the colour pattern is no longer distinct.

Eigenmann (1912, p. 72) recognized *K. guianensis* as a valid species but did not identify it in his Guyana collection. Instead he described, without comment, material from the Mahaica and Demerara as *Aequidens vittatus* (Heckel). The latter name applies, however, correctly to a very different species in the Paraguay and lower Paraná drainages (Kullander, in prep.), now in the genus *Bujurquina* (Kullander 1986).

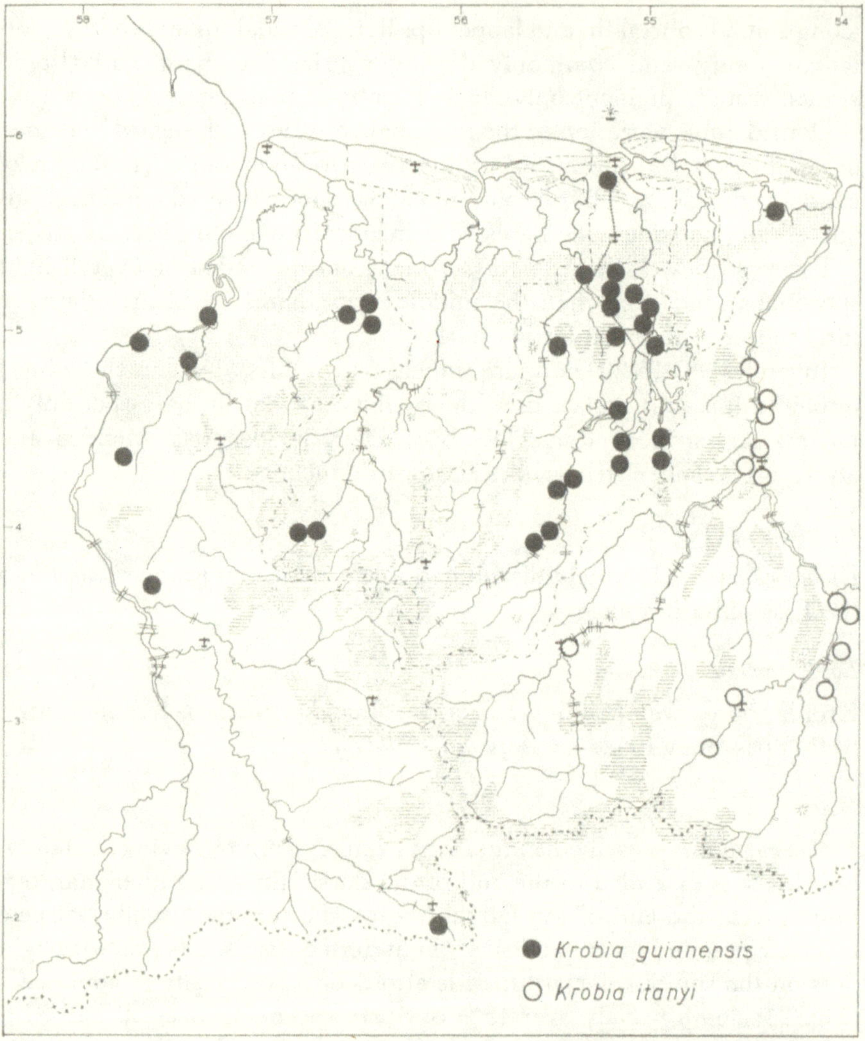

Fig. 90. Collecting localities of *Krobia* species in Surinam.

Lowe-McConnell (1969, p. 275) discussed the status of *K. guianensis* on basis of study of the holotype, concluding the distinctness of the species. She compared with Guianan *A. vittatus* (*sensu* Eigenmann) and separated *K. guianensis* as having scaly dorsal and anal fin bases, and '*A. vittatus*' as having naked dorsal and anal fins ('or with only one scale at base of each membrane in old specimens', citing Eigenmann (1912)).

Eigenmann's (most now in FMNH) and Lowe-McConnell's material (in BMNH) agree with the holotype of *K. guianensis* and with our

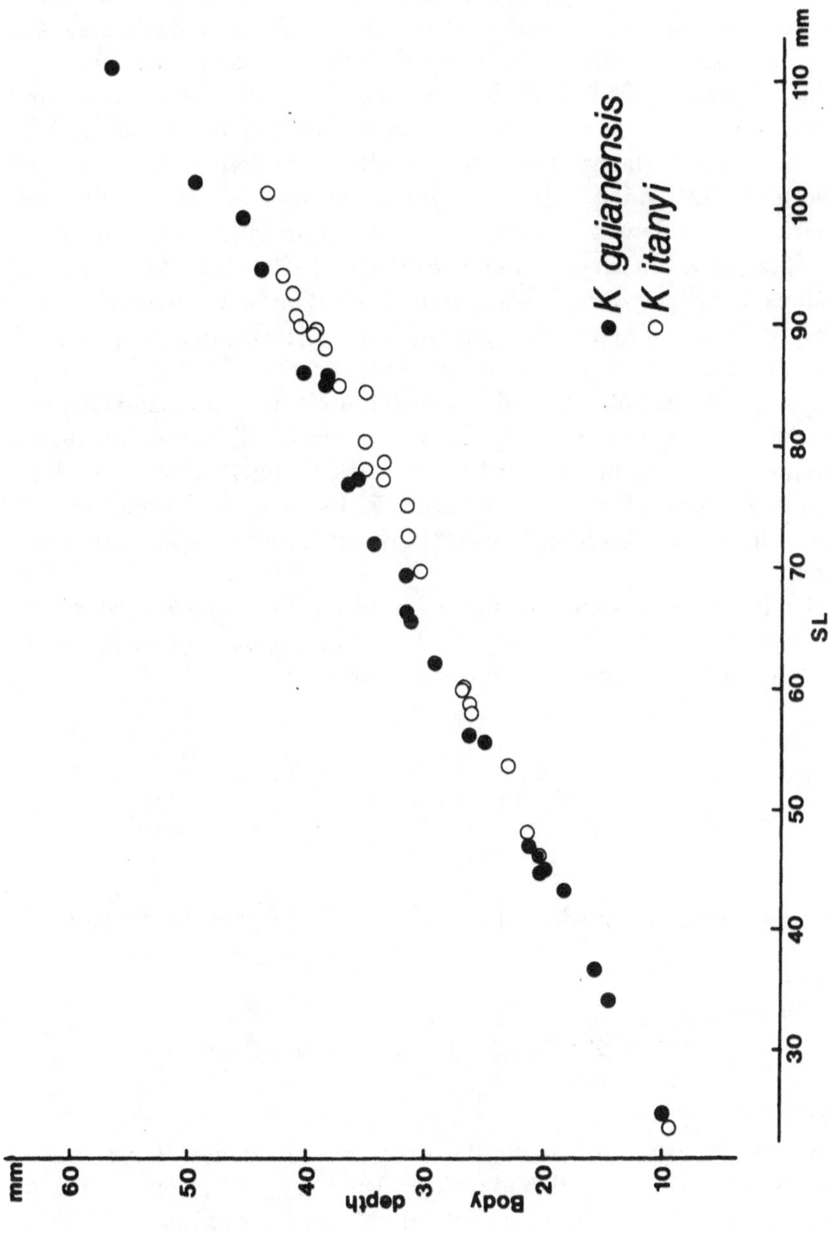

Fig. 91. *Krobia guianensis* and *K. itanyi*. Body depth plotted against SL, to show that *K. itanyi* tends to be the slenderer species.

Surinamese material, which is quite variable in shape and degree of fin squamation, and including specimens comparable to the type as well as rather elongate specimens and such with little scaled or naked dorsal and anal fins in single samples. Thus we identify it all as *K. guianensis*.

There is a superficial similarity to *Bujurquina vittata* which also features a stripe below the eye and a band from the orbit to the dorsal fin base end, but it has the lateral line canals on the dorsal lobe of the caudal fin between rays D1 and D2 and naked dorsal and anal fins. In overall aspect the species are different enough to prevent any mistakes in identification, but Eigenmann (in Eigenmann & Kennedy 1903) once described '*A.*' *vittatus* as *A. paraguayensis*, and apparently relied on Steindachner's (1875) description of a similar Amazonian species (*Bujurquina peregrinabunda* Kullander 1986) as *A. vittatus* for the identification.

Krobia guianensis thus has a wide distribution in Suriname and Guyana, from the lower Marowijne (replaced upstream by *K. itanyi*) west to the Demerara (cf. Eigenmann 1912, Lowe-McConnell 1969). A single specimen is reported from the Mazaruni R. by Lowe-McConnell (1969). The species is not present in Lowe-McConnell's extensive Rupununi collection.

Behavioural and some habitat data given for *Aequidens vittatus* by Keenleyside & Bietz (1981) are probably rather for *K. guianensis*, but we have not examined material from that study.

Krobia itanyi (Puyo)

(Figs 90-93)

Aequidens itanyi Puyo 1943, p. 146, fig. 4 (two syntypes from 'crique du haut Itany' [lost]).

Surinamese records

Acara itanyi Puyo 1949, p. 239, fig. 126 (descr., nearly as in Puyo 1943).

Neotype (Fig. 92)

ZMA 119.531. An adult male, 89.6 mm SL. Suriname, Distr. Marowijne, Marowijne R. system, kreek on left bank of Marowijne R., W Manbari Val, 6 km N Stoelmanseiland. 20 April 1967. Leg. H. Nijssen (HN 127).

Material

Over 500 specimens, all from the Marowijne R. drainage.

Fig. 92. *Krobia itanyi*, neotype, 89.6 mm SL, ZMA 119.531, from kreek on left bank of Marowijne R., W Manbari Val.

Description

Based on the neotype, 89.6 mm SL; counts from 25 specimens measured. Largest specimen examined 125 mm SL (ZMA 106.698).

Shape: moderately elongate, moderately compressed laterally. Dorsal and ventral outlines about equally, and little arched; dorsal fin base straight and little sloping save for strong terminal decurvature. In facial view outline narrowly elliptic with rounded nape and chest.

Head: moderately elongate with evenly and but slightly curved dorsal contour and straight, less steep ventral contour. Snout long, rounded off anteriorly. Interorbital flat except close to orbit, which little below frontal contour, in middle of head length, and in upper half of head. Mouth terminal, but below prominent postlabial skin fold; distinctly narrower than interorbital. Maxilla not reaching to orbit, little exposed; lower jaw articulation below anterior quarter of orbit; ascending processes of premaxilla not reaching to orbit. Jaws equal anteriorly.

Scales: Squ. long. 23(24), 24(1). Predorsal scales feebly ctenoid, except next to orbits; a median series of 7 scales, and an overlapping scale pair next to dorsal fin; anterior scales of almost flank scale size, smaller posteriorly. Chest scales of flank scale size. From between pelvic fin bases rostrad 12 scales in a median series; close to pelvic fin 3/4 size of flank scales, smaller rostrally. Cheek with 3(25) scale series; cycloid scales, upper 4/5 flank scale size, slightly smaller rostroventrally. Operculum with flank scale size cycloid scales. Suboperculum with 3 flank scale size scales, interoperculum with 2 cycloid scales, 4/5 size of flank scales.

Upper lateral line extending to below 2nd dorsal fin ray, overlapping lower by 2 scales. Lower lateral line reaching forwards to above first anal fin ray, continued by 2 canals on caudal fin base. Lateral line counts:

10 + 4 pores/5 pores + 4(1, 23.5 mm), 16/pore + 7(1), 13/9(1), 15/8(2), 15/9(11), 16/7(1), 16/8(3), 16/9(4), 17/8(1); 1-2 canals continuing lower line on caudal fin; 10 of 25 specimens with a canal between rays V4-5 or D3-4 or both.

Fin scales: soft dorsal fin with an interradial cycloid scale basally on each side, but a series of very small body scales along fin base. Third to 5th soft part membranes of anal fin with 1-2 cycloid interradial fin scales basally. Caudal fin basal scale layer with concave hind margin; interradial scales in single series, distal scales cycloid; 1/3 of fin scaled marginally.

Fins: first dorsal fin spine above first lateral line scale, a little longer than 1/3 length of last; spines subequal from 4th, last 2 again slightly increasing. Lappets short, free, anterior rounded, posterior truncate. Soft fin pointed, 4th ray longest, slightly produced, to past middle of caudal fin. D. XIII.10(2), XIII.11(2), XIV.9(2), XIV.10(18), XV.15(1). Anal fin in posterior 3rd of body. Soft fin pointed, 4th ray longest, not produced, not reaching middle of caudal fin. A. III.8(22), III.9(3). Caudal fin hind margin slightly emarginate, ray D5 by little the longest; dorsal lobe longer than ventral lobe. Pectoral fin subacuminate, reaching to above 3rd anal fin spine. P. 13(21), 14(4). Pelvic fin pointed, first ray slightly produced, reaching base of 3rd anal fin ray.

Jaw teeth: an outer series of stronger teeth, in upper jaw along anterior 3/4 of jaw rim, in lower jaw along about anterior half. Inner wide band of smaller teeth anteriorly, one series in upper along anterior 3/4 of jaw, in lower as far back as outer series. In 23 specimens 46-101 mm, 12-20/(extreme 14)17-24 (extreme 27) teeth in outer hemiseries of upper/lower jaw, an irregular increase in upper as only anterior half toothed in young, almost entire in large specimens, but large specimens with only 15-16 teeth occur; lower jaw number remaining about 18-20 with singular exceptions. In upper jaw anteriorly 2 inner series in young, 3, exceptionally 4, in large specimes; in lower jaw 3 series in young, 4-5 in large specimens.

Gill rakers: externally on first ceratobranchial 3(2), 4(18), 5(5). Gill rakers and lower pharyngeal tooth plate examined in a 77.0 mm specimen (ZMA 106.696). First gill arch (Fig. 93) externally with a papilliform epibranchial raker, a somewhat broader raker in the angle and 3 short blunt ceratobranchial rakers; all denticulated; medial soft swellings very reduced. Internally 2 + 1 + 8 denticulated short rakers, upper 3 papilliform, remainder more or less blunt, with moderately developed medial swellings. Epibranchial 1 slender, with little diverging medial processes. Pharyngobranchial 1 slender, rod like. Microbranchiospines always present externally along 2nd-4th gill arches and internally

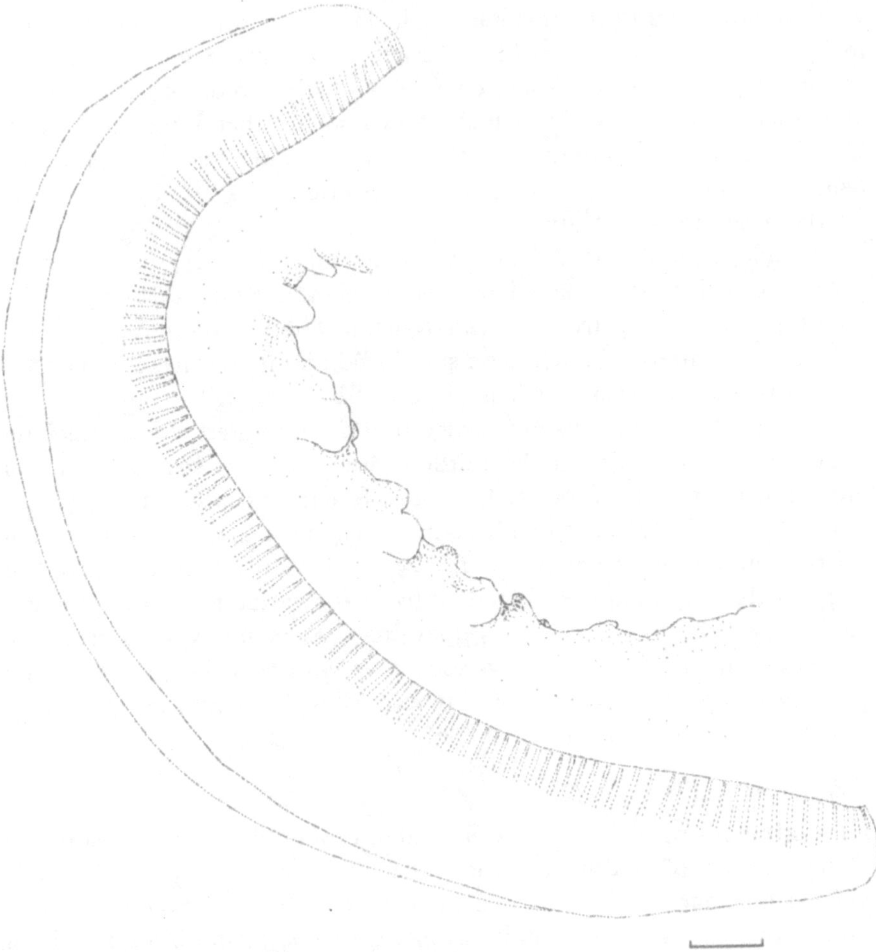

Fig. 93. *Krobia itanyi*. First gill arch in lateral aspect. Scale 1 mm. From ZMA 109.696, 77.0 mm SL.

on 4th arch, frequently also ventrally internally on 3rd, and occasionally, in large specimens, ventrally internally on 2nd arch.

Tooth plates: lower pharyngeal tooth plate and dentition as in *K. guianensis*. Two tooth plates with 7 and 6 pointed teeth on 4th ceratobranchial.

Vertebrae: 12 + 13(8).

Variation: little variation in body shape. Young with interorbital more flattened than large specimens and hence eye tangented by predorsal contour. Maxilla usually reaching to anterior margin of orbit in large, slightly beyond in small specimens. In the smallest specimens mouth width about equal to interorbital width. Head scales mostly cycloid also in large specimens; predorsal scales ctenoid from ca 60 mm; largest specimen (125 mm) with posterior cheek and all opercular scales ctenoid, and occasional ctenoid opercular scales also in other large specimens. Usually a median predorsal series of 8 scales, but irregularities occur, usually then in the replacement of the posterior 1-3 scales by scale pairs overlapping along midline.

Juvenile, 23 mm, with dorsal and anal fins naked; largest specimen (125 mm) with anal fin naked and few basal interradial dorsal fin scales; anal fin otherwise invariably with a few interradial scales basally on soft part, and 1-3 interradial scales on soft dorsal fin membranes (but last 3-5 membranes naked) from at least 35 mm SL.

Soft dorsal and anal fins in young roundish or pointed, dorsal reaching about 1/3 of caudal fin; anal fin little shorter. In large specimens dorsal and (shorter) anal fin produced, at most to end of caudal fin. Pelvic fin pointed, to 3rd anal fin spine base in young; produced, at most to anal fin base middle in large specimens. Pectoral fin reaching to above 3rd spine or first ray of anal fin. Caudal fin shape variable: feebly rounded, subtruncate or, usually, feebly emarginate as in neotype, dorsal lobe longer or lobes equal; many large specimens with rays D4 and D5 slightly prolonged (projection corresponding at most to 1/4 length of rest of fin). No sexual dimorphism in fin shape.

Colouration

Ground colour whitish preventrally and intermandibularly, more greyish along middle of abdomen, underside of head and lips; sides pale yellowish. Cheek and preorbital greyish whitish; gill cover tending to faint brownish posteriorly. Faint brown wide band from hind margin of orbit down to lower margin of preoperculum.

Dark brown sharply delimited stripe from orbit across snout tip below nostrils; another paralleling above nostril; a 3rd, widened medially, straight between orbits. Those stripes leaving greyish whitish prominent stripe between eye and nostril, otherwise interspaces grey. Nape brown, back pale brown.

Prominent blackish brown band from just behind orbit, caudad above operculum, below lateral line, to Bar 4, in squ. long. and superior scale series, 1-1 1/2 scales wide on side, narrower before and behind Bar 5; slightly lighter zone above band back to above Bar 5.

Five faint brownish vertical bars across sides, confluent dorsally, not reaching quite to ventral body margin: 1) posteriorly on caudal peduncle; 2) anteriorly on caudal peduncle; 3) above soft anal fin anteriorly; 4) above vent-anal fin origin; 5) distinguishable chiefly by pale interspace before Bar 4 and narrowed lateral band anteriorly and posteriorly, but merging with abdominal colour which a little more greyish than that of posterior vertical bars. Inconspicuous narrow dark margin on scales of dorsal sides; on abdominal sides and posteriorly indistinct dark vertical spot on each scale base.

Dorsal fin light greyish brown, lappet tips and margins of soft part clear; scattered clear dots on soft part basally. Anal fin similar, but lower margin dark brown; clear dots on last 3 membranes. Pelvic fin outer margin and rays greyish brown, lighter inwardly, membranes clear. Caudal fin with blackish brown large, somewhat ovate spot basally on dorsal lobe, not reaching upper fin margin or lower lateral line level; fin base below greyish brown, like rest of fin which, however, lightened by clear long dots on membranes save the 4 ventralmost, on outer half only membrane margins pigmented and hind margin of fin slightly smoky.

Other specimens similar, i.e. light with brownish back, faint vertical bars, prominent lowermost snout stripe, variably expressed, but not prominent cheek stripe and contrasting lateral band and caudal base spot. Caudal fin dotting always indistinct and mostly restricted to inner half of fin. Usually dorsal caudal fin margin with narrow black marginal and white submarginal line. The appearance of the lateral band varies; on side almost always broken up into more or less continuous blotches. There is one blotch before Bar 5; another is situated in Bar 5 and is the most prominent (= midlateral spot), roundish in young, oblong in large specimens; the third blotch is oblong and situated in Bar 4. The last (third) spot may be extended onto Bar 3 (onto the end of the upper lateral line), or, occasionally, a fourth spot forms in Bar 3 on the termination of the upper lateral line. Notably, the course of the band, or series of blotches, is directed towards the end of the dorsal fin base.

Juvenile, 23 mm, much like large specimens, but suborbital stripe more backwards slanting, snout dark with light stripe from eye to nostril indicated; caudal fin dusky, but caudal base spot distinct, black.

Distribution (Fig. 90)

Krobia itanyi is known only from the Marowijne drainage, from near the coast to highland tributaries.

Notes

Krobia itanyi was described (Puyo 1943) on the basis of two specimens, 140 and 70 mm total length from a creek in the upper R. Itany (= Litani R. in Surinam). The description was repeated in Puyo (1949) with the major addition of a mention of scaly dorsal and anal fin bases, and habitat data. The description and figure agree, save for some minor details, with the *Krobia* species identified by us from many localities in the Marowijne R. system.

The material used by Puyo (1943, 1948, 1949) for his studies on French Guianan freshwater fishes is not preserved. (cf. Puyo 1949, p. 7: "Les mensurations, les descriptions des genres et des espèces ont été prises et faites sur des spécimens vifs..."). As apparently no types exist, and as *K. itanyi* belongs to a group of several similar species, we propose a neotype (ZMA 119.531) from material apparently conspecific with Puyo's syntypes.

Krobia itanyi differs from *K. guianensis*, the most similar species, especially in the contrasted blotchy lateral band which ends in the bar behind the midlateral spot or, occasionally, in the next vertical bar, but is not continued to the dorsal fin base end. Facial stripes are usually well evident, and the spinous dorsal fin is immaculate. At least adults are slenderer than *K. guianensis*, depth usually less than 45% of SL instead of usually over 45% of SL in specimens over 50 mm SL (Fig. 91).

Puyo (1943) compared *K. itanyi* with *Aequidens vittatus* Heckel, but apparently the data compared are chiefly from Eigenmann's (1912) description of *K. guianensis* as *Aequidens vittatus*.

Meinken (1963) identified *Aequidens itani* (*sic*) in an aquarium fish import from Surinam; from his figure it appears rather to be *Krobia guianensis*, the established aquarium stock of which is handled as *Aequidens itanyi* and is included as such in standard aquarium fish works and many original articles.

CLEITHRACARA n. gen.

Type species

Acara maronii Steindachner, 1882.

Diagnosis

South American cichlids of small size (to 71 mm SL), with four dentary lateralis foramina, distinguished by characters in combination: preoperculum with two scales on horizontal limb; triserial predorsal squamation; extensively scaled unpaired soft fins; one supraneural; two palatine-lateral ethmoid articulations; lacking accessory caudal fin lateral lines, parhypurapophysis, interarcual cartilage, ceratobranchial 4 teeth and mesial foramen on second infraorbital. Similar to *Laetacara* in having a deep notch in the dorsal margin of the anterior ceratohyal; uniquely diagnosed by the much widened anterior arm of epibranchial 1 and a small ventral process on the quadrate approximately midway between its head and the level of the anterior end of the preoperculum.

Included species

Acara maronii Steindachner, 1882.

Etymology

The generic name alludes, with some sacrifice to brevity, to the trivial name given by aquarists to the only contained species, Keyhole cichlid, referring to the flank marking. It is derived from the Greek *kleithron*, lock, and the Tupi word for cichlids, *Acara*. It is to be treated as of feminine gender.

Distribution

The Guianas, from western Guyana to the Ouanary drainage in French Guiana.

Notes

The only known species of *Cleithracara* is a Guianan endemic that until now has been referred to the genus *Aequidens*. Kullander (1983a) recognized the distinctness of *C. maronii*, listing some diagnostic characters.

Cleithracara maronii may be most closely related to *Laetacara*, which also has a distinct notch in the dorsal margin of the anterior ceratohyal. These genera are similar in having a scaled preoperculum, triserial predorsal squamation, ribs on caudal vertebrae and mesethmoid separate from vomer, and in lacking accessory lateral line on the caudal fin, parhypurapophysis and ceratobranchial 4 teeth. *Laetacara* species, however, have only one (the anterior) palatoethmoid articulation, well-developed microbranchiospines, an interarcual cartilage and two supraneurals. Small species of *Laetacara* lack the middle foramen on the second infraorbital.

Nannacara is similar to *Cleithracara* in having a scaled preoperculum, a single supraneural, ribs on caudal vertebrae, separate mesethmoid and vomer, and in lacking both interarcual cartilage and ceratobranchial 4 dentition. *Nannacara* also has a prominent posterior palatoethmoid articulation. Distinguishing characters of *Nannacara* are listed on p. 194. There are no characters uniquely shared by *Nannacara* and *Cleithracara*.

The preopercular squamation is a distinctive character of *Nannacara*, *Cleithracara* and *Laetacara*, but not a synapomorphy of these genera. In one *Mesonauta* species, tentatively identified as *M. acora* (Castelnau), the cheek + preopercular scale pattern appears to be the same as in *Cleithracara*; with three scale series on the cheek and two scales on the lower limb of the preoperculum. Preopercular scales occur in several other cichlids: e.g., *Pterophyllum*, *Etroplus* Cuvier, *'Cichlasoma' centrarchus* (Gill & Bransford), infrequently in *Mesonauta insignis* (Heckel), *Chaetobranchopsis orbicularis* (Steindachner), and *Cichlasoma* species. The single preopercular scale occurring in *Cichlasoma* and *Mesonauta* species appears to be homologous with the posterior preopercular scale in *Cleithracara*. In *Nannacara* and *Laetacara* there are only two cheek scale series, which extend posteriorly over the vertical limb of the preoperculum; there are three preopercular scales, one ventrally on the vertical limb and two on the horizontal.

The shape of epibranchial 1 distinguishes *C. maronii* from all other *Cichlasoma*-like cichlids. *Pterophyllum* also has a wide epibranchial 1 but separate mesial expansion and pharyngobranchiad 1 arm. In geophagine cichlids epibranchial 1 has a ventral expansion separate from the pharyngobranchiad 1 arm and capped with connective tissue instead of with cartilage. Most cichlids, however, have a slender epibranchial 1.

There is some variation among cichlids in the design of the anteroventral corner of the quadrate, but no ventral process shielding the anguloarticular process as in *C. maronii* has been observed among other cichlids.

Several other characters are distinctive, not least the colour pattern, the shape of the medial extrascapular, the pronouncedly bicuspid pharyngeal dentition, and the shape of the distal postcleithrum.

Several loss characters characterize *Cleithracara* but are not unique. Ceratobranchial 4 tooth plates are missing also in, e.g., *Laetacara*, *Nannacara*, *Apistogramma* and *Satanoperca*, although such tooth plates are tentatively regarded as plesiomorphic for cichlids. All *Cichlasoma*-like cichlids except *Cleithracara*, *Nannacara*, and *'Nannacara' bimaculata*, have two supraneurals, but most geophagine cichlids have only one. The interarcual cartilage is absent in *Cleithracara*, *Nannacara*, and *Crenicichla*; it is reduced in size in *Cichlasoma*-like cichlids, and often absent unilaterally in these, although a well-developed rod-like or globular interarcual cartilage seems to be plesiomorphic for cichlids.

Cleithracara maronii (Steindachner)

(Figs 94-115)

Acara Maronii Steindachner 1882, p. 141, pl. II, fig. 4 (descr.; Maroni-Fluss in Guiana).

Surinamese records

Acara Maronii; Regan 1905b, p. 344 (descr.; material from "Surinam"). — Pellegrin 1904, p. 179 (descr.; material from Maroni).
Acara maronii; Puyo 1949, p. 238 (bibliogr.; brief descr.; criques ... Rivières affluents du Maroni, ... Maroni... Oyapock).
Aequidens maronii; Boeseman 1956, p. 196 (listed: "Surinam" [RMNH 21395]).
'Aequidens' maronii; Kullander 1983a, p. 103 (distribution), p. 280 (discussion of systematic position).
Astronotus (Acara) maronii; Eigenmann & Eigenmann 1891, p. 68 (listed).

Material

98 specimens, including holotype, from the Marowijne, Commewijne, Cottica, Suriname, Nickerie, and Corantijn river systems.

Fig. 94. *Cleithracara maronii*, 60.2 mm SL, RMNH 31141, from Saramacca Kreek.

Description

Chiefly from holotype and RMNH 31141 specimens (Fig. 94). Measurements from 25 specimens; counts below are from the same material. Largest specimen examined 70.9 mm SL (ZMA 107.022).

Shape: body deep, strongly compressed laterally; caudal peduncle short, deep and narrow. In frontal aspect a narrow oval with narrowly rounded chest and almost keeled nape. Dorsal and ventral contours about equally arched. Caudal peduncle with the ventral edge a little longer than the dorsal. Predorsal contour steep, with curved nape and straight front from above middle of orbit; prepelvic contour less steep, straight.

Head: short and deep, compressed. Snout short, rounded. Orbit situated below forehead contour (interorbital curved); all of it in upper half and most in anterior half of head. Maxilla little exposed, reaching to (young) or not (large) to orbital margin. Mouth terminal, narrower than interorbital; jaws isognath. Both lip folds interrupted anteriorly. Operculars and pectoral girdle bones entire.

Scales: Squ. long. 22(19), 23(6); circumpeduncular scale series 16; scales ctenoid except anteriormost cheek scales, uppermost subopercular, interopercular and anterior or both preopercular scales, in median and parallel prepelvic scale series, on pectoral axilla base, under gill cover and around vent and genital papilla. Predorsally three series of scales caudad from between orbits; marginal of eight or nine, medial (anteriorly) of three to five; as a rule posteriorly three scales of marginal series overlapping over midline, anteriorly three medial scales, but between these either medial scales and/or marginal scales overlapping over midline; anterior of flank scale size, like chest scales. Two ctenoid scales between pelvic fin bases; succeeded forwards by about eight or nine cycloid, posteriorly four-fifths size of anterior flank scales, rostrally smaller scales.

Cheek with 3(24) or 2(1) scale series, invariably two large scales on lower limb of preoperculum; two vertical series of scales on operculum, one scale series (three scales usually) on suboperculum and (of three scales) interoperculum (Fig. 95).

Upper lateral line 3-3 1/2 (initially) to 1 1/2-2 scales distance from dorsal fin, in third scale series above that containing the lower. Lateral lines 13/6(1), 15/6(4), 15/7(5), 16/5(1), 16/6(5), 16/7(4), 17/5(1), 17/6(4). Lower line continued on caudal fin by 1(20) or 2(4) canals except in one small specimen; no tubed lateral line scales on dorsal or ventral lobe of caudal fin.

Fin scales: dorsal, anal and caudal fins extensively scaled; pectoral fin base scaled in 70.9 mm specimen only. Spinous dorsal fin running in a

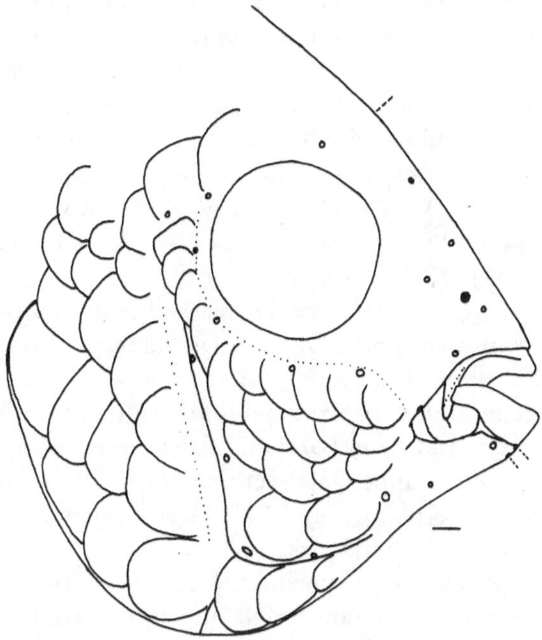

Fig. 95. *Cleithracara maronii.* Lateral aspect of head to show scale pattern and lateralis pores. Scale 1 mm. From RMNH 31141, 52.8 mm SL. Situation of coronalis pore and anterior two dentary pores which out of view, indicated by dash-lines.

shallow groove margined by projecting body scale edges; scaled sheath on soft fin continuous with body squamation, a basal band of broad scales covering rays, interradial scales distally smaller and cycloid, in single or double series along membranes; single interradial scales from posterior third of spinous fin, on soft fin. Four (small specimens) to 20 (large) scales (including basal) along middle membranes; width of scaled sheath at most equal to length of last dorsal fin spine. Anal fin squamation similar, but denser. Basal one-third to half of caudal fin densely scaled; hind edge of scaled layer about straight vertical; scales gradually smaller distally, ctenoid.

Fins: first dorsal fin spine inserted in advance of vertical from hind margin of operculum, its length half or more that of the last; spines subequal in length from fifth, last three or so again slightly increasing. Lappets anteriorly rounded, posteriorly truncate. Soft fin pointed, fifth and sixth rays produced, reaching end of caudal fin or beyond, may be nearly as long as body.

Anal fin origin marking posterior third of body; soft fin like soft dorsal fin, but slightly shorter. Pectoral fin asymmetrical, fourth ray longest, tip well-rounded, reaching to above first or second anal fin spine base. Pelvic fin pointed, first ray produced, with filamentous tip at most to caudal fin base. Caudal fin rounded, with 3 procurrent and 8 principal rays in each lobe. Young with shorter, not produced dorsal, anal and pelvic fins. D. XIV.12(2), XV.10(1), XV.11(10), XV.12(9), XV.13(1), XVI.12(1; first spine apparently divided), XVI.11(1). A. III.10(15), III.11(9), III.12(1). P. 12(1), 13(10), 14(14).

Jaw teeth: strong, conical, feebly recurved; anteriorly many worn (rounded or truncated tips). Outer series along entire jaw rim (only anterior third of upper jaw toothed in smallest specimens), inner (about two series) chiefly in disorganized band anteriorly; outer series of upper jaw well separated from inner band and size difference more distinct than in lower jaw; anterior upper jaw teeth somewhat procumbent. In eight large specimens, 48.6-66.3 mm, 17-23/14-20 (mean 20.0/17.3) teeth in upper/lower jaw outer hemiseries.

Gill rakers: from 66.3 mm specimen (Fig. 96): First gill arch with an epibranchial papilliform, a joint broad-tipped, and six subconical ceratobranchial rakers externally; upper ceratobranchial connected medially to a small soft swelling not reaching above median skin ridge, which gradually reduced in size on anterior rakers. Internal rakers 1 + 1 + 8 short, knoblike. All rakers edentulous. External ceratobranchial rakers 5(4), 6(18), or 7(3).

Lower pharyngeal tooth plate (Fig. 97) with a few subconical teeth rostrally, otherwise all teeth decidedly bicuspid with long antrorse posterior and smaller antrorse anterior cusp although the largest postermedial may have the anterior cusp reduced to a few minute nipples; marginal teeth slender, inner teeth laterally compressed.

Osteology.

Examination of a single alizarin-alcian blue preparation, NRM A83/1948392.3010, 36.9 mm SL, was complemented by partial dissection of alcohol specimens and radiographs to verify critical characters. As suggested by Figures 98-114, all from the 36.9 mm specimen, *C. maronii* is osteologically very similar to other *Cichlasoma*-like cichlids (cf. descriptions of *Cichlasoma* in Kullander 1983, *Tahuantinsuyua* and *Bujurquina* in Kullander 1986).

The neurocranium (Fig. 98) is high and wide, with somewhat foreshortened vomer. Median frontal crests contribute significantly to the height of the orbital region. The pharyngeal apophysis of the parasphenoid is well-developed. The posterior myodome has a relatively wide

Fig. 96. *Cleithracara maronii*. Outline of first gill arch in lateral aspect. Scale 1 mm. From NRM A83/1965144. 3011, 66.3 mm SL.

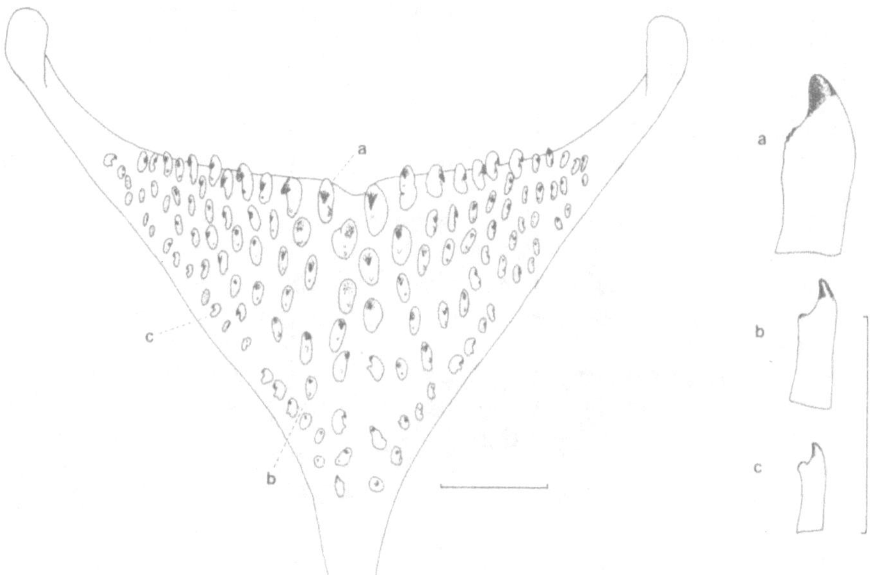

Fig. 97. *Cleithracara maronii*. Lower pharyngeal tooth plate in occlusal aspect and teeth (*a*, *b*, *c*) in medial aspect. Scales 1 mm. From NRM A83/1965144.3011, 66.3 mm SL.

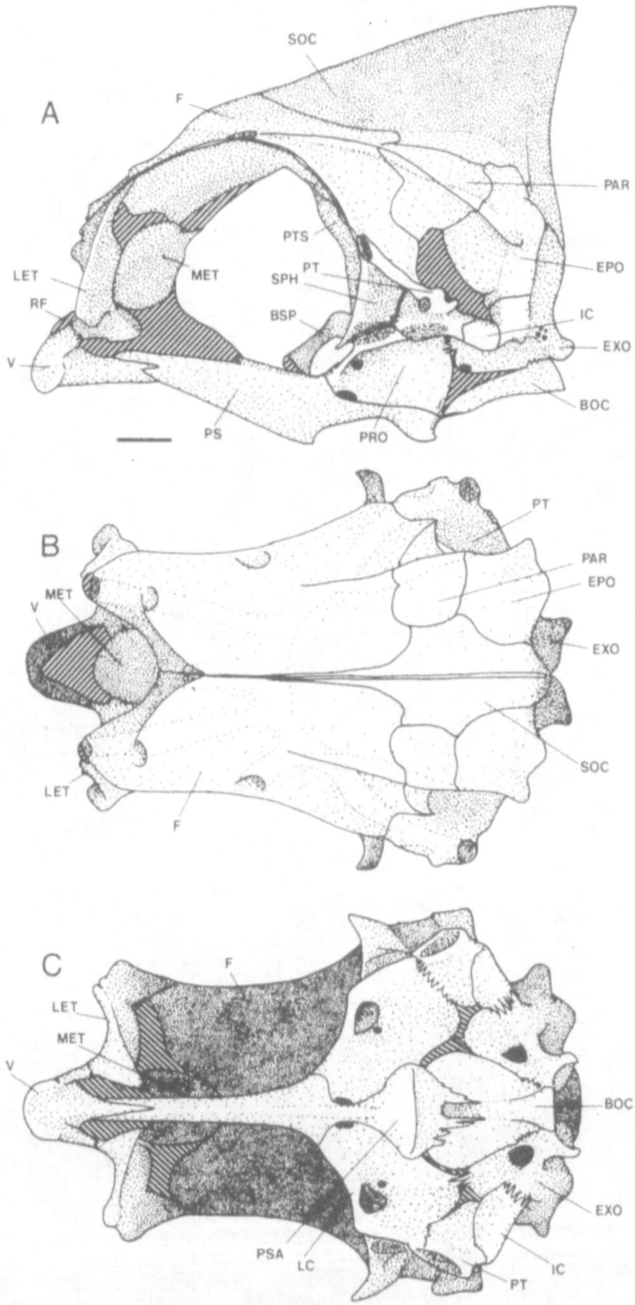

Fig. 98. *Cleithracara maronii.* *A* lateral, *B* dorsal, *C* ventral aspects of neurocranium. Cartilage shown as oblique lines. Scale 1 mm. From NRM A83/1948392.3010, 36.9 mm SL. *BOC* basioccipital, *BSP* basisphenoid, *EXO* exoccipital, *EPO* epioccipital, *F* frontal, *IC* intercalar, *LC* lateral commissure, *LET* lateral ethmoid, *MET* mesethmoid, *PAR* parietal, *PRO* prootic, *PS* parasphenoid, *PSA* pharyngobranchiad apophysis of parasphenoid, *PT* pterotic, *PTS* pterosphenoid, *RF* rostral fenestra (cartilage), *SOC* supraoccipital, *SPH* sphenotic, *V* vomer.

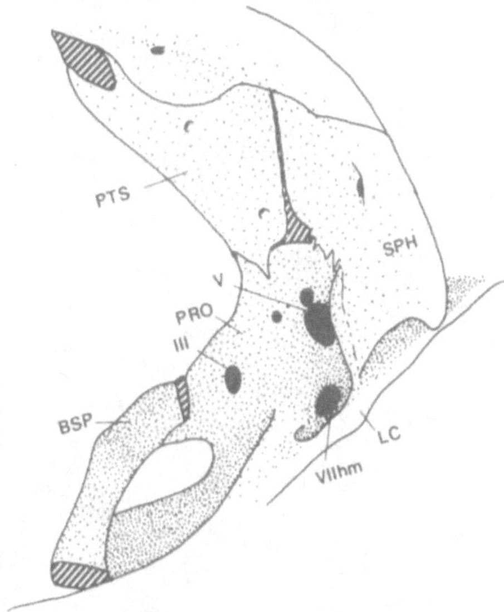

Fig. 99. *Cleithracara maronii*. About anterolateral aspect of pars jugularis. From NRM A83/1948392. 3010, 36.9 mm SL. Cartilage hatched. *BSP* basisphenoid, *LC* lateral commissure, *III* oculomotor nerve foramen, *PRO* prootic, *PTS* pterosphenoid, *SPH* sphenotic, *V* trigeminus nerve foramen, *VIIhm* hyomandibular trunk nerve foramen.

opening. The mesethmoid is sutured laterally to the lateral ethmoids, but free from the vomer, separated from it by a large rostral fenestra. The lateral commissure (Fig. 99) is rodlike with a characteristic mesial laterad bend. There is no anterior prootic bridge or prootic-pterosphenoid bridge.

The palatine (Figs 100-101) is short and wide and articulates with the lateral ethmoid both posterior to the lateral ethmoid-vomer suture and intraorbitally. The quadrate has a short blunt process extending ventrad from the lateral ventral edge, slightly posterior to the articulating head and at the level of the recess for the posterior dorsal anguloarticular process (Fig. 102).

The cephalic lateralis system is much like in other *Cichlasoma*-like cichlids, with the following noteworthy: The branches of the preopercular canal are narrow and distinct (Fig. 100); usually in *Cichlasoma*-like cichlids the 4th and 5th from anteriormost foramina are not separated from the main canal by recognizable branches. The coronalis foramen is elevated by prominent medial frontal crests and the anterior two frontal canal branches are not as relatively divergent as is usually the case in cichlids (Fig. 98B). There are four infraorbital ossicles (Fig. 103), the posteriormost tightly attached to the sphenotic; the anterior three have

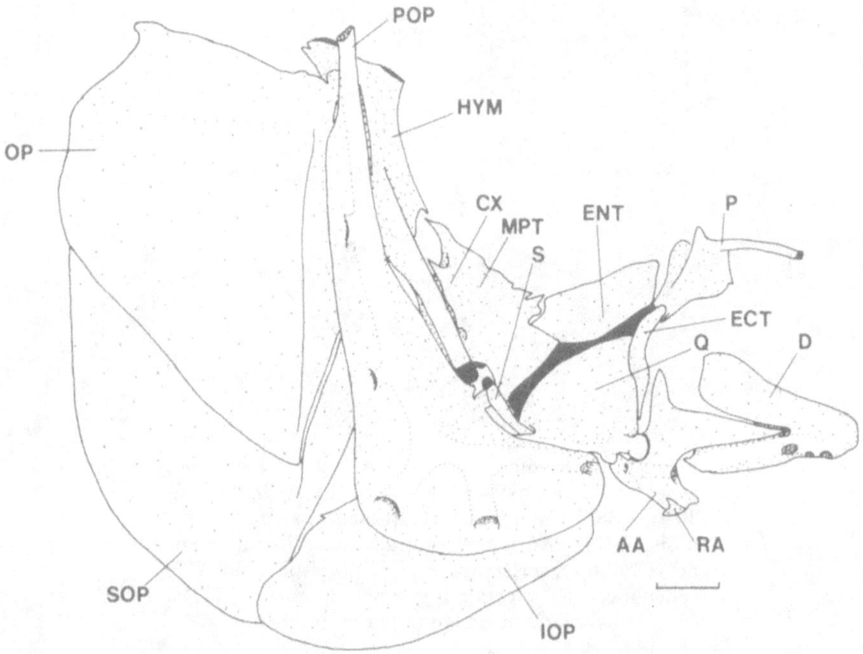

Fig. 100. *Cleithracara maronii*. Lateral aspect of suspensorium, lower jaw (teeth omitted) and operculars. Scale 1 mm. Cartilage black. From NRM A83/1948392.3010, 36.9 mm SL. *AA* anguloarticular, *CX* calyx, *D* dentary, *ECT* ectopterygoid, *ENT* entopterygoid, *HYM* hyomandibula, *IOP* interoperculum, *MPT* metapterygoid, *OP* operculum, *P* palatine, *POP* preoperculum, *Q* quadrate, *RA* retroarticular, *S* symplectic, *SOP* suboperculum.

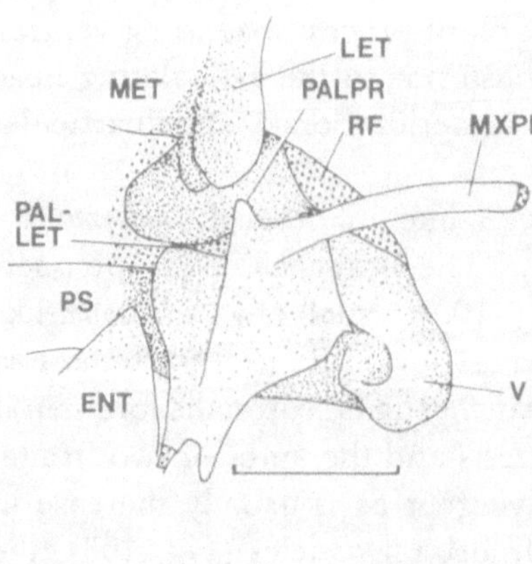

Fig. 101. *Cleithracara maronii*. Palatine and ethmovomerine block in lateral aspect. Coarse stippling cartilage. Scale 1 mm. From NRM A83/1948392.3010, 36.9 mm SL. *ENT* entopterygoid, *LET* lateral ethmoid, *MET* mesethmoid, *MXPR* maxillad process of palatine, *PAL-LET* posteromedial palatoethmoid articulation, *PALPR* anterodorsal process of palatine with palatopalatine ligament and medial palatoethmoid articulation, *PS* parasphenoid, *RF* rostral fenestra (cartilage plus), *V* vomer head.

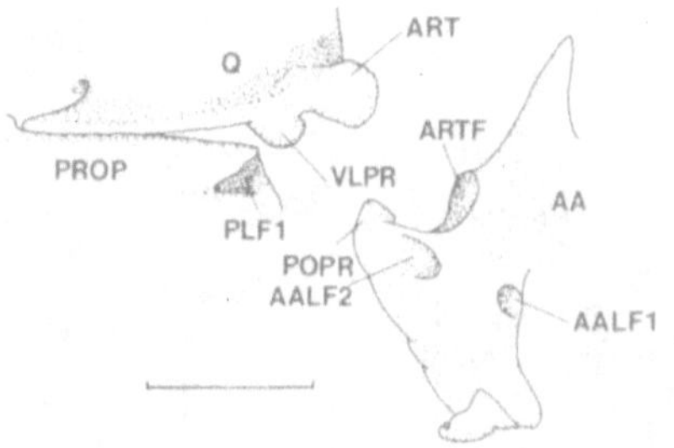

Fig. 102. *Cleithracara maronii*. Lateral aspect of disarticulated lower jaw articulation. Scale 1 mm. From NRM A83/1948392. 3010, 36.9 mm SL. To show particularly the ventrolateral process (*VLPR*) of the quadrate little posterior to the articulating head. ART articulating head of quadrate, *AA* anguloarticular, *ARTF* articulating facet of anguloarticular, *AALF1-2* lateralis canal foramina of anguloarticular, *PLF1* foramen of preopercular lateralis canal, *POPR* posterior process of anguloarticular, *PROP* preoperculum, *Q* quadrate.

ventral lamina, wide on the first two; the lacrimal and second infraorbital make contact on the medial side of the first infraorbital; each infraorbital has only terminal canal openings. The distribution of skin pores is depicted in Fig. 95.

In small specimens the medial extrascapular appears to have a single wide ventral opening whereas in large specimens there are two openings, one toward the lateral extrascapular the other to the body surface (Fig. 104). The dorsal tip of the distal postcleithrum (Fig. 105) has three pro-

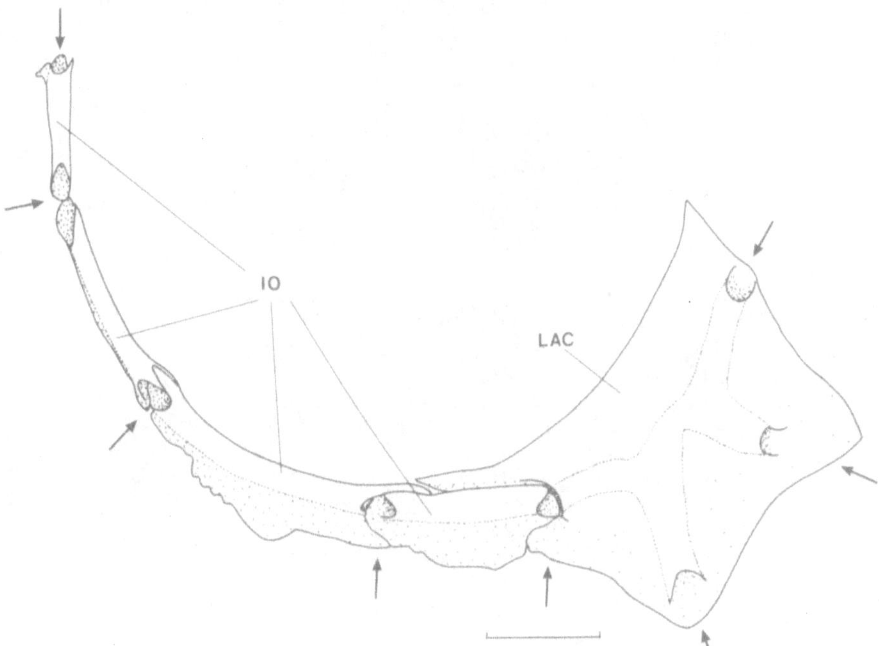

Fig. 103. *Cleithracara maronii*. Lateral aspect of suborbital series. Scale 1 mm. Arrows point to foramina of lateralis canal. From NRM A83/1948392.3010, 36.9 mm SL. *IO* infraorbitals, *LAC* lacrimal.

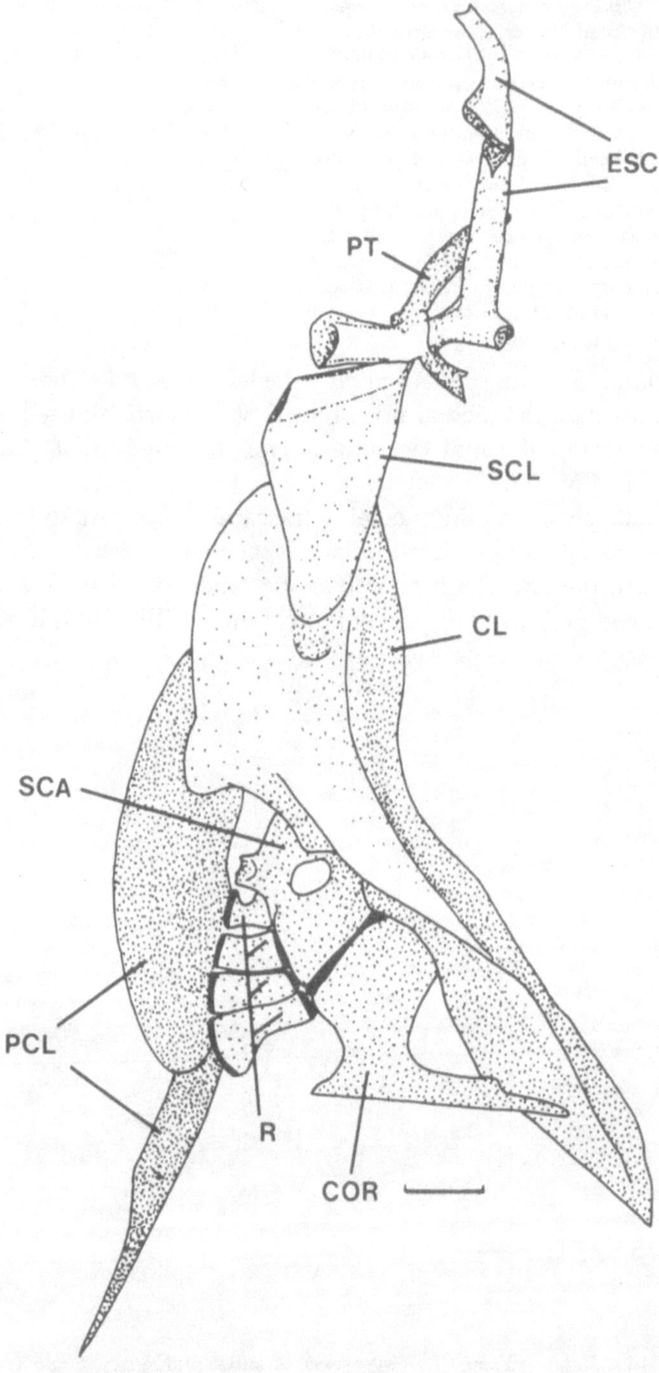

Fig. 104. *Cleithracara maronii*. Lateral aspect of right side pectoral girdle. Cartilage black. Scale 1 mm. From NRM A83/1948392.3010, 36.9 mm SL. *CL* cleithrum, *COR* coracoid, *ESC* extrascapulars, *PCL* postcleithra, *PT* posttemporal, *R* radials, *SCA* scapula, *SCL* supracleithrum.

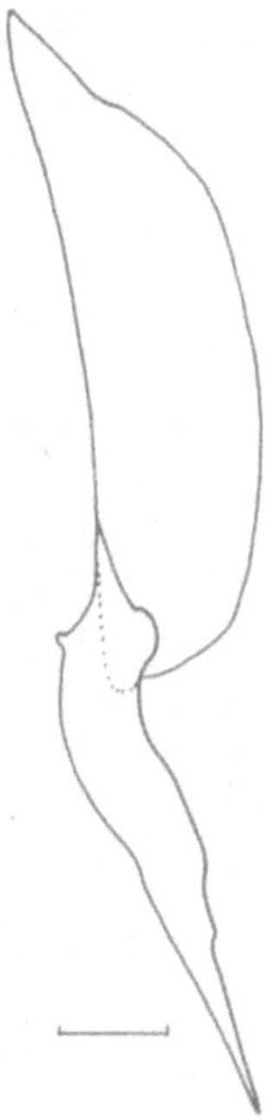

Fig. 105. *Cleithracara maronii*. Outline of right side postcleithra in medial aspect. Scale 1 mm. From NRM A83/1948392.3010, 36.9 mm SL. Obscured outline of proximal postcleithrum dotted.

jections; one pointed continuing the axis of the bone, one blunt posterior and one nipple-like process projecting rostrad unlike the other two which are overlapped by the proximal postcleithrum.

Most notable in the branchial skeleton (Figs 106-107) is the very wide, cartilage-lined anterior arm of the first epibranchial (Fig. 107). The

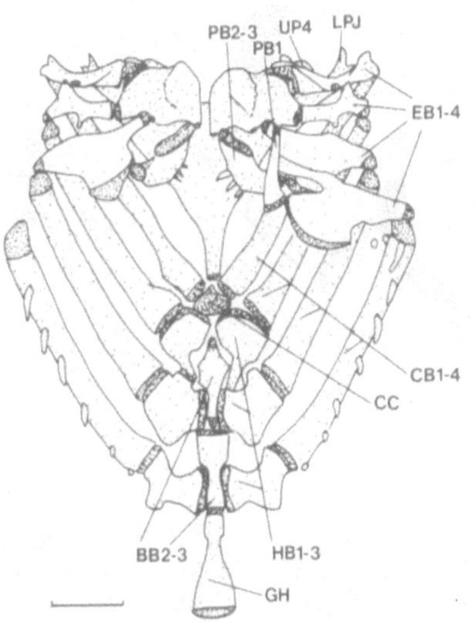

Fig. 106. *Cleithracara maronii*. Dorsal aspect of gill arch skeleton. Right side first epibranchial and pharyngobranchial removed to show epibranchial 2 - pharyngobranchial 2 contact. Denser stippling represents cartilage. Scale 1 mm. From NRM A83/1948392.3010, 36.9 mm SL. *CB1-4* ceratobranchials 1-4, *CC* central cartilage, *BB2-3* basibranchials 2-3 (basibranchial 1 covered by glossohyal), *EB1-4* epibranchials 1-4, *GH* glossohyal (basihyal), *HB1-3* hypobranchials 1-3, *LPJ* lower pharyngeal tooth plate (teeth not drawn), *PB1* pharyngobranchial 1, *PB2-3* pharyngobranchials 2-3, *UP4* fourth upper tooth plate.

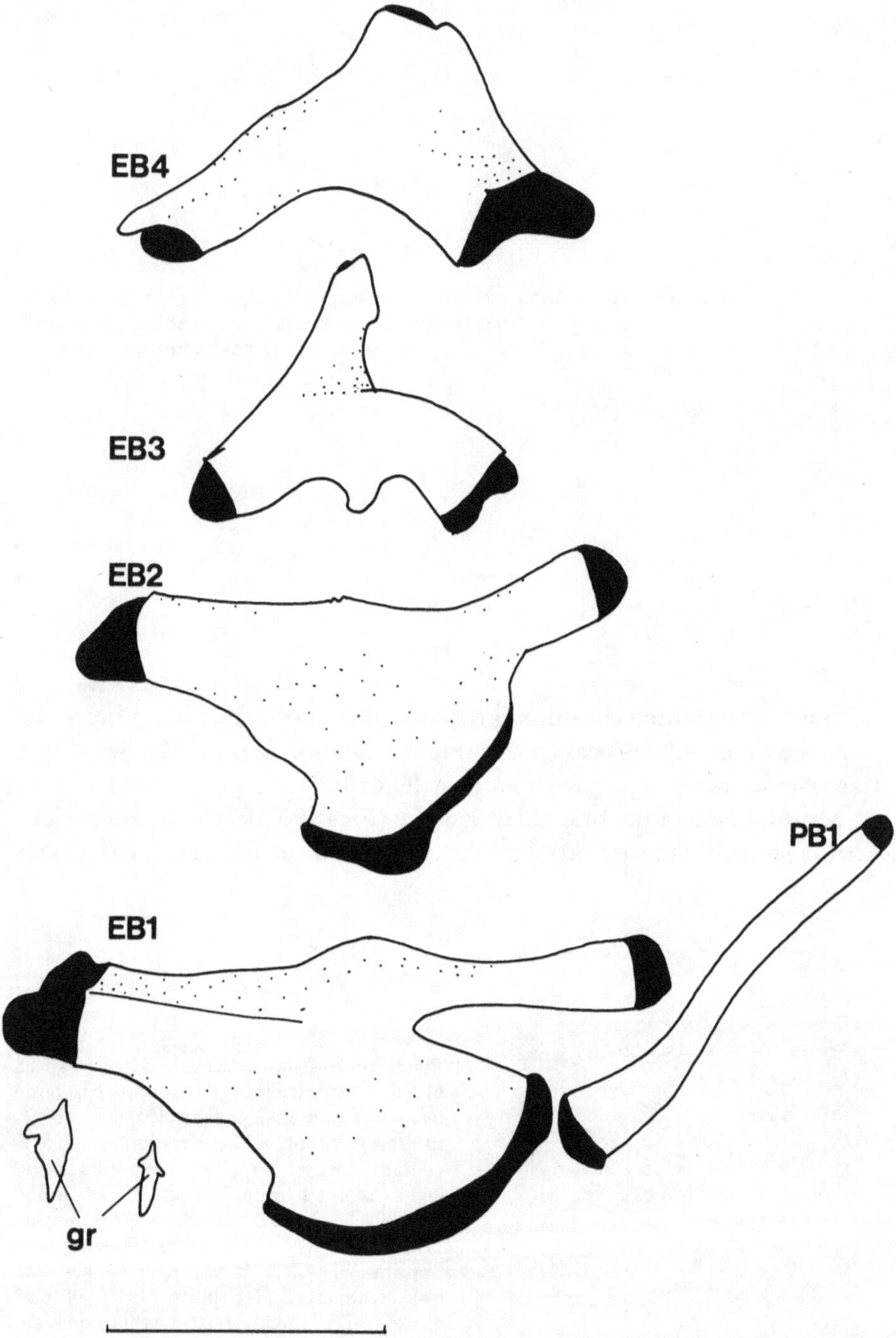

Fig. 107. *Cleithracara maronii*. Rostral aspects of right side epibranchials (*EB1-4*) and first pharyngobranchial (*PB1*). Cartilage black. Scale 1 mm. From NRM A83/1948392.3010, 36.9 mm SL. *gr* gill rakers associated with epibranchial 1.

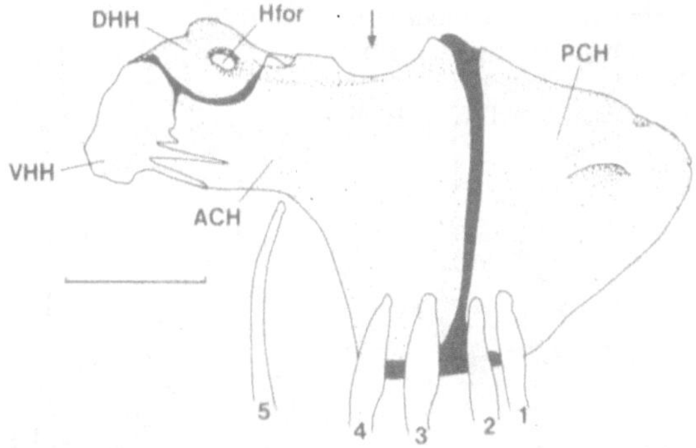

Fig. 108. *Cleithracara maronii*. Lateral aspect of left hyoid. Cartilage black. Scale 1 mm. From NRM A83/1948392.3010, 36.9 mm SL. *ACH* anterior ceratohyal, *DHH* dorsal hypohyal, *Hfor* hypohyal foramen, *PCH* posterior ceratohyal, *VHH* ventral hypohyal, *1-5* branchiostegal rays (cut). Arrow points to notch in dorsal contour of anterior ceratohyal.

uncinate process is long and there is no interarcual cartilage. Microbranchiospines are minute and restricted to the posterior part of the external side of ceratobranchials 2-4. In intact specimens microbranchiospines usually cannot be verified without dissection. There are no tooth plates on ceratobranchial 4. The hyoid (Fig. 108) is relatively deep with a notched dorsal edge of the anterior ceratohyal; the hyoid artery passes on the lateral face, in a shallow groove; the ceratohyal suture is covered by cartilage on the external face. The urohyal (Fig. 109) is deep in lateral aspect, corresponding to the head shape.

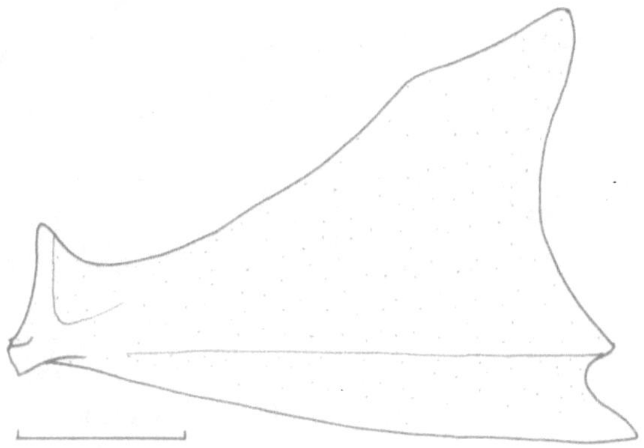

Fig. 109. *Cleithracara maronii*. Urohyal in left lateral aspect. Scale 1 mm. From NRM A83/1948392.3010, 36.9 mm SL.

Fig. 110. *Cleithracara maronii*. Premaxillae in facial aspect; dentition not complete. Scale 1 mm. From NRM A83/1948392.3010, 36.9 mm SL. Rostral cartilage distinguished by heavy dotting.

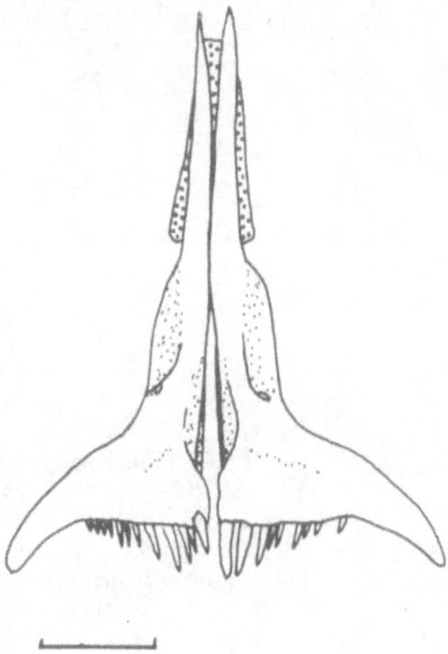

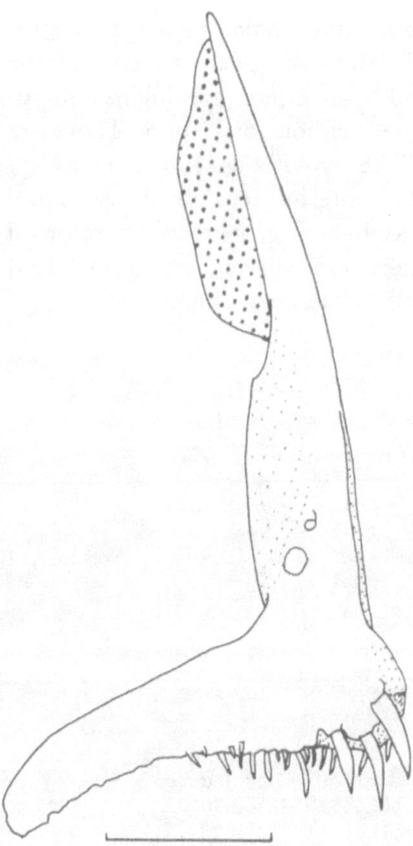

Fig. 111. *Cleithracara maronii*. Right premaxilla in rostrolateral aspect. Rostral cartilage indicated by heavy dotting. Dentition not complete. Scale 1 mm. From NRM A83/1948392.3010, 36.9 mm SL.

The alveolar ramus of the premaxilla (Figs 110-111) is relatively wide proximally; the ascending processes of the premaxilla are distinctly longer than the alveolar, there are two small foramina in the external face of the base of the articulating process. The maxilla (Fig. 112) has a large articulating head relative to its length.

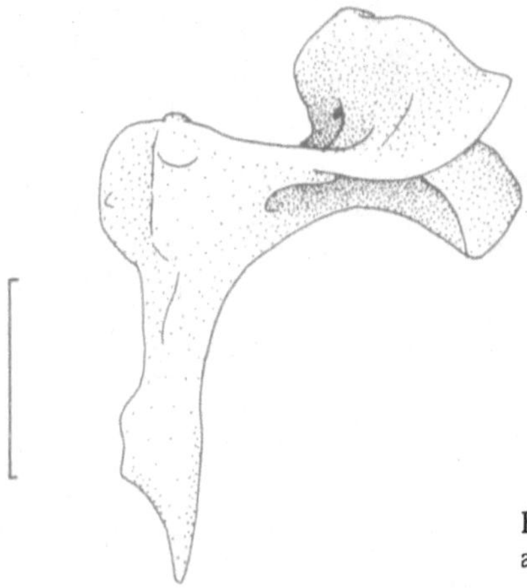

Fig. 112. *Cleithracara maronii*. Right maxilla in anterolateral aspect. Scale 1 mm. From NRM A83/1948392.3010, 36.9 mm SL.

Vertebrae: 24; the first two bear epineural ribs; the third has short paired hypapophyses and bears the first pleural and epipleural ribs although the basapophysis is not distinctly projecting. The last basapophysis and pleural rib are on the 12th vertebra, but epipleural ribs extend back to the 14th vertebra as intermuscular bones; the first hemal arch is on the 13th vertebra the hemal spine of which projects between the first two interhemals. Pleural ribs of the 11th and 12th vertebrae articulate with the first interhemal (Fig. 113). Among seven radiographed specimens six have 12 + 12 vertebrae, one has 12 + 11.

The single supraneural has a rounded dorsal tip in young; in large specimens it bears a prominent antrorse spine.

The caudal skeleton (Fig. 114) includes no complete vertebral centrum. The last three vertebrae contribute to the fin ray support. There is no parhypurapophysis. There are three distal radials ventrally, but only two dorsally, suggesting artifactual loss and/or co chondrification; a minute paired cartilage between the second epural and hypural 5 may be an abnormality.

Colouration

Large specimens (especially from RMNH 31141) with light, yellowish white ground colour on chest, preventrally, along abdominal midline,

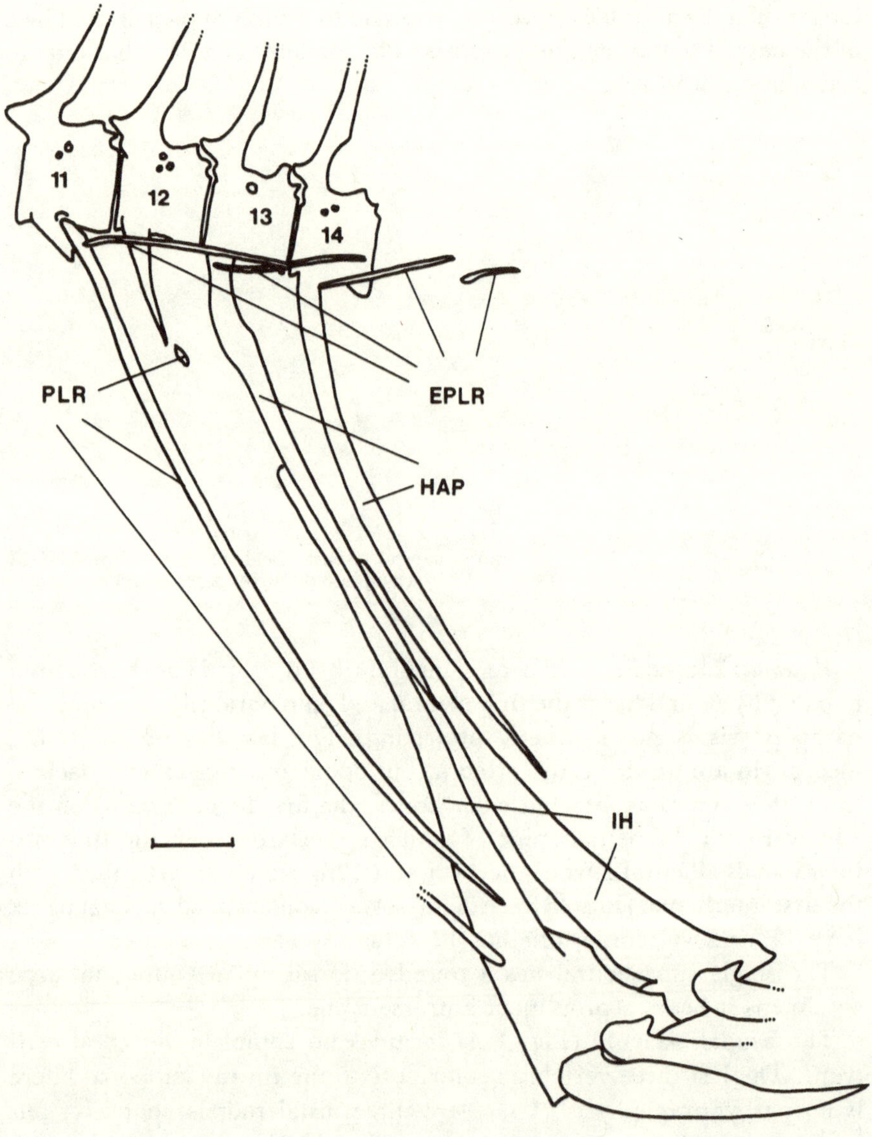

Fig. 113. *Cleithracara maronii*. Transitional region abdominal to caudal of vertebral column, to show epipleural rib configuration. Scale 1 mm. From NRM A83/1948392.3010, 36.9 mm SL. *EPLR* epipleural ribs of vertebrae 11-14 (those of 12th to 14th are free intramuscular elements), *HAP* hemal spines, *IH* interhemals, *PLR* pleural ribs (posterior a vestige), *11-14* 11th to 14th vertebral centra.

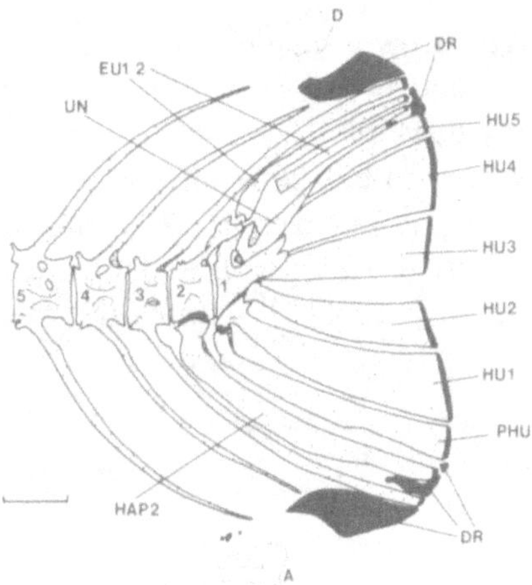

Fig. 114. *Cleithracara maronii.* Lateral aspect of caudal skeleton. Certilage black. Scale 1 mm. From A83/1948392.3010, 36.9 mm SL. Most rays disarticulated during maceration, and distal radials additional to those shown may have been lost. *A* outline of distal edge of last anal fin pterygiophore, *D* outline of distal edge of last dorsal fin pterygiophore, *DR* distal radials of caudal fin, *EU1-2* epurals 1-2, *HU1-5* hypurals 1-5, *HAP2* hemal spine of second preural centrum, *PHU* parhypural, *UN* uroneural, *1-5* preural centra.

anterior abdominal sides, branchiostegal membrane and between darker flank markings. Preorbital and nape grey; faint whitish stripe between eye and nostril. Lower jaw brownish grey, contrasting against lighter adjacent areas. Operculum dorsally greyish, ventrally lighter.

Wide, dark brown tear-shaped (ventrally widened to eye diameter) stripe from lower edge of orbit obliquely caudad-ventrad onto sub- and interopercular junction; cheek rostrally light, brownish yellow. Dark brown wide stripe from dorsal edge of orbit obliquely caudad-dorsad to nape midline, just anterior to dorsal fin origin, where meeting stripe from other side.

Back neutral to reddish brown, slightly lighter along lateral line. Below lateral line a brown band about one scale deep, slightly darker than back, from slightly behind gill cover caudad; sides below light. Below end of spinous dorsal fin, on, but mostly above, lateral line, a black round or oblong blotch; delimited anteriorly and posteriorly by light (yellowish white) deep spots, posterior of which continued as light bar down to anal fin base, anterior usually restricted, but may extend very faintly below lateral line. Above black blotch brown to dorsal fin, below continued as brown to dark brown bar to slightly above ventral edge of body, widened rostrad and continuous with horizontal band. Sides and caudal peduncle behind light vertical bar brown like back. Faint background pattern on sides, especially on posterior sides, of more or less zig-zagged lines as scale centers light, dorsal and ventral edges darker.

Unpaired fins immaculate; greyish brownish, caudal fin slightly lighter, dorsal corner narrowly white-margined. Pelvic fin with greyish brown outer edge, lighter, to white, inwardly. Often outer edge of pelvic fin white and anal fin with narrow black lower edge.

Young light, yellowish white with light brown zig-zag lines; slightly countershaded. Head stripes prominent, that below orbit not widened ventrally; horizontal band and bar below flank blotch hardly apparent. Flank blotch prominent. A shadow of a vertical bar between dorsal and anal fin ends, caudal peduncle lighter. Unpaired fins with dark bases, pale greyish brown; dorsal fin lappets, most of soft dorsal and anal fin and distal two thirds of caudal fin light, slightly smoky, but anal fin lower edge blackish.

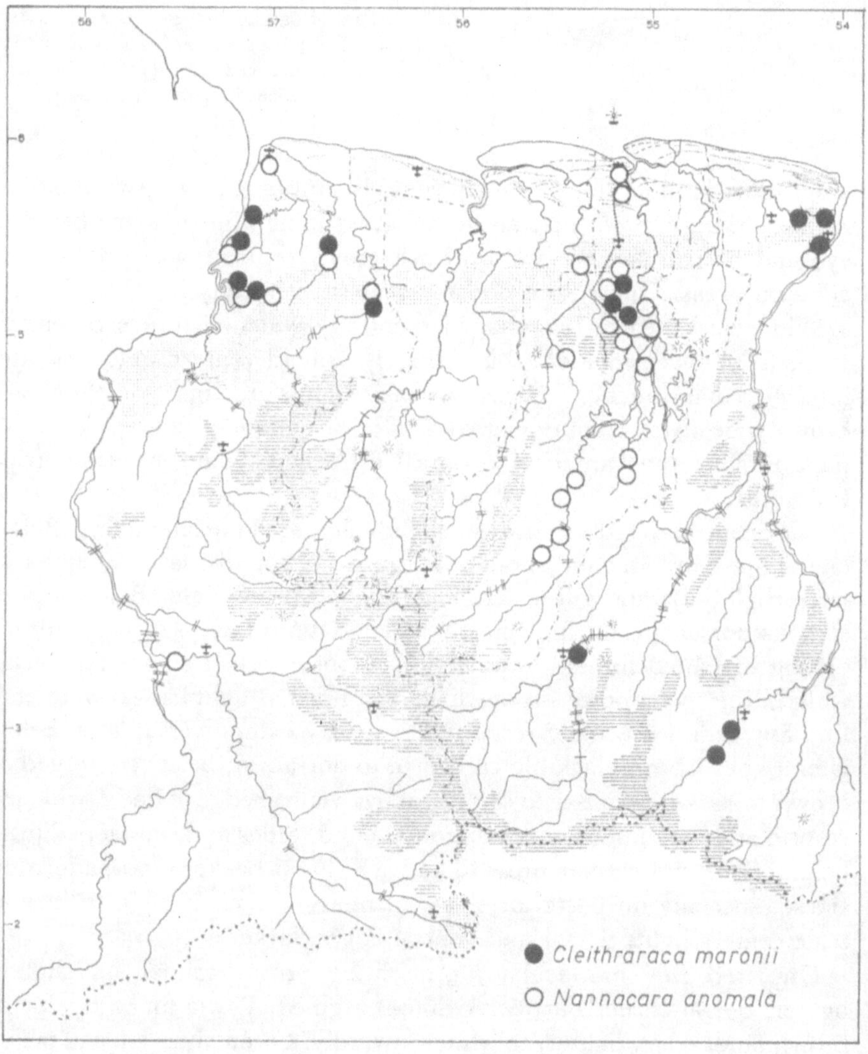

Fig. 115. Collecting localities of *Nannacara anomala* and *Cleithracara maronii* in Surinam.

Distribution (Fig. 115)

In Surinam, *C. maronii* occurs in the lower parts of the drainages from the Corantijn east to the Marowijne; only in the Marowijne drainage there are collection localities in southern tributaries (Paloemeu, Oelemari). The species ranges from the Barima R. in Guyana east to the Ouanary in French Guiana.

Notes

Steindachner (1882) gave an excellent description of the holotype, supplemented by a very fine figure. The species is distinct enough never to have caused any problem with its identification. It is easily distinguished from all other Surinamese cichlids by the combination of extensively scaled vertical fins and scaled preoperculum, and also by the distinctive colouration: light with contrasting black stripe from the nape through the eye over the cheek, and light-margined black blotch very high on the posterior sides.

Outside Surinam, *C. maronii* has been collected in small numbers, in Guyana in the lower Demerara R. system at Georgetown, Kumaka, Mackenzie, Wismar (Eigenmann 1912; Lowe-McConnell 1969); and in the Barima near the mouth of the Aruka R., as reported by Eigenmann (1912). From French Guiana there is material from Ouanary, and the Mana and Approuague systems. Notable is the apparent absence from the relatively well-collected Essequibo and Oyapock systems, and the rather scattered localities. However, *C. maronii* is obviously nowhere abundant and within the area of known distribution there are extensive areas that have not been adequately surveyed. Future collecting will probably produce a picture of a more continuous distribution along the coast of the Guianas.

Cleithracara maronii was recorded for Trinidad by Price (in Axelrod 1958). Evidently there was a population at Buenos Aires (Cedros Ward, S Trinidad), verified up to 1965, but not encountered in 1974 (Kenny 1979; but his reference to Price (1955) for the first record is unverifiable. SOK has confirmed the determination from a photograph of a young specimen provided by Kenny and Bruce.) This place is on the coast of the Boca de la Serpiente opposite mouths of the northern Orinoco Delta rivers. Kenny (1979) hypothesizes a colonization from the Orinoco taking into account then the presence of the species in NE South America. However, the nearest mainland locality is on the Barima R. in Guyana. The species is not collected in the Orinoco system or elsewhere in Venezuela.

NANNACARA Regan

Nannacara Regan 1905b, p. 344 (type species by monotypy *N. anomala* Regan).

Only two species are currently recognized in *Nannacara*, viz. *N. anomala* and *N. aureocephalus* Allgayer (1983). The geographic range of the genus includes the Guianas from the Aruka drainage east to the Oyapock drainage, Caripito and R. Morichal Largo in Venezuela, and the Belém area in Brazil. A record from Manaus (Regan 1912) is probably erroneous.

Nannacara was established originally to contain a single *Aequidens*-like species distinguished by having the upper lateral line separated from the spinous dorsal fin by only half a series of scales for most of its course, instead of 'well separated' (Regan 1905b). The relatively close proximity of the upper lateral line to the dorsal fin is probably correlated with the small size and still relatively large scales, and although of some use in identification, there are apparently other character states of greater diagnostic value.

From study of two alizarin-alcian blue specimens, partial dissection of alcohol material, and external morphology of remaining material of Surinamese *Nannacara*, the following distinctive characters were found (A-F). Characters A, C and F have been verified also in *N. aureocephalus*.

A. 14 (7 + 7) principal caudal fin rays (Fig. 116). All other cichlids so far studied have 16 (8 + 8) principal caudal fin rays. The plesiomorphic number for percoids is 17 (9 + 8). The caudal skeleton of *Nannacara* features somewhat approximated hypurals, especially hypurals 2 and 3, in distinction from other cichlids. Otherwise the caudal skeleton is similar to that of other *Aequidens*-like cichlids.

B. A fenestra in the wide lateral ledge of the quadrate, through which passes the tip of the posterior process of the anguloarticular when the lower jaw is maximally lowered. The posterior process of the anguloarticular is of modal size; the lateral ledge of the quadrate appears relatively wide, however. In other cichlids, the corresponding region of the quadrate features a simple recess which may or may not be indicated as a bulge on the laterodorsal aspect of the ledge.

C. Loss of the second from anteriormost frontal lateralis foramen (Fig. 117). This character may reflect a somewhat foreshortened ethmoid region, as well as a general reduction of the lateralis system, including reduced infraorbital series (character D), and largely pored flank lateral

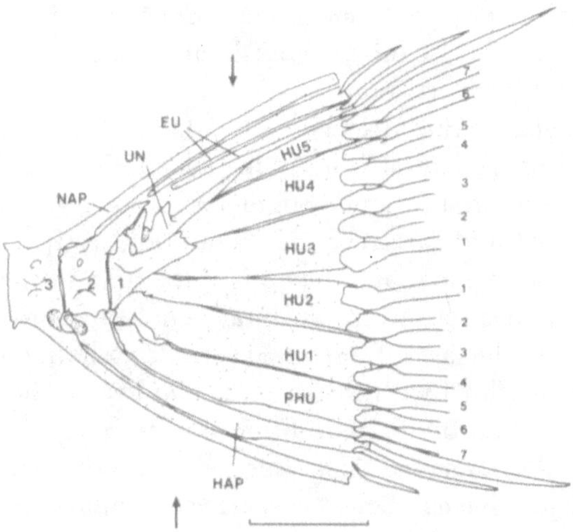

Fig. 116. *Nannacara anomala*. Caudal skeleton. Scale 1 mm. From NRM A82/1966503.3454, 27.1 mm SL. Distal radials not shown, principal rays cut. Arrows point to end of dorsal and anal fin bases, respectively. Vertebrae (*1-3*) and principal caudal fin rays (*1-7*, *1-7*) numbered. *EU* epurals, *HU1-5* hypurals 1-5, *HAP* last hemal spine, *NAP* neural spine, *PHY* parhypural, *UN* uroneural.

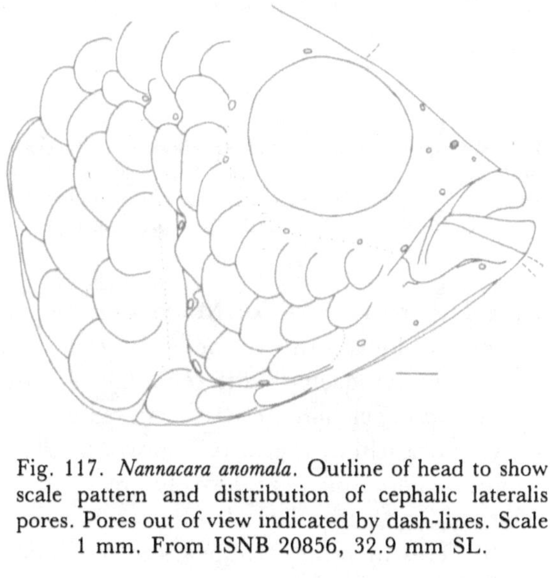

Fig. 117. *Nannacara anomala*. Outline of head to show scale pattern and distribution of cephalic lateralis pores. Pores out of view indicated by dash-lines. Scale 1 mm. From ISNB 20856, 32.9 mm SL.

lines. The only other cichlid in which the second frontal lateralis foramen has been reported lost, is *Taeniacara candidi* Myers. That species also has chiefly pored instead of tubed flank lateral lines; the cephalic lateralis system is more regressed than in *N. anomala*.

D. Infraorbitals reduced to two (Fig. 118); the posterior reaching to the ventral tip of the lateral wing of the sphenotic, penetrated by a canal posteriorly ending in a foramen slightly anterior to the end of the bone. As in *Cleithracara* and *Laetacara* the infraorbitals, especially the second, bear wide ventral lamina, and the ends of the first infraorbital are backed by the lacrimal and second infraorbital. The configuration in *Nannacara* represents loss of the last infraorbital (dermosphenotic), co-ossification of a remnant of the third with the second, and loss of the middle foramen on the second infraorbital. In addition the lacrimal is narrow compared to *Laetacara* and *Cleithracara*. A similar reduction of the infraorbital series is found in *Apistogramma*, *Apistogrammoides* (Kullander 1987a), and the smaller *Laetacara* species.

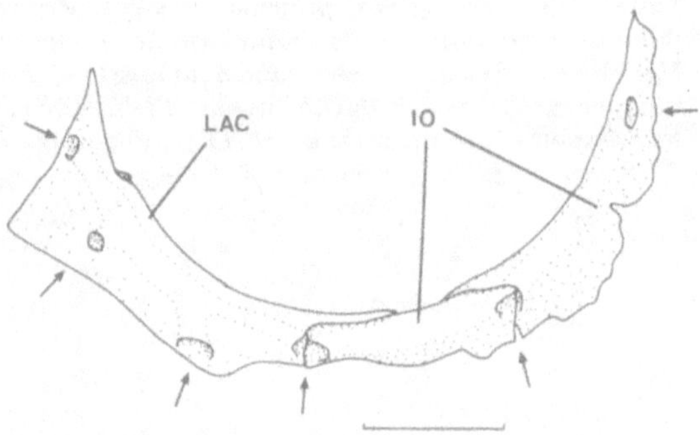

Fig. 118. *Nannacara anomala*. Left side suborbital series in lateral aspect. Scale 1 mm. From NRM A82/1966503.3454, 32.0 mm SL. Arrows point to lateralis canal openings. *IO* infraorbitals, *LAC* lacrimal.

E. Epibranchial 1 shape (Fig. 119): Moderately slender; pharyngobranchiad 2 arm about as long as the rest of the bone; pharyngobranchiad 1 arm very short. This extreme length difference between the medial arms appears to be unique among cichlids. *Hemichromis* features a similarly long and slender pharyngobranchiad 2 arm (Greenwood 1985a; pers. obs.), but the pharyngobranchiad 1 arm is relatively longer than in *Nannacara*.

F. A membrane uniting the last dorsal and anal fin ray with the caudal peduncle (Fig. 120). This membrane appears to be unique to *Nannacara* among cichlids.

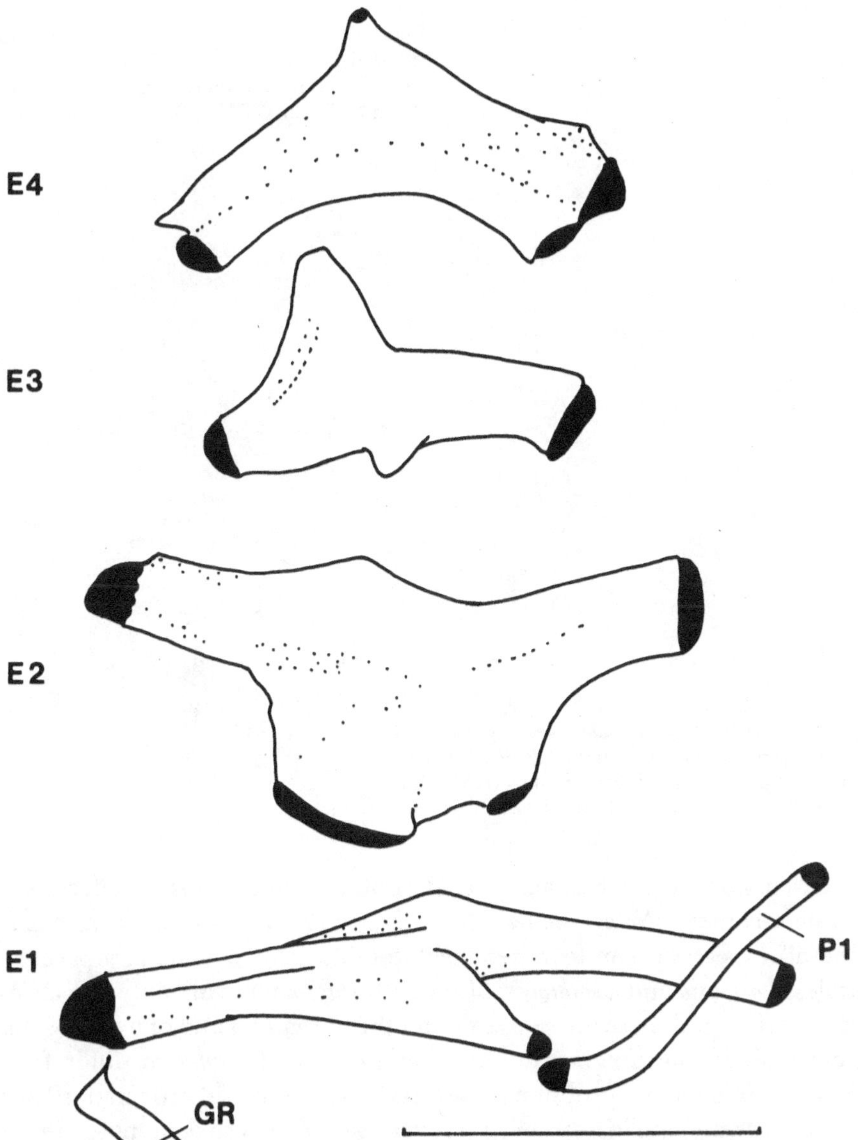

Fig. 119. *Nannacara anomala*. Right side upper gill arch elements in anterior view. Scale 1 mm. From NRM A82/1966503.3454, 32.0 mm SL. Cartilage black. *E1-4* epibranchials 1-4, *GR* gill raker in angle of first epi- and cerato-branchial, *P1* pharyngobranchial 1.

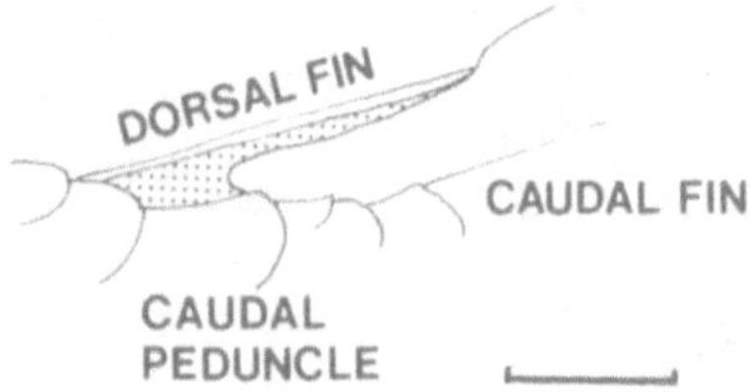

Fig. 120. *Nannacara anomala.* Membrane (shaded) uniting last ray of dorsal fin with caudal peduncle unique to the genus. Scale 1 mm. From ISBN 20854, 290 mm SL.

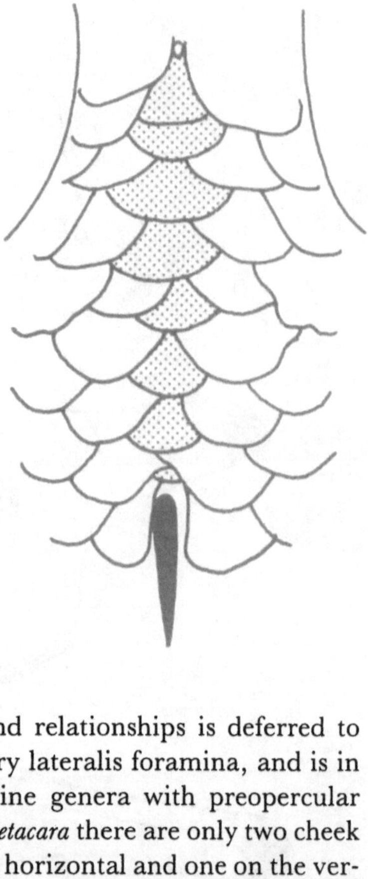

Fig. 121. *Nannacara anomala.* Scale pattern of nape, with predorsal midline scales shaded. First dorsal fin spine black. From NRM A82/ 1966503.3454, 41.0 mm SL.

Discussion of other anatomical data and relationships is deferred to another context. *Nannacara* has four dentary lateralis foramina, and is in overall aspect similar to other cichlasomine genera with preopercular scales, *Laetacara* and *Cleithracara*. Like in *Laetacara* there are only two cheek scale series, and there are two scales on the horizontal and one on the vertical limb of the preoperculum. *Nannacara* and *Cleithracara* differ from *Laetacara* in a having prominent posterior palatoethmoid articulation, and in having only one instead of two supraneurals. There is no notch in the dorsal margin of the anterior ceratohyal as in *Laetacara* and *Cleithracara* species, lip folds are continuous, and the predorsal squamation is uniserial. The predorsal scale pattern differs from the uniserial pattern in *Krobia* and *Bujurquina* as the posterior scales in the middle series are relatively small and never possess the notched margin characteristic of other South American cichlids with uniserial predorsal pattern (Fig.121).

Nannacara species are small; males reaching 56 mm Sl, females 39 mm SL in the wild. A number of losses or reductions may reflect the small adult size and are not unique characters: parhypurapophysis, microbranchiospines, ceratobranchial 4 tooth plates, interarcual cartilage, accessory caudal fin lateral lines lost; very short caudal peduncle not including any vertebrae; naked dorsal and anal fins; vertebrae 11 + 13 (epipleural ribs on the first two caudal vertebrae).

Ribeiro (1918) referred also *Acara dorsiger* Heckel, and *N. hoehnei* Ribeiro, to the genus *Nannacara*. The former is now referred to *Laetacara* (Kullander 1986). *Nannacara hoehnei* is a distinctive *'Aequidens'* species.

Nannacara bimaculata Eigenmann, in the Essequibo system, is similar to *N. anomala* and *N. aureocephalus* in predorsal and head squamation (but commonly three cheek scale series), and high dorsal fin count (D.XVI.8-9, XVII.7). It differs in having the full number of frontal lateralis foramina, 16 principal caudal fin rays, interrupted lower lip fold, and a completely different colour pattern (yellowish with brown vertical bars, high-set midlateral blotch and dark spot on caudal fin base; sexes isomorph but dark spots along dorsal fin base more intense in females). The phylogenetic relationships of *N. bimaculata* require further study.

Nannacara anomala Regan

(Figs 115-125)

Nannacara anomala Regan 1905b, p. 344 (R. Essequibo [syntypes BMNH 1864.1.21:27-28]).

Synonymy

?*Nannacara taenia* Regan 1912, p. 505 (the Amazon at Manaos).

Surinamese records

Nannacara anomala; Boeseman 1952, p. 197 (Listed: Zanderij I, savanna streamlet [RMNH 18369]; Makambi creek [RMNH 18370]; Kabelstation, Surinam River [RMNH 18371]; Paramaribo, in ditch [RMNH 18372]; plantation Kwatta, Paramaribo [RMNH 18449]). — Boeseman 1954, p. 23 (listed; neighbourhood of Paramaribo [RMNH 19792]; probably neighbourhood of Paramaribo [not found in RMNH]). — Boeseman 1956, p. 196 (listed: Surinam [RMNH 21385]).

Material

Over 550 specimens from the Marowijne, Suriname, Saramacca, Nickerie, and Corantijn river systems. Thirty-six specimens from the Nickerie, Corantijn, Marowijne and Suriname R. systems measured; counts from 143 specimens, except pectoral fin and lateral line counts, which are only from those measured. Description based chiefly on fine specimens from Zanderij and Carolina Kreek.

Fig. 122. *Nannacara anomala*, male, 44.9 mm SL, ISNB 20854, from near Zanderij.

Fig. 123. *Nannacara anomala*, female, 29.0 mm SL, ISNB 20854, from near Zanderij.

Description

Compound, but chiefly from 44.9 mm male from Zanderij (ISNB 20854, Fig. 122). Largest specimen recorded a male, 56.4 mm SL (ZMA 107.303); largest female recorded 39.1 mm SL (ZMA 105.344).

Shape: elongate, moderately compressed laterally, dorsal and ventral contours about equally arched; in frontal aspect elliptic, with narrowly rounded nape and chest. Caudal peduncle short, dorsal edge shorter than ventral.

Head: moderately elongate, rounded off rostrally; predorsal contour about straight, prepelvic contour curved. Orbit tangential; interorbital slightly curved. Mouth terminal, wider than interorbital; jaws equal anteriorly; tip of maxilla exposed, to anterior margin of orbit or little behind. Lips thick, folds continuous.

Scales: Squ. long. -(3), 21(48), 22(92). Circumpeduncular scale series 16. Flank scales ctenoid. Predorsal scales (Fig. 121) in 3 series, a median of usually 8 caudad smaller scales (anterior slightly smaller than anterior flank scales), about 8 marginal; cycloid, as also those next to orbit and a few along anteriormost part of dorsal fin. Chest scales ctenoid, of about flank scale size. About 8 prepelvic scales along midline, cycloid like those in parallel series anterior to pelvic fin bases; posterior 2/3 size of anterior flank scales, slightly smaller anteriorly.

Below eye 3 curved scale series, covering cheek and preoperculum (Fig. 117): lower cheek scale series continued over vertical limb of preoperculum posteriorly, lowermost series consisting of 2 scales on horizontal and 1 on vertical limb of preoperculum; lateralis pores on preoperculum exposed by incisions in scales or appearing between scales; posterosuperior cheek scales ctenoid, nearly of flank scale size, rest smaller, cycloid. Two vertical scale series on operculum; one series each on pre- and inter-operculum, all large, largest of flank scale size; dorsal opercular and subopercular scales ctenoid, rest cycloid. Cheek scale series (2)139, (3)4.

Upper lateral line commencing (by 1-3 scales) in 2nd scale series above that containing lower, at 2 1/2 scales distance from dorsal fin, continued in scales series above, posterior half at 1/2 scale distance from dorsal fin, to below about end of spinous dorsal fin, above anterior end of lower lateral line; largely of only pored scales. Lateral line total counts: upper, 13(1), 14(3), 15(8), 16(14), 17(9), 18(1); lower, 3(1), 4(9), 5(15), 6(8), 7(2), -(1); a canal on caudal fin base in 8 of 36 specimens checked.

Pectoral and pelvic fins naked; dorsal and anal fins naked, but adjacent body scales reaching out over bases of posterior spinous dorsal fin, soft dorsal fin and anal fin. Proximal about 1/3 or slightly less of caudal fin with an evenly wide layer of chiefly ctenoid scales.

Fins: dorsal fin origin above lateral line origin, first spine less than 1/3 length of last; spines increasing in length to last, but little from 5th; lappets free, truncate, reaching little beyond spine tips; soft fin pointed, 4th ray longest, produced in large males, to near caudal fin end, pointed, not produced in large females, to about middle of caudal fin. Membrane connecting last ray with caudal peduncle in 137 of 142 intact specimens, absent in 5. D. XV.8(5), XVI.6.i(4), XVI.7(21), XV.7.i(31), XVI.8(51), XVI.8.i(1), XVI.9(1), XVII.6.i(7), XVII.7(10), XVII.7.i(4), XVII.8(5), XVIII.7.i(1), -.7.i(1), -.-(1); "i" refers to a rudimentary terminal raylet, but last principal ray also commonly unbranched.

Soft anal fin pointed, 4th ray longest, slightly produced, to about 3/4 of caudal fin in large males; not produced, to about middle of caudal fin in large females. Membrane connecting last ray with caudal peduncle in 129 of 143 specimens, absent in 14. A. II.8(1, abnormal), III.6.i(1), III.7(5), III.7.i(25), III.8(85), III.8.i(15), III.9(11); "i" refers to a rudimentary terminal raylet, but last principal ray also commonly unbranched.

Pectoral fin elongately rounded, 5th ray longest, 6th subequal in length; extending to above second anal fin spine. P. 11 (2), 12 (33), 13 (1). Pelvic fin pointed, first ray longest; produced in large males, to along soft anal fin base; not produced in females, to above spinous anal fin base. Caudal fin elongately rounded; principal rays 6 + 7(1), 7 + 7(142).

Cephalic lateralis pores distributed as shown in Fig. 117.

Jaw teeth: 10 specimens 29.0-44.9 mm, with 19-25/17-23 (mean 21.2/20.9) teeth in upper/lower jaw outer hemiseries. Teeth conical, pointed, feebly recurved; those of outer series strong, evenly spaced, along entire jaw rim, not much enlarged anteriorly; inner teeth smaller, in 2-3 series forming wide band anteriorly, posteriorly a single series not reaching as far as outer series.

Gill rakers and lower pharyngeal tooth-plate. External first ceratobranchial gill rakers -(4), 5(45), 6(92), 7(2); no microbranchiospines, no gill rakers on lower pharyngeal tooth plate. In 39.8 mm specimen figured (Fig. 124)

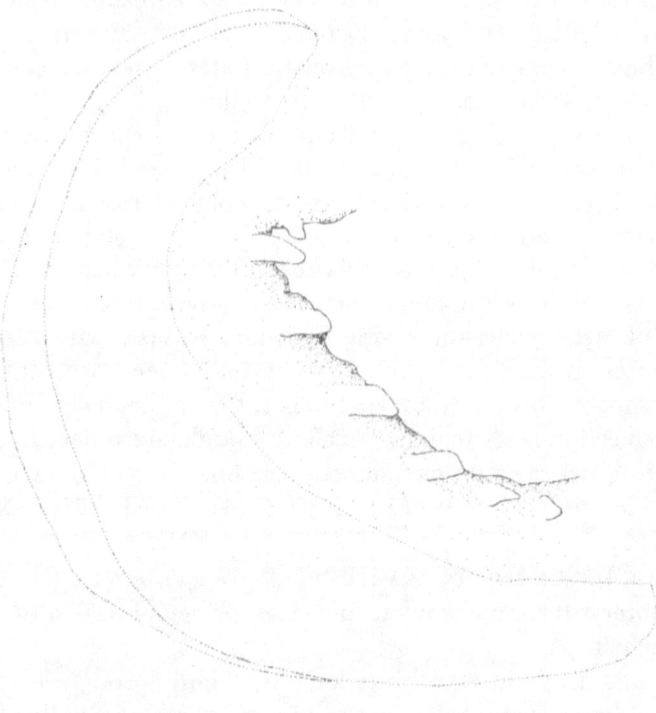

Fig. 124. *Nannacara anomala*. First gill arch in lateral aspect. Scale 1 mm. From ZMA 105.344, 39.8 mm SL.

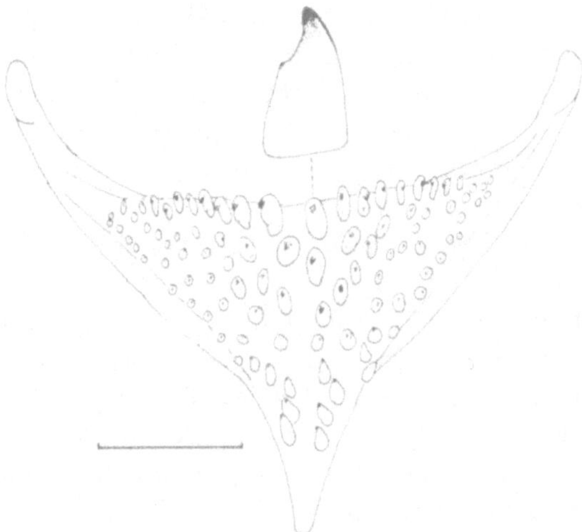

Fig. 125. *Nannacara anomala*. Lower pharyngeal tooth plate in occlusal aspect and posterior tooth in lateral aspect. Scale 1 mm (tooth plate), 0.5 mm (tooth). From ZMA 105.344, 39.8 mm SL.

first arch with short, edentulous external rakers, one epibranchial, longest raker in epi-/ceratobranchial joint angle, 5 rostrad successively shorter ceratobranchial rakers; internal rakers similar, 1 + 1 + 9. Lower pharyngeal tooth plate in same specimen (Fig. 125) relatively wide; rostrally subconical teeth, caudally teeth laterally compressed, and with more or less evident shelf on rostral edge.

Vertebrae: 11 + 13 (5).

Colouration

Specimen 44.9 mm, Fig. 122, representative of adult males, with yellowish ground colour on chest, belly, and, sparsely overlayered with dark pigment, on lower sides and in bar interspaces. Snout, preorbital, pre-, sub-, and inter-operculum, and lower jaw pale greyish. Cheek yellowish. Nape brown, lighter along midline.

Brown stripe from orbit slightly backwards inclined over cheek to lower edge of preoperculum. Faint light stripe between orbit and nostril, adjacent preorbital edge slightly darker than the rest. Brown horizontal band from orbit onto caudal fin base, slightly intensified where crossing vertical bars and slightly widened and intensified from 5th vertical bar caudad. Another, less distinct brownish horizontal band along back on and slightly below upper lateral line. Five dark vertical bars over sides (numbered from posteriormost forwards): 1) on caudal peduncle;

2) between ends of soft dorsal and anal fins; 3) above soft anal fin origin; 4) above spinous anal fin; 5) above middle of abdomen. Bars 3-5 from dorsal horizontal band to slightly above ventral edge of sides; Bars 3-4 continued to dorsal fin base, Bar 5 divided into 2 slightly diverging spots close to dorsal fin base. Bars strongest dorsally, fading out ventrally; areas enclosed by bars and bands standing out as light blotches. Area around dorsal fin origin dark, fainter pigmentation continued ventrad to lower horizontal band.

Dorsal fin light brown, rays light, greater part of spinous portion with minute light circles (or vermiculations); lappets clear, with dark tips; soft portion with clear margin. Caudal fin greyish brown, rays light, basally a slightly darker spot continuing lower horizontal band. Anal fin brownish grey, rays lighter, lower edge blackish. Pelvic fin outer edge whitish, otherwise brownish grey.

Smaller specimens which do not have such a dark nape show a brownish stripe from orbit obliquely caudad and dorsad, not reaching nape midline. Bar 5 commonly intensified, forming an indistinct spot, in lower horizontal lateral band. Often also another vertical bar, very faint, in advance of Bar 5, not below lower horizontal band. Appearance of lower horizontal band very variable, washed out or contrastingly dark brown, but always most prominent flank marking. Many large males extensively darkened, with subdued pattern, but dark dot on base of scales of lower sides. Vermiculations in dorsal fin best developed in large males. Adult females (Fig. 123) much like males, but lacking vermiculations in spinous dorsal fin. Instead two prominent, sharply delimited dark brown spots basally on dorsal fin above Bar 5 subdivisions.

Pelvic fin white. Young males commonly showing dorsal base spots but then not as intense as in females. Most larger females also with stripe below orbit continued onto interoperculum and cheek, gill cover and lower jaw with dark brown speckles, branchiostegal membrane also darkened. Juveniles similar to adults, but fins light, without light dorsal fin margin.

Distribution (Fig. 115)

Nannacara anomala is common in the lowlands of Surinam and Guyana, from the Aruka R. in Guyana (Eigenmann 1912) to the lower Marowijne. The absence from the Saramacca and Coppename drainages may be due to insufficient collecting in these drainages. The only upland Surinamese specimen in our material is a 30.0 mm male from the Paloemeu; it is somewhat stoutbodied (depth 38.3% SL).

Notes

Nannacara taenia Regan (1912) may be a synonym of *N. anomala*. It is based on a 31 mm female aquarium specimen in brooding pattern and in very bad condition, said to come from Manaus. The syntypes of *N. anomala* are two males in bad condition and the characters put forth by Regan (1912) as distinguishing the two species may be explained by sex differences and different living conditions. In any case, data from the holotype of *N. taenia* are within the variation shown by Surinamese *N. anomala*. According to Arnold (1912), the type of *N. taenia* arrived in a shipment of living ornamental fish from Manaus to Germany in 1911. It died in Arnold's aquarium in December 1911 and was sent to Regan in January 1912. As no *Nannacara* has been found in Central Amazonia since (except additional aquarium material supplied by Arnold) and as other of Arnold's localities are unreliable, it seems, despite the details on the origin of the fish, that some confusion may have occurred. A species distinct from both *N. anomala* and *N. aureocephalus* occurs near Belém, State of Para, Brazil, and should be compared with *N. taenia*.

Nannacara anomala was reported from Caripito, Venezuela, by Schultz (1949), and we have examined a *Nannacara* specimen from Estancia El Puente, R. Morichal Largo (MBUCV-V-15377pt., 21.7 mm SL). The latter is rather small and not well preserved, and Schultz material was not reexamined. It is plausible that the Venezuelan *Nannacara* is distinct from *N. anomala*, as both the Morichal Largo specimen, and Schultz's material (3 specimens) has four anal fin spines.

Nannacara aureocephalus is based on a series of overgrown, diseased and poorly preserved aquarium specimens originally collected in the Mana and Comté drainages (type locality a creek tributary to the Mana, 11 km from Saut Sabbat). We have reexamined the holotype and paratypes deposited in MNHN, additional aquarium material from the same source, and also single specimens and small series of *Nannacara* material from various localities in French Guiana.

Allgayer (1983) distinguished *N. aureocephalus* from *N. anomala* with reference to (1) deeper body, (2) shorter head, (3) smaller eye, and (4) bright yellow colour on cheek, postocular area, preoperculum and operculum. Allgayer's data on *N. anomala* are probably from Eigenmann (1912). Characters 2 and 3 likely reflect size allometry, and yellow head sides (character 4) are found also in *N. anomala* males.

French Guianan *Nannacara* do seem to be more deep-bodied than Surinamese *N. anomala*, but we note a difference between Mana + Orapu material (n = 3, 25.2-33.8 mm SL, depth 36.7-36.9% of SL), and Approuague (n = 6, 18.2-31.0 mm SL, depth 38.4-41.4% of SL) and Oyapock material (n = 5, 28.2-39.1 mm SL, depth 37.9-40.1% of SL).

The holotype of *N. aureocephalus* is slender (depth 34.0% of SL), but sunken.

Large males of *N. aureocephalus* are readily recognized on the spotted soft unpaired fins, however. Otherwise *N. aureocephalus* and *N. anomala* are very similar, and more study is required to find useful separating characters for females and young.

Some of Allgayer's data on *N. aureocephalus* indicate additional differences from *N. anomala*, but are rather due to different methods of counting. His count of 23-24 scales in a lateral series possibly includes scales on the caudal fin base; the holotype has only 21. There are 16 scale series around the caudal peduncle, not 11-12 as given by Allgayer.

CRENICICHLA Heckel

Crenicichla Heckel, 1840, p. 416-417 (type species *C. macrophthalma* Heckel, by subsequent designation in Eigenmann & Bray, 1894, p. 620).

Synonymy

Batrachops Heckel, 1840, p. 432 (type species, *B. reticulatus* Heckel, by subsequent designation in Eigenmann & Bray, 1894, p. 620).
Boggiania Perugia, 1897, p. 149 (type species, by monotypy, *B. ocellata* Perugia, 1897).

Local names

Datra fisi, Adoenko, Donfó (for all species).

Body elongate. Fleshy lips. Lower jaw prognathous. Orbit in upper half of head. Two well-developed lateral lines, continuing on caudal fin with 1-3 scales; dorsal part of head naked. Cycloid scales on cheek, operculum, suboperculum, anterior part of back, and belly; ctenoid scales on sides.

Caudal fin with basal and interradial scales. Pectoral, dorsal and anal fins without basal scales, although base sometimes covered with skin. Dorsal fin acute posteriorly. Anal fin origin just below first soft dorsal ray; anal fin posteriorly acute and with 3 anal spines. Ventral fin acute, V 1.5. Pectoral fin rounded. Caudal fin with 16 primary rays. Teeth conical and recurved, outer tooth series fixed, often considerably larger than inner tooth series. Gill rakers on first gill arch denticulated externally.

Crenicichla saxatilis (Linnaeus)

(Figs 126-127, 136)

Sparus saxatilis Linnaeus, 1758, p. 278 (Surinam).

Surinamese records

Perca saxatilis; Bloch, 1974, p. 79, fig. 309 (description; Surinam).
Crenicichla saxatilis; Heckel, 1840, p. 432 (Surinam). — Regan, 1905, p. 159 (Surinam). — Steindachner, 1915, p. 571-573 (description; Marowijne R.). — Wheeler, 1958, p. 226 (description; photograph of holotype of *Scarus pavoninus* Gronovius, 1854). — Nijssen, 1967b, p. 246 (Surinam). — Richter & Nijssen, 1980, p. 125 (Surinam, Brokopondo Reservoir). — Kullander, 1982, p. 655 (discussion on types). — Ploeg, 1986a, p. 47 (description; designation of lectotype and restriction of the type locality: Suriname R., Carolina Creek). — Ploeg, 1986b, p. 222 (Marowijne R. drainage).

Scarus pavoninus Gronovius, 1854, p. 63 (Surinam).
Crenicichla saxatilis albopunctata Pellegrin, 1904, p. 374 (in part: Surinam).
Crenicichla alta; (non Eigenmann, 1912) Nijssen, 1966, p. 175 (Brokopondo Reservoir). — Nijssen, 1967a, p. 236 (Suriname R. drainage).

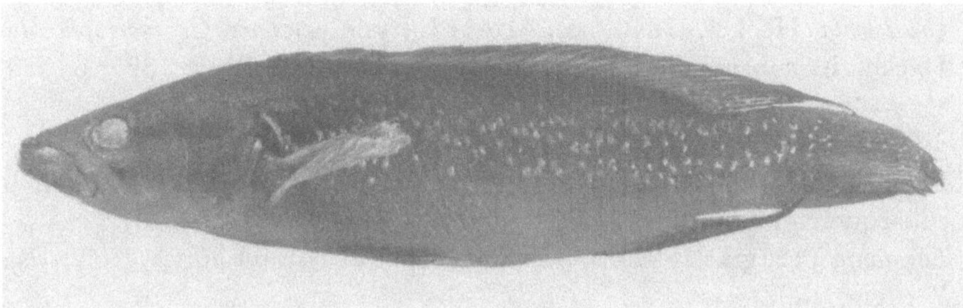

Fig. 126. *Crenicichla saxatilis*, male, 183 mm SL, ZMA 106.502A, from Para River, Suriname R. drainage.

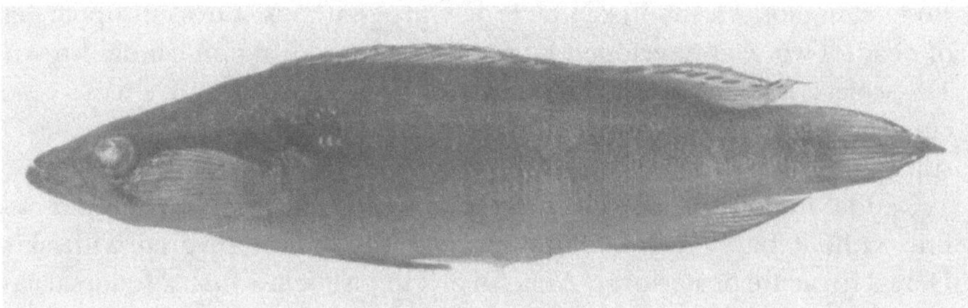

Fig. 127. *Crenicichla saxatilis*, female, 152 mm SL, ZMA 106.502A, from Para River, Suriname R. drainage.

Material

636 specimens, including lectotype, from Nickerie, Suriname, Commewijne and lower course of Marowijne R. drainages.

Description

Shape: elongate; trunk rectangular; laterally compressed.

Head: not flattened; snout rounded or pointed. Lips broad; nostrils situated just between postlabial skinfold and orbit; maxilla reaching to below orbit; caudal peduncle almost as deep as long; preoperculum serrated.

Teeth: conical, directed backwards with red crowns; 5 tooth series in lateral parts of upper jaw, 4 series in lateral part of lower jaw. In smaller specimens less tooth series; in specimens larger than about 80 mm SL always one more row in upper than in lower jaw. In specimens smaller than about 80 mm SL two series in both jaws. In upper jaw all teeth equal in length; in lower jaw outer teeth the largest; in both jaws only outer series fixed, inner series evertible.

Scales: ctenoid, except for cycloid scales present on cheek, operculum, and head (ventrally, and dorsally above anterior part of lateral line). Lateral line scales larger than adjacent scales; 2 (occasionally 3) scales in a vertical row between the lateral lines, 21-24 in anterior, and 9-11 in posterior lateral line. Two lateral line scales extend on caudal fin. From operculum to caudal fin base 52-60 scales in a row below lateral lines.

Gill rakers: usually 9 (8), dorsal 4-5 rakers bent, the remaining knob-shaped.

Fins: dorsal fin origin just beyond posterior edge of operculum; spines increasing in length to 7th (occasionally to 8th or 9th) posteriorly becoming equal in length; soft part of dorsal fin acute posteriorly, usually reaching to 4/5 of caudal fin, occasionally to beyond caudal fin. In females not extending to beyond 1/2 the caudal fin. In juveniles up to about 60 mm SL reaching to caudal fin base. D XVII-XIX.13-16. Anal fin acute posteriorly, reaching to caudal fin base, occasionally longer in males. A III.9-10. Ventral fin reaching to halfway the anus; second (soft) ray the longest; distally acute. V 1.5. Pectoral fin as long as ventral fin; distally rounded; P 15. Caudal fin rounded; basal scales present, most in the middle of the fin base (4 scale rows versus one scale row dorsally and ventrally). Interradial scales extend to 1/4 of caudal fin, also present in the smallest specimen examined; C 16.

Colouration

Males: brownish; dorsally darker, ventrally lighter. No conspicuous line between these areas. Anterior part of head and snout greyish; preventral area pale; cheek yellowish. Pectoral and ventral fins light grey, other fins darker. Dark blotch just above lateral line on caudal fin base, covering three rays. Dark humeral blotch covering 4 scales in a horizontal row, posterior to operculum, just *below*—in larger upstream specimens occasionally *on*—lateral line; humeral blotch irregularly round with a notch, always present at upper anterior side. Dark suborbital stripe; dark horizontal stripe from orbit to humeral blotch. Dark horizontal stripe tapering from orbit to distal tip of jaw. About 9 dark vertical bars above lateral lines. Many silvery dots (smaller than scale) occur in an irregular pattern on darker part of the sides, more numerous posteriorly. White dots (smaller than dots on sides) in posterior part of dorsal fin, anal fin, and in dorsal lobe of caudal fin. Humeral blotch and caudal fin blotch margined by silvery dots, giving the impression of an incomplete ocellus.

Females: ground colour as in male; humeral blotch only partly margined by silvery dots; no dots on sides, caudal fin blotch from hardly visible to conspicuous; faint dark lateral band; 13 black ocelli in dorsal fin, posterior to 11th spine; other females show occasionally ocelli a) from first spine to last soft ray, b) in last spines only, and c) no ocelli at all (intermediates occur). Dorsal fin margin above ocelli white. Caudal fin with dark dorsal margin, 5 ocelli below this margin; occasionally a black instead of a white margin.

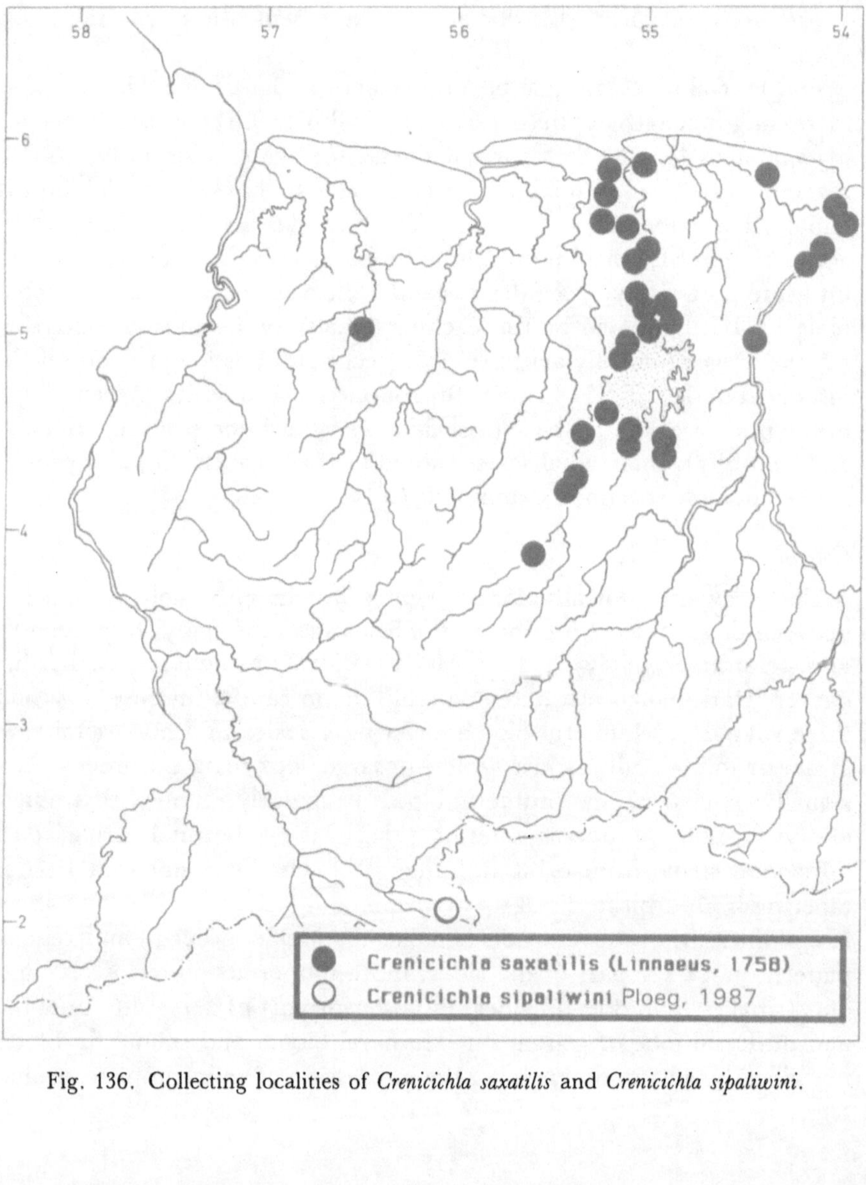

Fig. 136. Collecting localities of *Crenicichla saxatilis* and *Crenicichla sipaliwini*.

Juveniles: upper half of body dark, lower half light; dark band, running from tip of snout to caudal peduncle; lateral band disappearing with age. Humeral blotch round, on lateral band, below lateral line. Caudal fin blotch with light margin. Dorsal fin with white margin. No ocelli on fins; no dots on sides. Caudal fin dorsally with white margin and dark submargin, ventrally with white margin. Dorsal fin with white margin. No ocelli on fins; no dots on sides. Caudal fin dorsally with white margin and dark submargin, ventrally with white margin only. Shades of bars above lateral lines.

Distribution (Fig. 136)

Crenicichla saxatilis has been collected in the Nickerie, Suriname and Comewijne R. drainages, and in the Marowijne R. drainage downstreams of Tabbetje Hede. Outside Surinam it has been found in French Guiana; it possibly occurs in Guyana.

Crenicichla lugubris Heckel

(Figs 128, 137)

Crenicichla lugubris Heckel, 1840, p. 422 (Rio Negro).

Synonymy

Crenicichla funebris Heckel, 1840, p. 424 (Rio Guaporé).

Surinamese records

Crenicichla lugubris; Ploeg, 1987, p. 93 (Kabalebo R., Avanavero Fall).

Fig. 128. *Crenicichla lugubris*, 204 mm SL, ZMA 119.147, from Kabalebo River, Corantijn R. drainage.

Material

9 specimens from Corantijn R. drainage.

Description

Shape: predorsal contour slightly convex; preventral contour slightly convex anteriorly, straight below orbit.

Head: snout pointed in lateral view, rounded from above. Maxilla extending to anterior margin of orbit. Orbit large, diameter about half head depth at orbit level. Nostrils nearer to postlabial skinfold than to orbit.

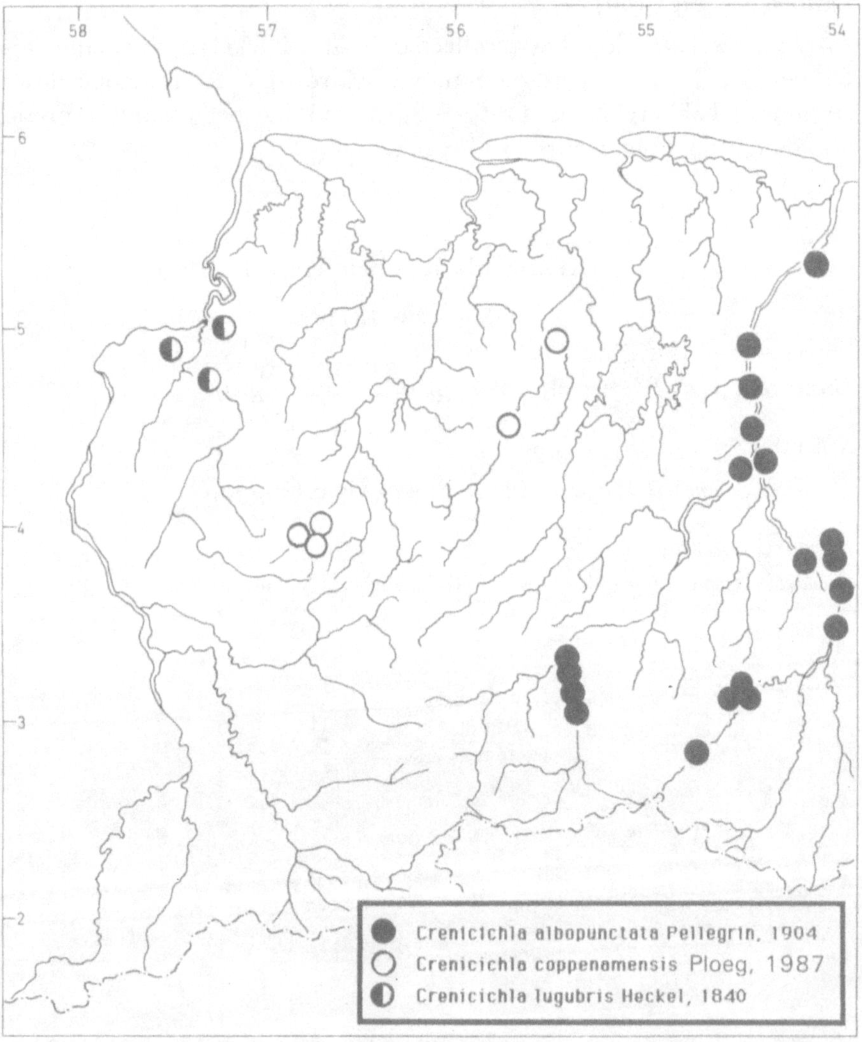

Fig. 137. Collecting localities of *Crenicichla albopunctata*, *C. coppenamensis*, and *C. lugubris*.

Fins: dorsal fin orgin just dorsal to hind edge of operculum. Dorsal fin spines increasing in length to 4th spine, the remaining equal in length; 5th from last ray longest, reaching to caudal fin base. Anal fin extending to caudal fin base; 4th ray from last longest. Ventral fin rounded; first soft ventral fin ray the longest. Caudal fin straight, with up to 6 basal scale rows.

Teeth: upper jaw with maximum 8 tooth series, lower jaw with maximum 5 tooth series. In both jaws outer teeth the largest.

Vertebrae number in one specimen (syntype NMW 61.148): 22 + 18.

Colouration

No sexual dimorphism in colour pattern was found. Ground colour of body brown, countershaded. No preorbital, postorbital or suborbital spot or stripe. Five bars between dorsal fin and upper lateral line. Dorsal, anal, ventral, and pectoral fins grey. Dorsal fin dark margined. Caudal fin with triangular black blotch, just above middle of base. Humeral spot just posterior to pectoral fin.

Distribution (Fig. 137)

C. lugubris has been collected in Surinam in the Corantijn R. drainage only. It is recorded from Guyana by Eigenmann (1912, Essequibo R. drainage). *C. lugubris* also occurs in northern Amazonian branches in Brazil.

Crenicichla multispinosa Pellegrin
(Figs 129-130, 138)

Crenicichla multispinosa Pellegrin, 1903, p. 124 (French Guiana).

Surinamese records

Crenicichla multispinosa; Pellegrin, 1904, p. 380 (in part: Surinam). — Regan, 1905, p. 164 (Surinam). — Ploeg, 1986b, p. 230 (restriction of type locality, Surinam, Marowijne R.). — Ploeg, 1987, p. 95 (Suriname and Marowijne R. drainages).
Crenicichla lugubris (non Heckel, 1840); Richter & Nijssen, 1980, p. 124 (Suriname R.).

Material

221 specimens from Marowijne and Suriname R. drainages.

Description

Shape: head depressed; predorsal and preventral contour straight.

Head: snout pointed in lateral view, rounded from above. Lower jaw slightly prognathous; maxilla not reaching anterior margin of orbit. Orbital diameter about half as deep as head at orbit level. Nostrils nearer to orbit than to postlabial skinfold.

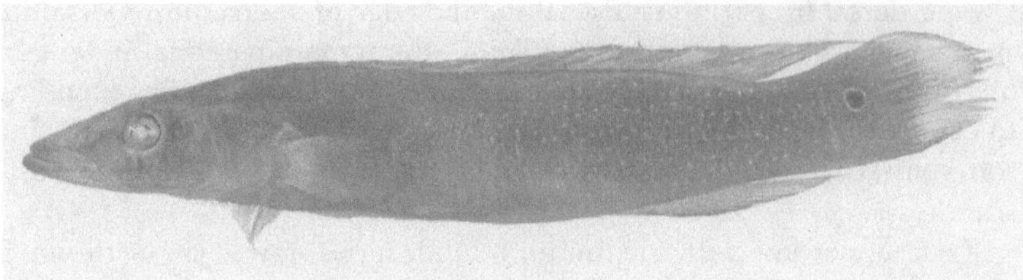

Fig. 129. *Crenicichla multispinosa*, male, 205 mm SL, ZMA 119.148, from Suriname River.

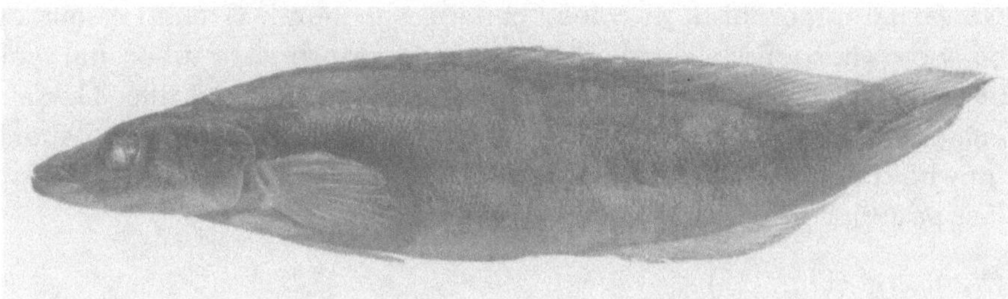

Fig. 130. *Crenicichla multispinosa*, female, 197 mm SL, ZMA 119.148, from Suriname River.

Fins: origin of dorsal fin just posterior to hind edge of operculum; spines increasing in length to about 9th, from there equal in length; rays reaching to 2/3 of caudal fin length in adults, to 1/5 of caudal fin length in juveniles; third from last soft ray the longest. Caudal fin with free ray tips; 5 basal scales at lateral line level, 2 ventrally and dorsally; interradial scales to 1/3 of caudal fin. Pectoral fin rounded.

Scales: cycloid on suboperculum, operculum, cheek, anterior part of back, and occasionally just posterior to base of pectoral fin, ctenoid in remaining areas. Lateral line scales larger than adjacent scales.

Teeth: upper jaw with maximum 7 tooth series, lower jaw with maximum 4 tooth series.

Vertebrae number in 3 specimens: 22 + 19/22 + 18/22 + 18.

Colouration

Males: brown, ventrally lighter, yellowish; cheek yellow; snout blue-greyish. Preventral area dirty white. A pattern of brown blotches on a bluish background below orbit (absent in some specimens, which then show a suborbital stripe). Posterior part of sides with many silvery dots, smaller than scale size. Dorsal fin whitish grey, scattered with whitish dots and dark margined. Anal fin whitish with few white dots, dark margined. Ventral and pectoral fins hyaline. Caudal fin with many silvery dots. Dark caudal spot, silvery ocellated.

Females: ground colour as in males. Dark lateral band running between lateral line scale rows. Neither suborbital stripe, nor a spot pattern below orbit. No dots on sides and on fins. Dorsal fin somewhat darker than in males, lacking the dark margins. Anal fin less whitish, dark margined. Ventral and pectoral fins as in males. Caudal fin with spot as in males, although less conspicuously ocellated. Dark band along middle; margins of this fin dorsally and ventrally dark.

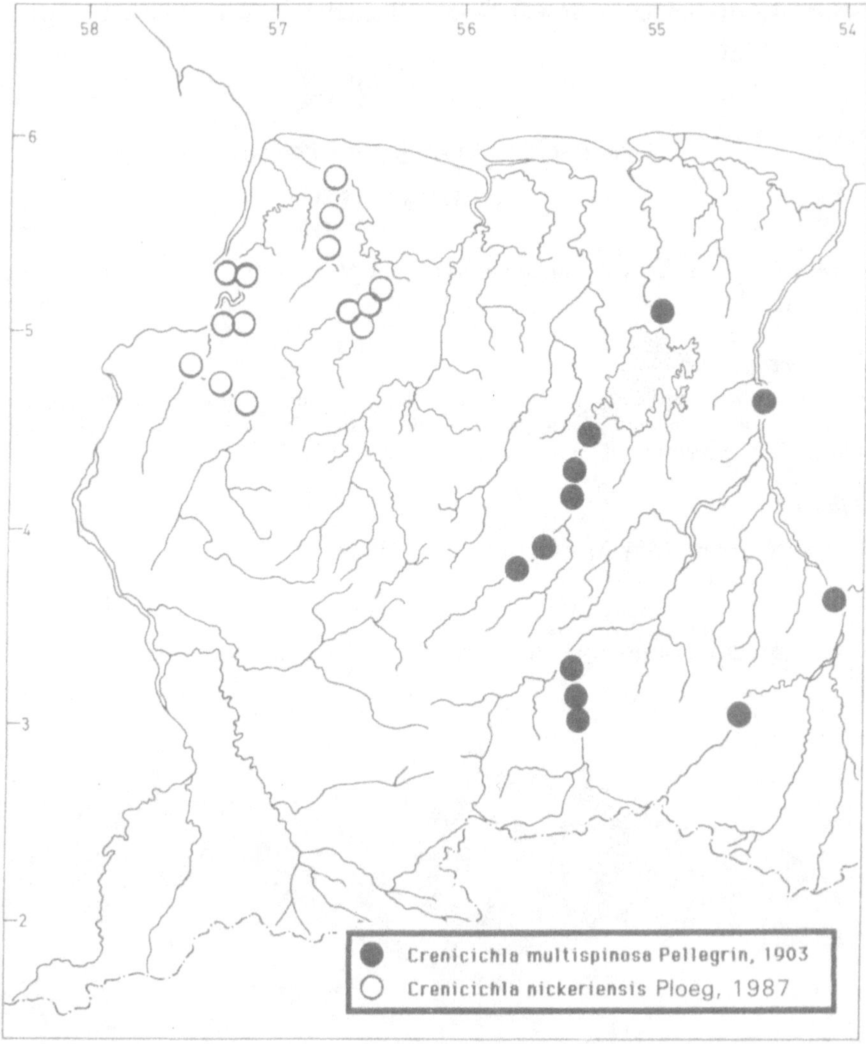

Fig. 138. Collecting localities of *Crenicichla multispinosa* and *Crenicichla nickeriensis*.

Juveniles: body showing 3 zones: a brownish upper zone, a darker zone from orbit to caudal peduncle, and a lighter ventral zone. No bluish part on head. Dots on flanks yellowish instead of silvery. Caudal spot ocellated with tan. Dark stripe from caudal to end of rays. Dorsal and anal fin dark margined. Larger juveniles look like adult males, except for the absence of a spot pattern below the orbit and the presence of yellowish coloured dots on the sides.

Distribution (Fig. 138)

Crenicichla multispinosa seems to be endemic in the Suriname and Marowijne R. drainages, where it was collected in the middle and upper courses only.

Crenicichla albopunctata Pellegrin

(Figs 131-132, 137)

Crenicichla saxatilis var. *albopunctata* Pellegrin, 1904, p. 374 (in part, Guyane).

Surinamese records

Crenicichla albopunctata; Ploeg, 1986b, p. 224 (restriction of type locality: Surinam, Marowijne R. drainage, Lawa R., Maka Creek; designation of the lectotype). — Ploèg, 1987, p. 75 (Marowijne R. drainage).

Material

253 specimens from Marowijne R. drainage.

Description

Shape: predorsal and preventral contour straight.

Fig. 131. *Crenicichla albopunctata*, male, 132 mm SL, ZMA 107.839, from Lawa River, Marowijne R. drainage.

Fig. 132. *Crenicichla albopunctata*, female, 124 mm SL, ZMA 107.840, from Lawa River, Marowijne R. drainage.

Head: snout somewhat rounded in lateral view, rounded from above. Maxilla extending to below orbit. Orbital diameter about half of head depth at orbit level. Nostrils slightly nearer to orbit than to postlabial skinfold.

Fins: dorsal fin origin just dorsal to pectoral fin base. Spines increasing in length to about 8th, from there equal in length. Fifth from last ray the longest, reaching to 4/5 of caudal fin in adult males, to 1/4 of caudal fin in adult females, to caudal fin base in juveniles. Tip of anal fin reaching to caudal fin base, 4th from last the longest. Caudal fin rounded with 2-3 basal scales at lateral line level, reduced to one dorsally and ventrally. Interradial scales to 1/4 of caudal fin; 4 scales between dorsal fin and most posterior scale of upper lateral line.

Teeth: upper jaw with maximum 5 tooth series in adults, lower jaw with maximum 4 tooth series. Fewer series in smaller specimens (2 series in smallest). Outer series approximately the same size as inner teeth. Teeth in upper jaw somewhat smaller than in lower jaw.

Vertebrae number in one specimen: 19 + 16.

Colouration

Males: dorsal half darker than caudal half; snout greyish. Preventral area dirty white-yellow; cheek yellowish; preoperculum greyish-brown. Preorbital stripe present; suborbital stripe faint, occasionally absent. Postorbital stripe sharply delimited. Black lateral band running from postorbital stripe to hind edge of caudal fin, becoming vague on sides, and very faint on caudal fin. Humeral spot on lateral band, just posterior to operculum. Silvery dots scattered on sides in dorsal 2/3 of body, on head, operculum and cheek. On the head these dots are present below and above postorbital stripe. On preoperculum they extend ventrally; humeral spot with dots. Dorsal fin grey with white dots, mainly posteriorly, smaller than those on sides of body. Anal fin grey with white dots only posteriorly. Ventral and pectoral fins light grey without dots. Caudal fin grey with dots like in dorsal fin, mainly dorsally. Caudal spot on lateral band at base of caudal fin, just dorsal to lateral line scales, surrounded by about 8 dots. Caudal fin margin white, except at end of lateral band.

Females: ground colour of body as in males. Dorsal fin with dark margin. Some females show 2 black, white margined ocelli over 11th and 12th spine; occasionally they are present from first dorsal fin spine to last ray, or are absent. Ocelli exclusively in females. Caudal fin dark margined, except in the middle. Anal fin with dark lower margin. Ventral fin with dark spine.

Juveniles: with conspicuous lateral band from tip of snout to end of caudal fin. Tannish zone above and below lateral line. Humeral spot on lateral band. Dark, ocellated caudal spot. Dorsal fin with small whitish ocelli. Yellow dots on lateral line scale rows at both sides of lateral band, in females larger than 100 mm SL (these dots are white in adult males). Caudal fin dorsally with dark submargin and white edges.

Distribution (Fig. 137)

Crenicichla albopunctata has been collected in the Marowijne R. drainage only. It is present almost everywhere in this drainage, e.g. Marowijne R., Tapanahony R., Paloemeu R., Oelemari R., and Lawa R. Outside Surinam it occurs in western French Guiana to Approuague R.

Crenicichla coppenamensis Ploeg

(Figs 133, 137)

Crenicichla coppenamensis Ploeg, 1987, p. 77 (description, Coppename R.).

Fig. 133. *Crenicichla coppenamensis*, male (holotype), 179 mm SL, ZMA 107.841, from Coppename River.

Surinamese records

Crenicichla coppenamensis; Ploeg, 1987, p. 77 (Coppename and Saramacca R.).
Crenicichla alta; (non Eigenmann, 1912), Boeseman, 1952, p. 198 (Saramacca R.).

Material

129 specimens from Coppename and Saramacca R. drainages.

Description

Shape: predorsal and preventral contour straight.

Head: snout somewhat rounded in lateral view, rounded from above. Maxilla reaching to below orbit. Orbit in upper part of head, about half as deep as head at orbit level. Nostrils nearer to orbit than to postlabial skinfold.

Fins: dorsal fin origin just posterior to hind edge of operculum. Spines increasing in length to about 9th, from where equal in length; 6th ray from last the longest, reaching to 1/3 of caudal fin, occasionally to end of fin in adult males, and to base of caudal fin in females and juveniles. Anal fin extending to caudal fin base with 5th ray from last the longest. Second ray of ventral fin the longest. Caudal fin rounded with 3-4 basal scales at lateral line level, diminishing to one dorsally and ventrally. Interradial scales to 1/4 of caudal fin length. Pectoral rounded.

Teeth: upper jaw with maximum 5 tooth series, all teeth equal in length. Lower jaw with maximum 4 tooth series, outer series with somewhat larger teeth. In specimens smaller than 55 mm SL, only 2 tooth series in both jaws.

Vertebrae number in one specimen: 20 + 14.

Colouration

Males: tan, dark dorsally, lighter ventrally. Head greyish; preventral area light; cheek yellowish. Black preorbital, postorbital, and suborbital stripe. Lateral band faint on sides, prominent on caudal fin, running between lateral line scale rows. A second band is occasionally present, running from below pectoral fin to 1/2 anal fin. Pale yellow dots—smaller than scales—are scattered on sides, more numerous along lateral lines. Black humeral spot posterior to operculum, 4 scales long, silvery ocellated. Spot in adults with centre on lateral line, in juveniles below lateral line. Dorsal fin grey, especially posteriorly scattered with white dots, smaller than dots on sides. Anal fin grey, showing scattered hyaline dots. Caudal fin grey with series of small white dots, arranged in lines. Black, ocellated spot dorsally on caudal fin, just posterior to basal scales. Many specimens show a small dark spot at the end of the caudal fin. Pectoral and ventral fins much lighter grey, almost white.

Females: somewhat lighter than males. Lateral band on head, and ring around humeral spot less conspicuous. No ring around caudal spot. Sometimes a horizontal band along abdomen, also present in some males. No yellow dots. Dorsal fin much lighter, dark margined and without dots. Anal fin lighter, without dots. Sexual dimorphism is apparent only in specimens larger than about 110 mm SL. Caudal fin without dots, dorsally dark margined, occasionally a white margin with a dark submargin instead. Often, a small dark spot at end of lateral band.

Juveniles: three horizontal zones: brown dorsally, a dark band from snout tip to caudal fin, and tan ventrally. Preventral area light. Humeral spot on lateral band well below lateral line, ocellated by a tan ring. Dorsal fin dark margined, without dots.

Distribution (Fig. 137)

Crenicichla coppenamensis was collected in the upper course of Coppename and Saramacca R. drainages.

Crenicichla nickeriensis Ploeg

(Figs 134, 138)

Crenicichla nickeriensis Ploeg, 1987, p. 81 (description; Nickerie R.).

Surinamese records

Crenicichla nickeriensis Ploeg, 1987, p. 81 (Nickerie R., Kabalebo R., Maratakka R. and Corantijn R.).

Material

183 specimens from Nickerie and Corantijn R. drainages.

Fig. 134. *Crenicichla nickeriensis*, male (holotype), 191 mm SL, ZMA 107.843, from Nickerie River.

Description

Shape: predorsal contour straight, preventral contour somewhat convex.

Head: snout rounded in lateral view and from above. Lower jaw prognathous; maxilla reaching to anterior margin of orbit. Orbital diameter about 1/3 of head depth at orbit level. Nostrils nearer to orbit than to postlabial skinfold.

Fins: dorsal fin origin just posterior to vertical from posterior edge of operculum. Spines increasing in length to about 10th, from where equal in length; 5th ray from last the longest, in males reaching to 1/3 of caudal fin length (occasionally halfway), in females to caudal fin spot, in juveniles to caudal fin base. Anal fin reaching to caudal fin base, 5th from last ray the longest. Caudal fin rounded with 2-4 basal scales at lateral line level, diminishing to one dorsally and ventrally. Interradial scales to 1/4 of caudal fin length. Pectoral fin rounded.

Teeth: upper jaw with maximum 6 tooth series, all teeth equal in length. Lower jaw with maximum 4 tooth series, labial series larger, inner teeth equal in length. External tooth series fixed, internal series evertible.

Vertebrae number in one specimen: 19 + 16.

Colouration

Males: brownish, dark dorsally, tan ventrally. No conspicuous axial separating line. Snout and head grey. Preventral area light, cheek tan, operculum grey. Faint preorbital stripe; black postorbital stripe. Chain of irregularly shaped black blotches along lateral lines from operculum to caudal fin spot; from here a faint stripe runs to end of caudal fin. Short, triangular suborbital stripe. Pink dots, irregularly distributed on dark part of sides, more concentrated posteriorly. Dots smaller than size of flank scales. Dorsal fin grey with pink dots, similar to those in posterior part of sides. Anal fin grey, scattered with dots, becoming smaller posteriorly. Dots arranged in horizontal rows. Ventral and pectoral fins pale grey, immaculate. Caudal fin grey with dots like those on sides; occasionally a small, dark spot at end of faint horizontal band. Ocellated caudal spot dorsally on caudal fin base, encircled by about 8 pink dots.

Females: ground colour of body and fins as in males. Preorbital and postorbital stripe, the latter conspicuous, dorsally and ventrally margined by a white zone. Humeral spot elongated horizontally, as long as eye, half as deep as long, situated on lateral line, ocellated by a tan zone. Chain of blotches along lateral lines absent. Dorsal fin lighter than in males; from edge of this fin an alternating pattern of dark, white, and dark spots between rays. Black or white spots in dorsal fin, from no spots at all to occasionally on every interradial membrane. Anal fin with dark margin; in last 3 rays 6 transverse lines, forming an alternating grey/ white pattern. Ventral and pectoral fins pale white. Caudal fin dorsally with dark margin and white submargin, ventrally a dark margin only; 3 faint bars on this fin; black caudal fin spot, ocellated by a pale ring, just above the extending lateral line scales.

Juveniles: dark lateral band running from snout tip to caudal peduncle, becoming more faint with age. Dark humeral spot, smaller than eye, on lateral band below lateral line. Dorsal fin grey with white margin and dark submargin, and with small white dots posteriorly, arranged in rows. Anal fin grey without dots. Ventral and pectoral fins much lighter than other fins. Caudal fin grey, dorsally with white margin and dark submargin, and with 3 white blotches in a row; dark caudal fin spot ocellated.

Distribution (Fig. 138)

Crenicichla nickeriensis has been collected in Nickerie, Kabalebo, Maratakka, and Corantijn Rivers.

Crenicichla sipaliwini Ploeg

(Figs 135-136)

Crenicichla sipaliwini Ploeg, 1987, p. 90 (description; Sipaliwini R.).

Surinamese records

Crenicichla sipaliwini; same as above.

Material

46 specimens from upper Corantijn R. drainage.

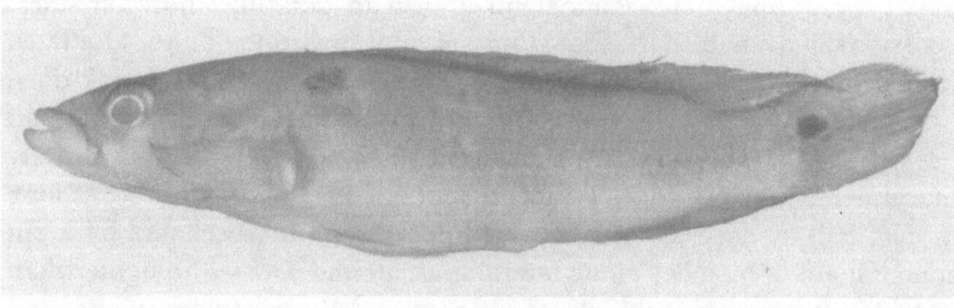

Fig. 135. *Crenicichla sipaliwini*, male? (holotype), 110 mm SL, RMNH 30561, from Sipaliwini River, Corantijn R. drainage.

Description

Shape: predorsal and preventral contour straight.

Head: snout pointed in lateral view, rounded from above. Maxilla not or just reaching orbit. Orbit about 1/3 as deep as head at orbit level. Scales on cheek much smaller than on operculum. Nostrils nearer to orbit than to postlabial skinfold.

Fins: dorsal fin origin just anterior to hind edge of operculum; spines increasing in length to about 9th, from where equal in length; 6th ray from last the longest, reaching to 1/3 of caudal fin length, occasionally to end of this fin in adults, and to caudal fin base in juveniles. Anal fin extending to caudal fin base; 5th ray from last the longest. Ventral fin with first soft ray the longest. Caudal fin rounded with 3 basal scales at lateral line level, diminishing to one dorsally and ventrally. Interradial scales to 1/4 of caudal fin length. Pectoral fin rounded.

Teeth: upper jaw with maximum 4 tooth series; all teeth equal in length. Lower jaw with maximum 4 tooth series; outer series with somewhat larger teeth.

Colouration

No conspicuous sexual dimorphism. Juveniles only differ in the position and size of the humeral spot. Belly yellow-grey, dorsally greyer; dorsal part of head dark grey, ventrally paler. Preventral area and cheeks yellowish. Conspicuous black preorbital and postorbital stripes; no suborbital stripe. Vague lateral band between lateral line scale rows. Black humeral spot, occasionally with dorsoanterior notch. Humeral spot in adults with its centre on lateral line; juveniles show a larger humeral spot with its centre below lateral line. Dorsal fin grey, mostly without black margin. Sometimes an ocellus in anterior part of dorsal fin, posteriorly vague dots arranged in rows. Anal fin grey with vague dots arranged in rows. Caudal fin with vague vertical bars. Black, ocellated spot dorsally on caudal fin, just posterior to basal scales. Small, dark spot on the distal parts of the middle caudal fin rays. Pectoral fins hyaline, ventral fins white.

Distribution (Fig. 136)

Crenicichla sipaliwini is known only from the Sipaliwini R., upper Corantijn drainage.

MATERIAL EXAMINED

The material is listed in the same order as the descriptions; for each species grouping is according to drainage from west to east. Data given are catalogue number, number of specimens, SL in mm (unless otherwise specified), district for Surinamese localities, collecting site, date, collector and, when known, station number in parenthesis.

Cichla ocellaris

ORIGIN UNKNOWN

ZMB 2839. Holotype, 183.4 mm SL. Coll. M.E. Bloch.

SURINAME

Corantijn R. drainage

USNM 225840. 1, 98.7. Distr. Nickerie, Corantijn R., River km 90. 7 Sep 1980. R.P. Vari (RPV 80-14).
USNM 226019. 116 juvs. Distr. Nickerie, Corantijn R. at 5°00'N 57°17'W. 17 May 1980. H.M. Madaire (M 80-17).
RMNH 31156. 2. Distr. Nickerie, Lucie R. 15 Feb 1975. M.S. Hoogmoed (594).
RMNH 31157. 1. Distr. Nickerie, Corantijn R. near Wakay Camp. 23 Mar 1971. Forestry Service.
RMNH 17242. 1, ca 370 (distorted). Distr. Nickerie, Lucie R. Nov 1910. J.F. Hulk.
USNM 226212. 2, 208.0-236.4. Distr. Nickerie, Mataway Kreek, ca 8 km from its mouth. 11 Sep 1980. R.P. Vari (RPV 80-25).
USNM 226018. 18, 33.5-145.5. Distr. Nickerie, Corantijn R. at Km 180, side channel of main river along Surinamese shore. 8 Sep 1980. R.P. Vari (RPV 80-19).

Nickerie R. drainage

ZMA 105.761. 1, ca 125. Distr. Nickerie, Stondansie val. 5 Apr 1967. H. Nijssen (HN 121).

Saramacca R. drainage

ZMA 105.572. 1, 324.1. Distr. Brokopondo, Toeboeka Kreek, at left bank of Saramacca R., ca 5 km N mouth of Kleine Saramacca R. 2 Mar 1967. H. Nijssen (HN 110).
ZMA 105.566. 2, 170->290. Distr. Brokopondo, Kleine Saramacca R., rapids ca 14 km ESE (geographically) of the mouth. 28 Feb 1967. H. Nijssen (HN 108).

Suriname R. drainage

NRM 11310. 2, 92.4-95.0. Distr. Brokopondo, Berg en Dal? No date. A. Klinckowström?
ZMA 105.007. 3, 205-253.4. Distr. Brokopondo, Brokopondo Meer E of the dam, shore of second and third auxiliary dam. 18-19 May 1966. H. Nijssen (HN 15).
ZMA 107.832. 1, 298.6. Same data as ZMA 105.007.
NRM A82/1966203.3422. 1, 222. Same data as ZMA 105.007.
ZMA 105.002. 6, ca 215.8 mm-ca 23 cm (distorted). Distr. Brokopondo, Brokopondo Meer, shore ca 1 km S Afobaka dam. 28 Jun 1966. H. Nijssen (HN 35).

ZMA 105.035. 1, 71. Distr. Brokopondo, Brokopondo Meer, ca 3.5 km SE Brownsweg, lake shore at railway Brownsweg-Kabelstation. 13 May 1966. H. Nijssen (HN 13).
ZMA 105.011. 1, >220. Distr. Brokopondo, Brokopondo Meer, at fourth auxiliary dam, ca 3 km E Afobaka dam. 13 Jun 1966. H. Nijssen (HN 27).
ZMA 105.058. 8, 108.8-160.7. Distr. Brokopondo, Brokopondo Meer, ca 300 m E eastern edge of the dam. 30 Aug 1966. H. Nijssen (HN 45).
ZMA 105.054. 1, 268.0. Distr. Brokopondo, Brokopondo Meer, ca 3.5 km SE Brownsweg. 31 Aug 1966. H. Nijssen (HN 46).
ZMA 105.071. 1, 77. Distr. Brokopondo, Brokopondo Meer, shore, ca 1 km S Afobaka dam. 28 Jun 1966. H. Nijssen (HN 34).
ZMA 105.059. 25, 70-125. Data as for ZMA 105.054.
NRM A82/1966353.3426. 3, 71-106. Data as ZMA 105.059. 1 specimen cleared and stained.
ZMA 105.083. 6, 47.6-65.7. Distr. Brokopondo, Brokopondo Meer, shore between first and second eastern auxiliary dams. 9 Sep 1966. H. Nijssen (HN 57)
ZMA 105.082. 1, 124. Distr. Brokopondo, Brokopondo Meer at Moengofoetoe on Sara Kreek, ca 3 km S reservoir dam. 9 Sep 1966. H. Nijssen (HN 56).
ZMA 105.133. 1, ca 245. Distr. Brokopondo, Koesikamba (= Apresina) val, ca 5 km NE Pokiegron. 22 Sep 1966. H. Nijssen (HN 64).
ZMA 105.086. 2, 235.6-267.8. Distr. Brokopondo, Brokopondo Meer, ca 500 m S western point of the reservoir dam. 7 Sep 1966. H. Nijssen (HN 53).
ZMA 105.165. 4, 44.6-62.8. Distr. Brokopondo, Sopo val (= Mama Dam). 22 Sep 1966. H. Nijssen (HN 65).
ZMA 105.161. 8, 50.9-87. Distr. Brokopondo, Brokopondo Meer, at third auxiliary dam. 24 Aug 1966. H. Nijssen (HN 41).
NRM A82/1966343.3425. 1, 50.0. Data as ZMA 105.161.
ZMA 105.194. 1, 414.0, >485.0 mm TL. Distr. Brokopondo, Brokopondo Meer, ca 1 km S dam. 8 Oct 1966. H. Nijssen (HN 70).
ZMA 105.483. 3, ca 73-ca 90. Distr. Brokopondo, Brokopondo Meer along northern shore at Afobaka. 15 Feb 1967. H. Nijssen (HN 103).
ZMA 105.405. 1, 52.8. Distr. Brokopondo, Gran Mau Kreek at right bank of Gran Rio ca 1 km NE Dombaai (= Bendi Watra). 30 Jan 1967. H. Nijssen (HN 99).
ZMA 105.732. 1, 130. Distr. Brokopondo, Awara kreek on right bank of Suriname R., ca 1.5 km S Botopasi. 18 Mar 1967. H. Nijssen (HN 111).
ZMA 105.669. 3, 52.3-252.8. Distr. Brokopondo, Suriname R., rapids 1 km S Botopasi. 22 Mar 1967. H. Nijssen (HN 117).
ZMA 105.804. 1, 70. Distr. Brokopondo, Parwapa (= Paba) kreek on left bank of Suriname R., 2.5 km N Botopasi at Foetoenakaba (= Voetoekaba). 20 Mar 1967. H. Nijssen (HN 113).
ZMA 107.575. 6, 53.0-ca 56.2. Distr. Brokopondo, Brokopondo Meer, 1 km E former Mamadam rapids. 17-18 Dec 1978. C.J.J. Richter.
ZMA 106.435. 1, ca 280. Distr. Brokopondo, Witte Kreek, 12 km S Brownsweg. 23 Nov 1966. Leg. H. Nijssen (HN 84).
ZMA 107.598. 6, 75-199.7. Distr. Brokopondo, Brokopondo Meer, 2 km SSW former Komkomjee Island. 14 Dec 1978. C.J.J. Richter.

GUYANE FRANÇAISE

R. Maroni drainage

ISNB 17569pt. 1, ca 30. R. Marouini downstream of Epoia. 22 Nov 1969. J.-P. Gosse & Leopold III (Sta. 9).
ISNB 19767. 1, 37.9. R. Ouaqui at Saut Bali. 18 Nov 1969. J.-P. Gosse & Léopold III (Sta 6).
MNHN 95-134-135. 2, ca 80.7-ca 110.0. Maroni. No date. E. Ducaux.

Drainage unknown or uncertain

MNHN A.9484. 1, 299.8. 'Cayenne'. No date. Mélinon.

Chaetobranchus flavescens

SURINAME

Corantijn R. drainage

RMNH 18802. 1, 161.8. Distr. Nickerie, Nanni Kreek, Doksipan. 3 Sep 1941. D.C. Geijskes.

Saramacca R. drainage

RMNH 18323. 1, 93.2. (Upper?) Saramacca R. 1902. Saramacca Expedition 1902.

Suriname R. drainage

RMNH 18166. 2, 86.4-93.5. Distr. Para, Coropina Kreek, Republiek. Oct 1946. D.C. Geijskes.

Geophagus surinamensis

SURINAME

Drainage unknown or uncertain

ZMB 2825. Holotype. 78.5 mm SL. Surinam. Before 1792. Coll. M.E. Bloch.

Saramacca R. drainage

ZMA 105.658. 5 juvs. Distr. Brokopondo, tributary kreek on right bank of Kleine Saramacca R., ca 11 km (geographically) ESE mouth. 27 Feb 1967. H. Nijssen (HN 106).

Suriname R. drainage

ZMA 105.687. 18, 41-119.9. Distr. Brokopondo, Jenjeekreek on right bank of Suriname R., 7.5 km N Botopasi. 21 Mar 1967. H. Nijssen (HN 114).

ZMA 105.745. 19, 52-143.1. Distr. Brokopondo, Parwapa Kreek (= Paba Kreek) on left bank of Suriname R., 2.5 km N Botopasi. 20 Mar 1967. H. Nijssen (HN 113).

NRM A87/19678121.3942. 3, 90-110. Same data as ZMA 105.745.

ZMA 105.123. 8. Distr. Brokopondo, Sopo Val (= Mama Dam) in the Suriname R. 22 Sep 1966. H. Nijssen (HN 65).

ZMA 105.272. 5. Distr. Brokopondo, Marowijne (= Gran Kreek), ca 63 km S Afobaka barrage. 20 Oct 1966. H. Nijssen (HN 79).

ZMA 105.216. 2. Distr. Brokopondo, Sara Kreek ca 27 km S Dam. 14 Oct 1966. H. Nijssen (HN 76).

ZMA 105.399. 39. Distr. Brokopondo, tributary kreek (= Gran Mau) on right bank of Gran Rio ca 1 km NE Dombaai (= Bendi Watra), geographically ca 13 km SW Djoemoe. 30 Jan 1967. H. Nijssen (HN 99).

ZMA 105.384. 65. Distr. Brokopondo, tributary kreek on right bank of Gran Rio, ca 4 km NE NE part of Awadam (= Awaradam), rapids. 31 Jan 1967. H. Nijssen (HN 101).

ZMA 105.436. 10. Distr. Brokopondo, tributary kreek on left bank of Gran Rio, ca 3 km NE NE part of Awadam (= Awaradam), rapids. 1 Feb 1967. H. Nijssen (HN 102).

ZMA 105.542. 7. Distr. Brokopondo, Gran Rio at NE part of Awadam (= Awaradam). 28 Jan 1967. H. Nijssen (HN 97).

ZMA 105.716. 6. Distr. Brokopondo, Suriname R. 1 km S Botopasi, rapids. 22 Mar 1967. H. Nijssen (HN 117).
ZMA 105.726. 22, largest 148.0. Distr. Brokopondo, Awara Kreek on right bank of Suriname R. ca 1.5 km S Botopasi. 18 Mar 1967. Leg. H. Nijssen (HN 111).
ZMA 107.584. 3. Distr. Brokopondo, Brokopondomeer, 1 km E former Mamadam rapids. 17-18 Dec 1978. C.J.J. Richter.
ZMA 107.597. 4. Distr. Brokopondo, Brokopondomeer, 2 km SSW former Komkomjee island. 14 Dec 1978. C.J.J. Richter.

Marowijne R. drainage

ZMA 106.197. 9, 76.5-100.0. Distr. Marowijne, Soea-Kisi-Kreek on left bank of Tapanahony R., 12 km SW center of Stoelmanseiland. 22 Apr 1967. H. Nijssen (HN 129).
ZMA 105.038. 88. Distr. Marowijne, Marowijne R. ca 25 km S Albina. 17 Jun 1966. H. Nijssen (HN 31).
ZMA 106.192. 2. Distr. Marowijne, Marowijne R. at Mopikondre (= Maria's Hoop), ca 30 km S Albina. 17 Jun 1966. H. Nijssen.
ZMA 106.194. 1. Distr. Marowijne, kreek on left bank of Marowijne R. W of Manbarie Val (6 km N Stoelmanseiland). 20 Apr 1967. H. Nijssen (HN 127).

GUYANE FRANÇAISE

R. Maroni drainage

NRM A84/1979415.3052. 2, 95.3-103.1. Saut Gostou. 12 Oct 1979. P. Planquette.
ZMA 106.198. 12. Kamaloea or Saloea Kreek on right bank of Marowijne R., 9 km SE mouth of Gran Kreek. 24 Apr 1967. H. Nijssen (HN 131).
ISNB 17409. 30. R. Ouaqui, right bank tributary of R. Tampok, Saut Bali. 18 Nov 1969. Léopold III & J.-P. Gosse (Sta. 6).
ISNB 18591. 3 juvs. R. Marouini, downstream of Epoia. 22 Nov 1969. Léopold III & J.-P. Gosse (Sta. 9).

Geophagus brokopondo

SURINAME

Suriname R. drainage

ZMA 119.530. Holotype. 107.7 mm SL. Distr. Brokopondo, Brokopondomeer along N shore at Afobaka. 15 Feb 1967. H. Nijssen (HN 103).
ZMA 105.480. 4, 70.0-123.2. Same data as holotype. Paratypes.
NRM A87/1967073.3941. 1,96.6. Same data as holotype. Paratype.
ZMA 105.081. 32, 30.7-58.7. Distr. Brokopondo, Brokopondomeer shore between first and second auxiliary dams. 9 Sep 1966. H. Nijssen (HN 57). Paratypes.

Geophagus brachybranchus

SURINAME

Corantijn R. drainage

RMNH 30972. 7, 41.1-91.5. Distr. Nickerie, right bank tributary of Kabalebo R., just below Avanavero Vallen. 6 Apr 1971. M. Boeseman. Paratypes.
USNM 226211. 73. Distr. Nickerie, Corantijn R. at km 180, side channel of main river along Surinamese shore. 8 Sep 1980. R.P. Vari (RPV 80-19). Paratypes.

USNM 225838. 1. Distr. Nickerie, Matapi Kreek, ca 1 km from mouth. 9 Sep 1980. R.P. Vari (RPV 80-21). Paratype.
USNM 226046. 2. Distr. Nickerie, Mataway Kreek, ca 8 km from mouth. 11 Sep 1980. R.P. Vari (RPV 80-25). Paratypes.
USNM 226047. 22. Distr. Nickerie, pool in front of Camp Hydro. 17 Sep 1980. R.P. Vari (RPV 80-36). Paratypes.
USNM 225839. 2. Distr. Nickerie, Matapi Kreek. 17 May 1980. H.M. Madaire (80-7). Paratypes.
USNM 289513. 2. Distr. Nickerie, small pool on side of Mataway Kreek, ca 2 km from its mouth. 8 Dec 1979. R.P. Vari (RPV 79-13). Paratypes.
USNM 225837. 19. Distr. Nickerie, river km 90. 7 Sep 1980. R.P. Vari (RPV 80-14). Paratypes.

Nickerie R. drainage

RMNH 30974. Holotype. 107.4 mm SL. Distr. Nickerie, rocky pool just above Blanche Marie Vallen. 13 Jan 1971. M. Boeseman.
RMNH 30975. 10, 73.7-101.2. Same data as holotype. Paratypes.
NRM A83/1971023.3017. 2, 49.5-115.6. Same data as holotype. 1 specimen cleared and stained. Paratypes.
RMNH 30976. 2, 84.9-102.3. Distr. Nickerie, left bank tributary kreek of upper Nickerie R., above Camp Stondansie. 30 Jan 1971. M. Boeseman. Paratypes.
RMNH 30973. 10, 65.6-138.4. Distr. Nickerie, right bank tributary of Nickerie R., ca 1/2 h below Blanche Marie Vallen. 11 Feb 1971. M. Boeseman. Paratypes.
NRM A83/1971064.3016. 3, 74.0-91.5. Same data as RMNH 30973. Paratypes.
ZMA 106.193. 19, 40.1-87.8. Distr. Nickerie, right bank tributary kreek of Nickerie R., 12 km WSW Stondansie Val. 5 Apr 1967. H. Nijssen (HN 120). Paratypes.

Geophagus harreri

SURINAME

Marowijne R. drainage

RMNH 30977. 17, 28.8-42.9. Distr. Marowijne, Tapanahony, pool 2.5 km downstreams of airfield. 30 Nov 1965. G.F. Mees.
NRM A83/1965482.3012. 5, 42.8-51.9. Same data as RMNH 30977. 1 specimen cleared and stained.
ZMA 107.751. 9, to 173.9. Distr. Marowijne, Tapanahony R., 2 km downstream of mouth of Paloemeu R. 17 Nov 1965. G.F. Mees.
ZMA 106.195. 2. Distr. Marowijne, Maka Kreek, on left bank of Lawa R., 10 km S centre of Stoelmanseiland. 21 Apr 1967. H. Nijssen (HN 128). Paratypes.
ZMA 106.940. 1 paratype + 8 specimens. Distr. Marowijne, Soeakisi Kreek on left bank of Tapanahony R., 12 km WSW of centre of Stoelmanseiland. 22 Apr 1967. H. Nijssen (HN 129). [Paratype listed as ZMA 106.197 by Gosse (1976).]
ZMA 106.939. 1. Distr. Marowijne, sandy bank of Marowijne R., 25 km S Albina. 17 Jun 1966. H. Nijssen (HN 31). [Listed as ZMA 105.038 by Gosse (1976).] Paratype.
ISNB 17564. 3. Distr. Marowijne, rapids of Paloemeu R. upstream of mouth of Waloemeroe Kreek. 3 Nov 1966. J.-P. Gosse & Léopold III (Sta. 155).
ISNB 17565. 6. Distr. Marowijne, rapids of Lawa R. downstream of Anapaikekondre. 18 Nov 1966. J.-P. Gosse & Léopold III (Sta. 169). [1 specimen given to H. Harrer.]

GUYANE FRANÇAISE

R. Maroni drainage

ISNB 20849. 2, 51.5-90.8. R. Marouini, first rapids downstream. 23 Nov 1969. J.-P. Gosse & Léopold III (Sta. 10).

ISNB 17570. 22, 28.6-56.8. R. Marouini downstreams of Epoia. 22 Nov 1969. J.-P. Gosse & Léopold III (Sta. 9).
ISNB 17569. 8, 88.3-141.6. R. Marouini downstreams of Epoia. 22 Nov 1969. J.-P. Gosse & Léopold III (Sta. 9).
ISNB 17566. 3. R. Inini at Saut Sonnelle. 16 Nov 1969. J.-P. Gosse & Léopold III (Sta. 3).
ISNB 17567. 1. Crique Bois Blanc, right bank tributary to the R. Inini. 16 Nov 1969. J.-P. Gosse & Léopold III (Sta. 4).
ISNB 17568. 1. Small tributary on right bank of R. Ouaqui upstream of Saut Bali. 18 Nov 1969. J.-P. Gosse & Léopold III (Sta. 7).
ISNB 17571. 1 paratype + 3 specimens, 150-183. Same data as ISNB 20849.
ISNB 17572. 29. Same data as ISNB 17571.
ZMA 106.941. 17. Crique Kamaloea (= Saloea), 9 km SE mouth of Gran Kreek. 24 Apr 1967. H. Nijssen (HN 131). [Listed as ZMA 106.198 by Gosse (1976).] Paratypes.
ZMA 107.833. 1. Same data as ZMA 106.941.
NRM A84/1979415.3051. 4, 25.4-172.9. Saut Gostou. 12 Oct 1979. P. Planquette.

Satanoperca leucosticta

SURINAME

Nickerie R. drainage

ZMA 106.196. 2, 60.7-98.2. Distr. Nickerie, Nickerie R., Stondansi Val. 5 Apr 1967. H. Nijssen (HN 121).
ZMA 105.818. 3, 56.1-89.4. Distr. Nickerie, tributary kreek on right bank of Nickerie R., 12 km WSW Stondansi Val. 5 Apr 1967. H. Nijssen (HN 120).

Apistogramma steindachneri

SURINAME

Drainage unknown or uncertain

RMNH 19334. 1, 31.5. Marowijne gebied?. 1951. E.C. Stol.
RMNH 19355. 2, 43.6-52.7. Marowijne gebied. Summer 1951. E.C. Stol.
RMNH 21402. 2, ca 26.7-31.0. Suriname. December 1951. E.C. Stol.

Corantijn R. drainage

RMNH 18470. 1, 40.5. Distr. Nickerie, Lucie R. 1 Dec 1910. K.M. Hulk.
RMNH 18471. 1, 30.3. Distr. Nickerie, upper Corantijn R.? 1910-1911. K.M. Hulk.
RMNH 30978. 3, 30.5-32.2. Distr. Nickerie, right bank tributary kreek to Kabalebo R., ca 8 km below Avanavero Vallen. 7 Apr 1971. M. Boeseman.
RMNH 30979. 1, 32.1. Distr. Nickerie, 20 km N Lucie R. 11 Feb 1975. M.S. Hoogmoed.
RMNH 30980. 1, 37.8. Distr. Nickerie, upper (SE) Kaboeri Kreek. 29 Mar 1971. M. Boeseman.
RMNH 30981. 7, 12.5-32.4. Distr. Nickerie, Avanavero Camp, a few km inland, in jungle kreek. 12 Sep 1965. G.F. Mees.
RMNH 30982. 4, 23.8-36.2. Distr. Nickerie, Avanavero R. 18 Sep 1965. G.F. Mees.
RMNH 30983. 3, 15.6-20.7. Distr. Nickerie, jungle kreek NE airstrip Kayser Mountains. 10 Aug 1968. M.S. Hoogmoed.
RMNH 30984. 1, 38.8. Distr. Nickerie, right bank tributary kreek of Kabalebo R., just below Avanavero Vallen. 6 Apr 1971. M. Boeseman.
RMNH 30985. 1, 37.5. Distr. Nickerie, right bank tributary to Kabalebo R., below Avanavero Vallen (Camp BWKW). 10 Apr 1971. M. Boeseman.

RMNH 30986. 3, 32.4-38.2. Distr. Nickerie, right bank tributary kreek of Kabalebo R., ca 8 km below Avanavero Vallen. 7 Apr 1971. M. Boeseman.
USNM 227331. 9, 12.6-29.0. Distr. Nickerie, small kreek entering Corantijn R. at Km 440, ca 20 km N Lucie R. mouth. 11 Dec 1979. R.P. Vari (RPV 79-18).
USNM 225811. 4. Distr. Nickerie, kreek opposite Guyanese logging camp, 2 1/4 h S Matapi, ca 2 km downstream of Cow Falls. 11 Sep 80. R.P. Vari (RPV 80-24).
USNM 225826. 3. Distr. Nickerie, Mataway Kreek, ca 8 km from its intersection with Corantijn R. 11 Sep 1980. R.P. Vari (RPV 80-25).
USNM 225812. 1. Distr. Nickerie, woodland stream about 1/2 km inland of Camp Mataway. 12 Sep 1980. R.P. Vari (RPV 80-26).
USNM 225813. 1. Distr. Nickerie, Kamp Kreek 100 m N turnoff to Camp Geology. 13 Sep 1980. R.P. Vari (RPV 80-27).
USNM 225814. 5. Distr. Nickerie, Steen Kreek where crossed by Amotopo-Paramaribo road. 13 Sep 1980. R.P. Vari (RPV 80-28).
USNM 225815. 34. Distr. Nickerie, small stream behind Kamp Infrastructure (opposite Avanavero airfield). 13 Sep 1980. R.P. Vari (RPV 80-29).
USNM 289715. 1. Distr. Nickerie, small stream behind Kamp Infrastructure (opposite Avanavero airfield). 13 Sep 1980. R.P. Vari (RPV 80-29).
USNM 225816. 9. Distr. Nickerie, small stream entering stream located about 200 m upriver of Dalbana Kreek mouth. 14 Sep 1980. R.P. Vari (RPV 80-30).
USNM 225817. 30. Distr. Nickerie, tributary to Dalbana Kreek ca 3 km upstream from where Dalbana Kreek is crossed by Amotopo to Camp Geologie road. 15 Sep 1980. R.P. Vari (RPV 80-32).
USNM 225818. 3. Distr. Nickerie, stream entering Corantijn R. at ca Km 385 slightly N Tiger Falls. 16 Sep 1980. R.P. Vari (RPV 80-33B).
USNM 225819. 4. Distr. Nickerie, small kreek entering Lucie R. ca 3 km upstream from junction of Lucie and Corantijn R. 18 Sep 1980. R.P. Vari (RPV 80-37).
USNM 225820. 10. Distr. Nickerie, stream entering south side of Lucie R. about 200 m downstream of ferry crossing. 18 Sep 1980. R.P. Vari (RPV 80-38).
USNM 289716. 6. Distr. Nickerie, stream on south bank of Lucie R, ca 6 km upstream of ferry crossing on Amotopo-Camp Geologie Road. 18 Sep 1980. R.P. Vari (RPV 80-39).
USNM 225821. 8. Distr. Nickerie, stream on south bank of Lucie R, ca 6 km upstream of ferry crossing on Amotopo-Camp Geologie Road. 18 Sep 1980. R.P. Vari (RPV 80-39).
USNM 225822. 1. Distr. Nickerie, small creek on E side of Corantijn R. ca 2 km upstream of mouth of Moskita Kreek. 19 Sep 1980. R.P. Vari (RPV 80-40).
USNM 225823. 19. Distr. Nickerie, small kreek entering Corantijn R. on E bank approximately 350 m below base of Wilhelm II Falls. 19 Sep 1980. R.P. Vari (RPV 80-41).
USNM 225824. 13. Distr. Nickerie, small kreek entering Corantijn R. on E side approximately 300 m N of Amotopo boat landing. 19 Sep 1980. R.P. Vari (RPV 80-42).
USNM 225825. 8. Distr. Nickerie, tributary to Sisa Kreek, N side approximately 700 m downstream of crossing of Amotopo to Kamp Geologie road. 20 Sep 1980. R.P. Vari (RPV 80-43).

Nickerie R. drainage

ZMA 107.017. 29. Distr. Nickerie, right bank tributary of Nickerie R., 12.5 km WSW Stondansie Val. 5 Apr 1967. H. Nijssen (HN 120) [measured: 10, 25.1-49.3].
RMNH 30987. 2, 51.2-52.8. Distr. Nickerie, left bank tributary to Nickerie R., below Blanche Marie Vallen. 16 Feb 1971. M. Boeseman.
RMNH 30988. 19, 27.3-48.0. Distr. Nickerie, forest kreek tributary to Nickerie R., ca 1 km behind Camp Waterloopk., below Stondansie Val. 3 Feb 1971. M. Boeseman.
RMNH 30989. 3, 21.2-30.3. Distr. Nickerie, kreek at second savannah, right bank tributary to middle Maratakka R. at woodcutters camp. 6 Mar 1971. M. Boeseman.

RMNH 30990. 5, 23.1-26.7. Distr. Nickerie, shortly before bivouac at middle course of Maratakka R. 1 Mar 1971. M. Boeseman.
RMNH 30991. 8, 28.3-46.9. Distr. Nickerie, left bank tributary of Nickerie R. below Blanche Marie Vallen. 16 Feb 1971. M. Boeseman.
RMNH 30992. 1, 46.1. Data as 30989.
RMNH 30993. 40, 18.9-48.0. Distr. Nickerie, left bank tributary to Nickerie R., 3 km behind Camp Waterloopk., below Stondansie Val. 1 Feb 1971. M. Boeseman.
RMNH 30994. 7, 23.3-29.2. Distr. Nickerie, right bank tributary to Fallawatra R., ca 35 min by boat (corial) from outlet of Fallawatra R. 2 Feb 1971. M. Boeseman.
RMNH 30995. 4, 21.3-48.0. Distr. Nickerie, left bank tributary to upper Nickerie R. above Camp Stondansie. 27 Apr 1971. M. Boeseman.
RMNH 30996. 12, 16.5-48.7. Distr. Nickerie, right bank tributary to Fallawatra R., ca 35 min from outlet. 2 Feb 1971. M. Boeseman.

Coppename R. drainage

ZMA 107.019. 10. Distr. Saramacca, left bank tributary kreek of Coppename R. in Wilhelminagebergte. 18 May 1967. H. Nijssen (HN 135).
ZMA 107.018. 23. Distr. Saramacca, left bank tributary kreek of Linker Coppename R. in Wilhelminagebergte (03°54′N 56°46′W). 9 May 1967. H. Nijssen (HN 132). [10 measured: 27.6-49.0.]
ZMA 107.020. 21. Distr. Saramacca, right bank tributary kreek of Coppename R. in Wilhelminagebergte. 19 May 1967. H. Nijssen (HN 136).
RMNH 30997. 2, 46.9-47.9. Distr. Saramacca, Coppename R., Raleigh Vallen. 20 Mar 1972. G.F. Mees.
RMNH 30998. 6, 26.6-ca 43.0. Distr. Saramacca, forest stream at Raleigh Vallen. 20 Mar 1972. G.F. Mees.

Saramacca R. drainage

RMNH 30999. 1, 32.9. Distr. Brokopondo, Goj Kreek, tributary to Saramacca R., above Posoegroenoe. 7 Apr 1964. M. Boeseman.
RMNH 31000. 6, 22.4-31.0. Distr. Brokopondo, Saramacca R., pool below Feddiprati soela. 9 Apr 1964. M. Boeseman.

Suriname R. drainage

ZMA 107.016. 6. Distr. Brokopondo, Kwati Watra Kreek, at left bank of Suriname R., 500 m S Botopasi. 19 Mar 1967. H. Nijssen (HN 112).
ZMA 107.009. 1. Distr. Brokopondo, Marowijne (= Gran) Kreek, 50.5 km S dam at Afobaka. 19 Oct 1966. H. Nijssen (HN 78).
ZMA 107.008. Male, 65.3. Distr. Brokopondo, Dabikwen (= Baling) Kreek near Brokobaka village ca 15 km N Afobaka. 19 Sep 1966. H. Nijssen (HN 61).
ZMA 107.011. 15. Distr. Brokopondo, kreek at right bank of Gran Rio, 600 m SW Ligorio. 27 Jan 1967. H. Nijssen (HN 95). [10 measured: 30.8-48.1.]
ZMA 107.010. 3. Distr. Para, Carolina or Malasie Kreek, 10 km SE Zanderij Airfield. 14 Dec 1966. H. Nijssen (HN 92).
ZMA 107.001. 4. Distr. Brokopondo, Marowijne (= Gran) Kreek, 50.5 km S Afobaka. 6 May 1966. H. Nijssen (HN 7).
ZMA 105.751. 14. Distr. Brokopondo, Parwapa (= Paba) Kreek on left bank of Suriname R., 2.5 km N Botopasi at Foetoenakaba (= Voetoekaba). 20 Mar 1967. H. Nijssen (HN 113).
ZMA 107.003. 2. Distr. Brokopondo, Marowijne (= Gran) Kreek, 53.5 km S Afobaka. 25 May 1966. H. Nijssen (HN 17).
ZMA 107.002. 2. Distr. Brokopondo, Marowijne (= Gran) Kreek, 55 km S Afobaka. 24 May 1966. H. Nijssen (HN 16).
ZMA 105.360. 1. Distr. Brokopondo, Pasa Kreek (= Dateke Kreek), ca 1.5 km S Brokopondo on riverway Brokopondo-Afobaka. 1 Sep 1966. H. Nijssen (HN 48).

ZMA 105.343. 22. Distr. Suriname, Carolina (= Malasie) Kreek, ca 10 km SE Zanderij airfield on road Zanderij-Kraka. 14 Dec 1966. H. Nijssen (HN 92). [10 measured: 20.5-62.8.]
ZMA 105.701. 62. Distr. Brokopondo, Jenjee Kreek on right bank of Suriname R., 7.5 km N Botopasi. 21 Mar 1967. H. Nijssen (HN 114).
NRM A82/1967122.3242. 6. Same data as ZMA 105.701.
ZMA 105.361. 1. Distr. Brokopondo, Marowijne (= Gran) Kreek, stagnant, geographically ca 51 km S Afobaka dam. 27 May 1966. H. Nijssen (HN 19).
ZMA 107.005. 1. Distr. Brokopondo, Marowijne (= Gran) Kreek, 58 km S Afobaka. 9 Jun 1966. H. Nijssen (HN 25).
ZMA 107.004. 6. Distr. Brokopondo, Marowijne (= Gran) Kreek, 51 km S Afobaka. 27 May 1966. H. Nijssen (HN 19).
ZMA 107.007. 5. Distr. Brokopondo, Tapoeripa Kreek at Brokopondo. 15 Sep 1966. H. Nijssen (HN 60).
ZMA 107.006. 15. Distr. Brokopondo, Tapoeripa Kreek at Hermansdorp, 1 km N Brokopondo. 25 Aug 1966. H. Nijssen (HN 43).
ZMA 106.013. 3. Distr. Brokopondo, right bank tributary kreek of Gran Rio, 1 km NE Dombaai or Bendi Watra. 30 Jan 1967. H. Nijssen (HN 99).
ZMA 107.012. 23. Distr. Brokopondo, tributary kreek on right bank of Gran Rio, 2.5 km NE Awaradam. 28 Jan 1967. H. Nijssen (HN 96).
ZMA 107.015. 18. Distr. Brokopondo, tributary kreek on left bank of Gran Rio, 3 km NE Awaradam. 1 Feb 1967. H. Nijssen (HN 102).
ZMA 107.014. 2. Distr. Brokopondo, tributary kreek on right bank of Gran Rio, 4 km NE Awaradam. 21 Jan 1967. H. Nijssen (HN 101).
ZMA 107.012. 1. Distr. Brokopondo, kreek at Brownsweg, railway Km 114. Oct 1960. Pijpers.
ZMA 112.552. 2. Distr. Para, tributary of Coropina Kreek near Republiek. 2 Nov 1972. P.J.H. van Bree.
RMNH 31001. 29, 7.9-43.4. Distr. Brokopondo, forest kreekje above Asidonhopo, tributary to Pikien Rio. 13 Aug 1964. M. Boeseman.
RMNH 31002. 3, 22.3-26.2. Distr. Brokopondo, margin of Brokopondomeer near Brownsweg. 19 Dec 1971. V.J. Mees.
RMNH 31003. 2, 24.5-28.3. Distr. Brokopondo, tributary kreek of Gran Kreek below Bofroedédé. 5 Mar 1964. M. Boeseman.
RMNH 31004. 1, ca 21.3. Distr. Para, Saramacca Kreek. 27 Apr 1965. G.F. Mees.
MNH 31005. 2, 24.5-33.2. Distr. Brokopondo, Brokopondomeer, railway Kabel-Brownsweg, at road to gold mine. 30 Mar 1965. G.F.
Mees. RMNH 31006. 2, 22.4-24.2. Distr. Para, Republiek. 2 Jul 1965. G.F. Mees.
RMNH 31007. 1, 13.9. Distr. Brokopondo, right bank tributary kreek of middle Gran Kreek, near present-day Brokopondomeer shore. 30 Jul 1964. M. Boeseman.
RMNH 31008. 6, 24.9-39.2. Distr. Brokopondo, kreek near Kajana, below Ligorio. 16 Jul 1965. G.F. Mees.
RMNH 31009. 1, 31.7. Distr. Brokopondo, right bank tributary kreek of middle Gran Kreek near present-day Brokopondomeer shore. 30 Jul 1964. M. Boeseman.
RMNH 31010. 3, 15.7-37.9. Distr. Brokopondo. tributary kreek of Gran Rio between Ligorio and Awaradam Vallen. 15 Aug 1964. M. Boeseman.
RMNH 31011. 1, 33.3. Distr. Brokopondo, Gansee, Brokopondomeer. 8 Apr 1965. G.F. Mees.
RMNH 31012. 1, 22.2. Distr. Brokopondo, Gansee. 22 Jun 1965. G.F. Mees.
RMNH 31013. 1, 34.7. Distr. Para, Saramacca Kreek. 7 Apr 1965. G.F. Mees.
RMNH 31014. 11, 22.9-38.2. Distr. Brokopondo, ca 2 km E Kabel along submerging railway to Abontjema. 10 Jul 1964. M. Boeseman.
RMNH 31015. 24, 13.7-44.3. Distr. Brokopondo, tributary kreeks of Gran Rio, between Ligorio and Awaradam Vallen. 15 Aug 1964. M. Boeseman.
RMNH 31016. 4, 9.8-61.9. Distr. Brokopondo, ca 2 km E Kabel along railway Kabel-

Abontjema, flooded area near shore of developing Brokopondomeer. 7 Jul 1964. M. Boeseman.
RMNH 20410. 1, 22.7. Distr. Para, Republiek, Coropina Kreek. 3 Sep 1955. P.W. Hummelinck.
RMNH 20393. 1, 33.2. Distr. Brokopondo, Kabel, ditch. 2 Sep 1955. P.W. Hummelinck.
RMNH 18820. 1, 36.5. Distr. Para, Vierkinderen, Republiek, swamp kreek. 5 Sep 1948. D.C. Geijskes, P.H. Creutzberg.
RMNH 31017. 2, 33.2-56.2. Distr. Brokopondo, Tapoeripa Kreek. 8 Mar 1966. G.F. Mees.
RMNH 31018. 1, 34.6. Distr. Brokopondo, Marowijne (= Gran) Kreek, ca 41 km (geogr.) S Afobaka barrage. 8 Jun 1966. W. Vervoort.
RMNH 31019. 4, 21.0-36.4. Distr. Para, Saramacca Kreek. 22 Apr 1965. G.F. Mees.
RMNH 31020. 9, 13.2-40.1. Distr. Para, Saramacca Kreek. 20 Apr 1965. G.F. Mees.
RMNH 31021. 1, 35.8. Distr. Para, Para (= Saramacca) Kreek. 29 Jul 1965. G.F. Mees.
ISNB 18604. 2, 21.0-39.7. Distr. Brokopondo, second brook on the new road Zanderij-Matta. 29 Nov 1966. J.-P. Gosse (Sta 174).

Apistogramma ortmanni

SURINAME

All from Distr. Nickerie, Corantijn R. drainage

RMNH 31022. 1, 37.2. Tributary of Kabalebo R. below Avanavero Vallen (Camp BWKW). 10 Apr 1971. M. Boeseman
RMNH 31023. 22, 23.9-37.5. Tributary of Kabalebo R. just below Avanavero Vallen. 6 Apr 1971. M. Boeseman.
RMNH 31024. 22, 19.2-39.9. Avanavero Vallen. Sep 1965. G.F. Mees.
RMNH 31025. 1, 28.2. Kabalebo R., Avanavero Vallen. 9 Apr 1971. M. Boeseman.
RMNH 31026. 4, 10.7-29.1. Avanavero, mouth of stream. 16 Sep 1965. G.F. Mees.
RMNH 31027. 11, 21.9-32.6. Avanavero. 18 Sep 1965. G.F. Mees.
USNM 227332. 206, 12.0-29.7. 10.0-29.9. Small kreek entering Corantijn R. at Km 440 about 20 km N Lucie R. mouth. 11 Dec 1979. R.P. Vari (RPV 79-18).
USNM 225827. 8. Woodland stream about 1/2 km inland of Camp Mataway. 12 Sep 1980. R.P. Vari (RPV 80-26).
USNM 225828. 31. Small stream entering stream located about 200 m upriver of Dalbana Kreek mouth. 14 Sep 1980. R.P. Vari (RPV 80-30).
USNM 225829. 1. Stream near Camp Anjoemara. 14 Sep 1980. R.P. Vari (RPV 80-31).
USNM 225830. 14. Stream entering Corantijn R. at approximately Km 385 slightly N of Tiger Falls. 16 Sep 1980. R.P. Vari (RPV 80-33B).
USNM 225831. 3, largest 41.2. Small kreek entering Lucie R. about 3 km upstream from junction of Lucie and Corantijn R. 18 Sep 1980. R.P. Vari (RPV 80-37).
USNM 225832. 13. Stream entering south side of Lucie R. about 200 m downstream of ferry crossing. 18 Sep 1980. R.P. Vari (RPV 80-38).
USNM 225833. 15. Stream on south bank of Lucie R., ca 6 km upstream of ferry crossing on Amotopo-Camp Geologie Road. 18 Sep 1980. R.P. Vari (RPV 80-39).
USNM 225834. 5. Small kreek on east side of Corantijn R. about 2 km upstream of mouth of Moskita Kreek. 19 Sep 1980. R.P. Vari (RPV 80-40).
USNM 225835. 2. Small kreek entering Corantijn R. on east side ca 300 m N of Amotopo boat landing. 19 Sep 1980. R.P. Vari (RPV 80-42).
USNM 225836. 2. Tributary to Sisa Kreek, N side ca 700 m downstream of crossing of Amotopo to Camp Geologie Road. 20 Sep 1980. R.P. Vari (RPV 80-43).

USNM 289923. 7. Distr. Nickerie, tributary to Dalbana Kreek ca 3 km upstream from where Dalbana Kreek is crossed by Amotopo to Camp Geologie road. 15 Sep 1980. R.P. Vari (RPV 80-32).

Guianacara owroewefi

SURINAME

Drainage unknown

NRM ZISP506. 2, 66.6-76.7. Surinam. [1891?]. A. Klinckowström. Paratypes.

Coppename R. drainage

ZMA 106.741. 24. Distr. Saramacca, kreek on right bank of Linker Coppename R. in Wilhelminagebergte (03°51'N 56°45'W). 10 May 1967. H. Nijssen (HN 133). Paratypes.
ZMA 106.740. 95. Distr. Saramacca, kreek on left bank of Linker Coppename R. in Wilhelminagebergte (03°54'N 56°46'W). 9 May 1967. H. Nijssen (HN 132). Paratypes.
NRM A82/1967192.3419. 31, 41.9-78.8. Same data as ZMA 106.740. Paratypes.
ZMA 106.743. 3. Distr. Saramacca, right bank tributary kreek of Coppename R. in Wilhelminagebergte (03°52'N 56°53'W). 19 May 1967. H. Nijssen (HN 136). Paratypes.
ZMA 106.742. 13. Distr. Saramacca, left bank tributary kreek of Coppename R. in Wilhelminagebergte (03°52'N 56°55'W). 18 May 1967. H. Nijssen (HN 135). Paratypes.

Saramacca R. drainage

ZMA 106.730. 39. Distr. Brokopondo, tributary kreek on left bank of Kleine Saramacca R., ca 13 km (geogr.) ESE mouth. 28 Feb 1967. H. Nijssen (HN 107). Paratypes.
ZMA 106.729. 81. Distr. Brokopondo, tributary kreek on right bank of Kleine Saramacca R., ca 11 km (geogr.) ESE mouth. 27 Feb 1967. H. Nijssen (HN 106). Paratypes.
NRM A82/1967091.3418. 31, 25.8-86.0. Same data as ZMA 106.729. Paratypes.
ZMA 106.731. 10. Distr. Brokopondo, Kleine Saramacca R. 14 km ESE of junction with Saramacca R. 28 Feb 1967. H. Nijssen (HN 108). Paratypes.
RMNH 31028. 34, 22-65. Distr. Brokopondo, pool below Feddiprati soela. 9 Apr 1964. M. Boeseman (169). Paratypes.
RMNH 31029. 83, 20-47. Distr. Brokopondo, pool below Feddiprati soela. 9 Apr 1964. M. Boeseman (169). Paratypes.
RMNH 18494. 10. Distr. Brokopondo, Jan Basi Gado. Dec 1902. P.J. de Kock. Paratypes.

Suriname R. drainage

RMNH 31030. 1, 53. Distr. Brokopondo, soela near bivouac on middle Gran Kreek, 31 Jul 1964. M. Boeseman. Paratype.
RMNH 31031. 7, 48-60. Distr. Brokopondo, soela on lower Gran Kreek, ca 4 km from mouth. 6 May 1964. M. Boeseman. Paratypes.
RMNH 31032. 1, 23. Distr. Brokopondo, Amanipari or Gamboekoe Kreek, tributary to Sara Kreek between mouth of Lokus Kreek and Dam. 24 Feb 1964. M. Boeseman. Paratype.
RMNH 31033. 22, 30-61. Distr. Brokopondo, Awaradam. 17 Jul 1965. G.F. Mees. Paratypes.
RMNH 31034. 3, 52-90. Distr. Brokopondo, kreek near Brokopondo. 5 Jan 1964. M. Boeseman. Paratypes.

RMNH 31035. 6, 45-68. Distr. Brokopondo, forest kreek tributary to Pikien Rio above Asidonhopo. 13 Aug 1964. M. Boeseman. Paratypes.
RMNH 31036. 1, 60. Distr. Brokopondo, Marchall Kreek. 28 Dec. G. F. Mees. Paratype.
RMNH 31037. 2, 24-26. Distr. Brokopondo, Marchall Kreek. 28 Dec. G. F. Mees. Paratypes.
RMNH 31038. 26, 48-72. Distr. Brokopondo, Kajana. 16 Jul 1965. G. F. Mees. Paratypes.
RMNH 31039. 1, 38. Distr. Brokopondo, Mama Kreek at Berg en Dal. 12 Mar 1966. G. F. Mees. Paratype.
RMNH 31040. 4, 67-88. Distr. Brokopondo, Compagnie Kreek. 15 Mar 1966. G. F. Mees. Paratypes.
RMNH 31041. 5, 62.3-75.9. Distr. Brokopondo, tributary kreek of Gran Kreek ca 12 km from mouth of Gran Kreek. 21 Jul 1964. M. Boeseman. Paratypes.
NRM A83/1964302.3014. 5, 62.3-69.8. Same data as RMNH 31041. Paratypes.
RMNH 31042. 60, 26-71. Distr. Brokopondo, tributary kreek of Gran Kreek ca 12 km from mouth of Gran Kreek. 21 Jul 1964. M. Boeseman. Paratypes.
RMNH 31043. 1, 72.8. Distr. Brokopondo, kreek 1/4 h upstream from Ligorio. 16 Jul 1965. G. F. Mees. Paratype.
RMNH 31044. 8, 41-55. Distr. Brokopondo, Grandam. 18 Jul 1965. G. F. Mees. Paratypes.
ZMA 107.577. 1. Distr. Brokopondo, Brokopondomeer, 1 km E of former Mamadam Rapids. 17-18 Dec 1978. C.J.J. Richter. Paratype.
ZMA 107.606. 7. Distr. Brokopondo, Brokopondomeer, 2 km SSW of former Komkomjee Islands. 14 Dec 1978. C.J.J. Richter. Paratypes.
ZMA 106.712. 1. Distr. Brokopondo, Mama Kreek on road Paranam-Afobaka, 1.5 km N divertion to Berg en Dal. 17 May 1966. H. Nijssen (HN 14). Paratype.
ZMA 106.715. 3. Distr. Brokopondo, Mama kreek, 1.5 km N Berg en Dal on road Paranam-Afobaka. 2 Sep 1966. H. Nijssen (HN 50). Paratypes.
ZMA 106.716. 6. Distr. Brokopondo, Dabikwen (= Baling) Kreek at Brokobaka, ca 5 km N Afobaka. 19 Sep 1966. H. Nijssen (HN 61). Paratypes.
ZMA 106.717. 6. Distr. Brokopondo, Dabikwen (= Baling) Kreek at Brokobaka, ca 5 km N Afobaka. 19 Sep 1966. H. Nijssen (HN 61). Paratypes.
ZMA 106.709. 18. Distr. Brokopondo, Sara Kreek, 27 km S Dam village. 14 Oct 1966. H. Nijssen (HN 76). Paratypes.
ZMA 106.720. 4. Distr. Brokopondo, Sara Kreek, ca 31 km S Dam village. 12 Oct 1966. H. Nijssen (HN 73). Paratypes.
ZMA 106.721. 121. Distr. Brokopondo, Witte Kreek, 12 km S Brownsweg. 23 Nov 1966. H. Nijssen (HN 84). Paratypes.
ZMA 106.711. 57. Distr. Brokopondo, Marowijne (= Gran) Kreek, ca 63 km S Afobaka reservoir. 20 Oct 1966. H. Nijssen (HN 79). Paratypes.
ZMA 106.723. 30. Distr. Brokopondo, tributary kreek at right bank of Gran Rio, ca 600 m SW Ligorio. 27 Jan 1967. H. Nijssen (HN 95). Paratypes.
ZMA 106.722. 5. Distr. Brokopondo, Marchall Kreek, 1.5 km N Marchall village. 8 Dec 1966. H. Nijssen (HN 87). Paratypes.
ZMA 106.726. 52. Distr. Brokopondo, Gran Mau Kreek, right bank tributary to Gran Rio, ca 1 km NE Dombaai (= Bendi Watra). 30 Jan 1967. H. Nijssen (HN 99). Paratypes.
ZMA 106.724. 60. Distr. Brokopondo, tributary kreek at right bank of Gran Rio, ca 2.5 km NE of NE part of Awadam (= Awaradam). 28 Jan 1967. H. Nijssen (HN 96). Paratypes.
ZMA 106.728. 104. Distr. Brokopondo, tributary keek on left bank of Gran Rio, ca 3 km NE of NE part of Awadam (= Awaradam). 1 Feb 1967. H. Nijssen (HN 102). Paratypes.
ZMA 106.727. 227. Distr. Brokopondo, tributary keek on right bank of Gran Rio, ca 4 km NE of NE part of Awadam (= Awaradam). 31 Jan 1967. H. Nijssen (HN 101). Paratypes.

NRM A82/1967052.3420. 30, 39.2-79.2. Same data as ZMA 106.727. Paratypes.
ZMA 106.732. 1. Distr. Brokopondo, Kwati Watra Kreek on left bank of Suriname R., 500 m S Botopasi. 19 Mar 1967. H. Nijssen (HN 112). Paratype.
ZMA 106.733. 46. Distr. Brokopondo, Jenjee Kreek on right bank of Suriname R., 7.5 km N Botopasi. 21 Mar 1967. H. Nijssen (HN 114). Paratypes.
ZMA 106.706. 1. Distr. Brokopondo, Compagnie Kreek at road Paranam-Afobaka. 9 May 1966. H. Nijssen (HN 8). Paratype.
ZMA 106.705. 4. Distr. Brokopondo, Marowijne (= Gran) Kreek, 50.5 km S Afobaka along bank. 6 May 1966. H. Nijssen (HN 7). Paratypes.
ZMA 106.714. 1. Distr. Brokopondo, Marowijne (= Gran) Kreek, 60 km S Afobaka. 8 Jun 1966. H. Nijssen (HN 24). Paratype.
ZMA 106.713. 5. Distr. Brokopondo, Marowijne (= Gran) Kreek, 51 km S Afobaka. 27 May 1966. H. Nijssen (HN 19). Paratypes.
ZMA 106.707. 1. Distr. Brokopondo, kreek at right bank of Sara Kreek, 31 km S Dam village. 12 Oct 1966. H. Nijssen (HN 74). Paratype.
ZMA 106.708. 3. Distr. Brokopondo, kreek at left bank of Sara Kreek, 19 km S Dam village. 13 Oct 1966. H. Nijssen (HN 75). Paratypes.
ZMA 105.747. 7. Distr. Brokopondo, Parwapa (= Paba) Kreek on left bank of Suriname R., 2.5 km N Botopasi at Foetoenakaba (= Voetoekaba). 20 Mar 1967. H. Nijssen (HN 113). Paratypes.
ZMA 106.710. 3. Distr Brokopondo, Marowijne (= Gran Kreek), ca 50 km S Afobaka barrage. 19 Oct 1966. H. Nijssen (HN 78). Paratypes.
ZMA 106.718. 4. Distr. Brokopondo, Sopo Val (= Mama Dam) in Suriname R. 22 Sep 1966. H. Nijssen (HN 65). Paratypes.
ZMA 106.719. 1. Distr. Brokopondo, Sara Kreek, ca 21 km S Dam. 12-13 Oct 1966. H. Nijssen (HN 71). Paratype.
ZMA 106.725. 2. Distr. Brokopondo, Gran Rio at NE part of Awadam (= Awaradam). 28 Jan 1967. H. Nijssen (HN 2?). Paratypes.
ZMA 106.734. 1. Distr. Brokopondo, Suriname R. 1 km S Botopasi, rapids. 22 Mar 1967. H. Nijssen (HN 117). Paratype.
RMNH 21421. 1. Distr. Brokopondo, Gansee. Aug 1950. H. Heyde. Paratype.
ZMA 105.081pt. 1. Distr. Brokopondo, Brokopondomeer shore between first and second E auxiliary dams. 9 Sep 1966. H. Nijssen (HN 57). Paratype.

Marowijne R. drainage

ZMA 106.737. 8. Distr. Marowijne, Maka Kreek on left bank of Lawa R., 10 km S Stoelmanseiland. 21 Apr 1967. H. Nijssen (HN 128). Paratypes.
ZMA 106.738. 4, 34-125. Distr. Marowijne, Soea Kisi Kreek on left bank of Tapanahony R., 12 km SW Stoelmanseiland. 22 Apr 1967. H. Nijssen (HN 129). Paratypes.
ZMA 107.752. 19. Distr. Marowijne, Tapanahony R. ca 2 km downstreams mouth of Paloemeu R. 17 Nov 1965. G.F. Mees. Paratypes.
ZMA 106.736. 22. Distr. Marowijne, kreek on left bank of Marowijne R., W Manbari Val, 6 km N Stoelmanseiland. 20 Apr 1967. H. Nijssen (HN 127). Paratypes.
RMNH 31045. 27, 18-66. Distr. Marowijne, R. Tapanahony, 2.5 km downstreams of airfield. 30 Nov 1965. G.F. Mees. Paratypes.
RMNH 31046. 3, 41-71. Distr. Marowijne, Tapanahony R. behind Paloemeu airfield. 27 Nov 1965. G.F. Mees. Paratypes.
RMNH 31158. Several. Distr. Marowijne, Tapanahony. Nov 1965. G.F. Mees. Paratypes.
ISNB 738. 17. Distr. Marowijne, R. Paloemeu at Papadron soela. 1 Nov 1966. Léopold III & J.-P. Gosse. (Sta 153). Paratypes.
ISNB 739. 58. Distr. Marowijne, rapids of R. Paloemeu above confluence with Waloemeroe Kreek. 3 Nov 1966. Léopold III & J.-P. Gosse. (Sta 155). Paratypes.

ISNB 740. 7. Distr. Marowijne, small river left bank tributary of Marowijne R. at Tabbetje Hede. 24 Nov 1966. Léopold III & J.-P. Gosse. (Sta 171). Paratypes.
ISNB 741. 1. Distr. Marowijne, right bank tributary of Lawa R. at Anapaïkekondre. 18 Nov 1966. Léopold III & J.-P. Gosse. (Sta 168). Paratype.
ISNB 742. 2. Distr. Marowijne, small left bank tributary of the Oelemari R. upstreams of Oelemari airfield. 10 Nov 1966. Léopold III & J.-P. Gosse. (Sta 159). Paratypes.
ISNB 743. 3. Distr. Marowijne, R. Tapanahony at Paloemeu airfield. 25 Oct 1966. Léopold III & J.-P. Gosse. (Sta 151). Paratypes.
RMNH 31159. 3. Distr. Marowijne, Benzdorp. Oct 1965. Bekker. Paratypes.

GUYANE FRANÇAISE

R. Maroni drainage

ISNB 744. 9. Crique Bois-blanc, right bank tributary of R. Inini. 16 Nov 1969. J.-P. Gosse & Léopold III (Sta. 4). Paratypes.
ISNB 745. 11. R. Ouaqui, right bank tributary of R. Tampok, Saut Bali. 18 Nov 1969. J.-P. Gosse & Léopold III (Sta. 6). Paratypes.
ISNB 746. 17. R. Marouini, below village Epoia. 22 Nov 1969. J.-P. Gosse & Léopold III (Sta. 9). Paratypes.
ISNB 747. 76. R. Inini, Saut Sonnelle. 16 Nov 1969. J.-P. Gosse & Léopold III (Sta. 3). Paratypes.
ZMA 106.735. 8. Crique on right bank of R. Maroni E Mooisantitabbetje (36 km N Stoelmanseiland). 19 Apr 1967. H. Nijssen (HN 125). Paratypes.
ZMA 106.739. 20. Kamaloea or Saloea Kreek on right bank of R. Maroni, 9 km SE mouth of Gran Kreek. 24 Apr 1967. H. Nijssen (HN 131). Paratypes.
ISNB 749. Holotype, 58.7 mm SL. R. Marouini below first rapids. 23 Nov 1969. Léopold III & J.-P. Gosse (Sta 10).
ISNB 748. 10, 35.1-69.8. Same data as holotype. Paratypes.

Guianacara sphenozona

SURINAME

Corantijn R. drainage (all Distr. Nickerie)

RMNH 31047. Holotype, male, 73.4 mm SL. Sipaliwini. 23 Jan 1966. G.F. Mees.
RMNH 31048 (28) + NRM A83/1966037.3015 (10). 43-85. Sipaliwini. 23 Jan 1966. G.F. Mees. Paratypes.
RMNH 31049. 32, 22-62. Sipaliwini. 23 Jan 1966. G.F. Mees. Paratypes.
RMNH 31050. 76, 13-40. Sipaliwini. 23 Jan 1966. G.F. Mees. Paratypes.
RMNH 31051. 61, 16-68. Avanavero Vallen. 24 Sep 1965. G.F. Mees. Paratypes.
RMNH 31052. 13, 31-70. Zandvallen. 23 Sep 1965. G.F. Mees. Paratypes.
RMNH 31053. 3, 65.1-81.1. Right bank tributary of Kabalebo R., below Avanavero Vallen (Camp BWKW). 10 Apr 1971. M. Boeseman. Paratypes.
ZMA 115.753. 2. Lucie R. ca 10 km from confluence with Corantijn R. 19 Feb 1976. M. Panday. Paratypes.
RMNH 31054. 4, 36-40. Avanavero, kreek at camp. 20 Sep 1965. G.F. Mees. Paratypes.
RMNH 31055. 8, 28-58. Avanavero Vallen. 9 Apr 1971. M. Boeseman. Paratypes.
RMNH 31056. 1, 19. Lucie R. 15 Feb 1975. M.S. Hoogmoed. Paratype.
USNM 227335. 32, 39-76. Stream entering Kabalebo R. at Camp Avanavero, ca 3 km downstream of DeVis Falls. 7 Dec 1979. R.P. Vari (RPV 79-10). Paratypes.
RMNH 31057. 90, 18-56. Sipaliwini. 9 Feb 1966. G.F. Mees. Paratypes.
RMNH 31058. 8, 24-68. Sipaliwini. 9 Feb 1966. G.F. Mees. Paratypes.
RMNH 31059. 5, 26-40. Sipaliwini. 4 Feb 1966. G.F. Mees. Paratypes.

RMNH 31060. 1, 55. Sipaliwini. 28 Mar 1966. G.F. Mees. Paratype.
RMNH 31061. 10, 24-53. Sipaliwini. 6 Feb 1966. G.F. Mees. Paratypes.
RMNH 18468. Several. Lucie R. 1 Dec 1910. K.M. Hulk. Paratypes.
USNM 225868. 1. Corantijn R., 5°00′N 57°17′W. 17 May 1980. H.M. Madaire (M 80-17). Paratype.
USNM 225869. 1. Small kreek close to Mataway (4°58.5′N 57°42′W). 18 May 1980. H.M. Madaire (M 80-20). Paratype.
RMNH 31062. 2, 53.4-66.5. Right bank tributary kreek of Kabalebo R., just below Avanavero Vallen. 6 Apr 1971. M. Boeseman. Paratypes.
USNM 225860. 2. Stream near Camp Anjoemara. 14 Sep 80. R.P. Vari (RPV 80-31). Paratypes.
USNM 225861. 7. Stream entering Corantijn R. at ca Km 385 slightly N Tijger Vallen. 16 Sep 1980. R.P. Vari (RPV 80-33B). Paratypes.
USNM 225862. 8. Interconnected rock pools near Camp Hydro. 17 Sep 1980. R.P. Vari (RPV 80-35). Paratypes.
USNM 225863. 10. Pool in front of Camp Hydro. 17 Sep 1980. R.P. Vari (RPV 80-36). Paratypes.
USNM 225864. 8. Stream on south bank of Lucie R, ca 6 km upstream of ferry crossing on Amatopo-Camp Geologie Road. 18 Sep 1980. R.P. Vari (RPV 80-39). Paratypes.
USNM 225865. 1. Small creek on E side of Corantijn R. ca 2 km upstream of mouth of Moskita Kreek. 19 Sep 1980. R.P. Vari (RPV 80-40). Paratype.
USNM 225866. 1. Small kreek entering Corantijn R. on E side approximately 300 m N of Amatopo boat landing. 19 Sep 1980. R.P. Vari (RPV 80-42). Paratype.
USNM 225867. 7. Tributary to Sisa Kreek, N side approximately 700 m downstream of crossing of Amatopo to Kamp Geologie road. 20 Sep 1980. R.P. Vari. (RPV 80-43). Paratypes.

Guianacara oelemariensis

SURINAME

Marowijne R. drainage

ISNB 750. Holotype. Male, 81.0 mm SL. Distr. Marowijne, small right bank tributary to the upper Oelemari R. 14 Nov 1966. J.-P. Gosse (Mission Suriname 1966, Sta. 163).
ISNB 751. 13, 45.0-74.9. Same data as holotype. Paratypes.
NRM A87/1966461.3944. 2, 59.5-71.1. Same data as holotype. Paratypes.
ISNB 752. 1, 73.6. Distr. Marowijne, small brook tributary to the upper Oelemari R. 14 Nov 1966. J.-P. Gosse (Mission Suriname 1966, Sta. 164). Paratype.
ISNB 753. 1, 46.3. Distr. Marowijne, small left bank tributary to the Oelemari R. upstream of Oelemari airfield. 10 Nov 1966. J.-P. Gosse (Mission Suriname 1966, Sta. 159). Paratype.
ISNB 754. 6, 33.7-47.6. Distr. Marowijne, soela of the Oelemari R., one day's boat journey upstream of the Oelemari airfield. 15 Nov 1966. J.-P. Gosse (Mission Suriname 1966, Sta. 166). Paratypes.

Cichlasoma bimaculatum

See Kullander (1983a) for full locality data.

SURINAME

Isolated localities: Paramaribo; Coronie.

Corantijn R. drainage: Nanni Kreek, Doksen Savanne; Nanni Kreek, Van Wouw; Avanavero; Lucie R.

Nickerie R. drainage: Wageningen; Cupido; Nieuw Nickerie.

Suriname R. drainage: Afobaka; Berg en Dal; Marchall Kreek; Gansee; Pasa (= Dateke) Kreek; Gran Kreek 50.5, 51, 53.5, 55, 58, 60 km S Afobaka; Tout lui Faut canal, 12 km S Paramaribo; 3.5 km SE Brownsweg; Dabikwen (= Baling) Kreek; Sara Kreek, 19, 31 km S Dam; Compagnie Kreek; Klaas Kreek; Brownsweg; Kwatta; Hermansdorp; Domburg.

Commewijne R. drainage: Tributary of the upper Commewijne R. Cottica R. drainage: Wia-Wia line, km 7.75; Wia-Wia; Charlottenburg.

Marowijne R. drainage: Albina.

Aequidens tetramerus

SURINAME

Corantijn R. drainage

RMNH 31063. 2, 18.7-86.5. Distr. Nickerie, right tributary kreek to Kaboeri Kreek, below Fajastik Kreek. 2 Apr 1971. M. Boeseman.
USNM 225877. 1. Distr. Nickerie, small kreek close to Mataway. 18 May 1980. H.M. Madaire (80-20).
USNM 289717. 2. Distr. Nickerie, small stream entering stream located about 200 m upriver of Dalbana Kreek mouth. 14 Sep 1980. R.P. Vari (RPV 80-30).
USNM 225873. 6. Distr. Nickerie, tributary to Dalbana Kreek ca 3 km upstream from where Dalbana Kreek is crossed by Amotopo to Camp Geologie road. 15 Sep 1980. R.P. Vari (RPV 80-32).
USNM 225874. 5. Distr. Nickerie, small kreek entering Lucie R. about 3 km upstream from junction of Lucie and Corantijn rivers. 18 Sep 1980. R.P. Vari (RPV 80-37).
USNM 226014. 8. Distr. Nickerie, stream entering south side of Lucie R., about 200 m downstream of ferry crossing. 18 Sep 1980. R.P. Vari (RPV 80-38).
USNM 226015. 8. Distr. Nickerie, small kreek on E side of Corantijn R. about 2 km upstream of mouth of Moskita Kreek. 19 Sep 1980. R.P. Vari (RPV 80-40).
USNM 225875. 1. Distr. Nickerie, small kreek entering Corantijn R., on east bank approximately 350 m below base of Wilhelm II Falls. 19 Sep 1980. R.P. Vari (RPV 80-41).
USNM 225876. 9. Distr. Nickerie, small kreek entering Corantijn R. on E side approximately 300 m N Amotopo boat landing. 19 September 1980. R.P. Vari (RPV 80-42).

Nickerie R. drainage

RMNH 31064. 1, 59.2. Distr. Nickerie, kreek at second savannah, right tributary to middle Maretakka R., at woodcutters camp. 6 Mar 1971. M. Boeseman.
ZMA 107.291. 1, 81.8. Distr. Nickerie, tributary kreek to Nickerie R., 12 km WSW Stondansi Val. 5 Apr 1967. H. Nijssen (HN 120).

Coppename R. drainage

ZMA 107.289. 4, 76.2-91.9. Distr. Saramacca, kreek at right bank of Coppename R. (03°52′N 56°53′W). 19 May 1967. H. Nijssen (HN 136).

Suriname R. drainage

ZMA 107.285. 1, 96.3. Distr. Brokopondo, Marowijne (= Gran) Kreek, ca 51 km S Afobaka dam. 27 May 1966. H. Nijssen (HN 19).
ZMA 107.286. 1, ca 110. Distr. Brokopondo, Marowijne (= Gran) Kreek, ca 58 km S Afobaka dam. 9 Jun 1966. H. Nijssen (HN 25).

Marowijne R. drainage

ISNB 19508. 6. Distr. Marowijne, small brook tributary to the upper Oelemari R. 14 Nov 1966. J.-P. Gosse (Sta. 164).
ISNB 19507. 7. Distr. Marowijne, small brook tributary to the upper Oelemari R. 14 Nov 1966. J.-P. Gosse (Sta. 163).
ISNB 19505. 4. Distr. Marowijne, small river on right bank of Tapanahony R., en aval of confluence with Paloemeu R. at Paloemeu airfield. 9 Nov 1966. J.-P. Gosse (Sta. 158).
ISNB 19514. 3. Distr. Marowijne, right bank affluent of Lawa R., at Anapaikekondre. 18 Nov 1966. J.-P. Gosse (Sta. 168).
RMNH 31065. 2, 72.5-94.4. Distr. Marowijne, Tapanahony R. behind Paloemeu airfield. 27 Nov 1965. G.F. Mees.
RMNH 31066. 1, 47.7. Distr. Marowijne, Oelemari airstrip. 29 Jan (or Jul) 1975. M.S. Hoogmoed, W.N. Polder.
ZMA 107.293. 1, 57.1. Distr. Marowijne, kreek on left bank of Marowijne R. W Manbari Val, 6 km N Stoelmanseiland. 20 Apr 1967. H. Nijssen (HN 127).
ZMA 197.295. 2, 35.5-84.5. Distr. Marowijne, Soeakisikreek on left bank of Tapanahony R., 12 km SW Stoelmanseiland. 22 Apr 1967. H. Nijssen (HN 129).
ZMA 107.294. 4, 52.9-76.9. Distr. Marowijne, Maka Kreek on left bank of Lawa R., 10 km S center of Stoelmanseiland. 21 Apr 1967. H. Nijssen (HN 128).

GUYANE FRANÇAISE

R. Maroni drainage

ZMA 107.292. 5, 76.8-125.0. Crique on right bank of R. Maroni, E Mooisantitabbetje, 36 km N Stoelmanseiland. 19 Apr 1967. H. Nijssen (HN 125).

Aequidens paloemeuensis

SURINAME

Marowijne R. drainage

ISNB 755. Holotype. Adult male, 95.3 mm SL. Distr. Marowijne, small brook tributary to right bank of R. Paloemeu on road between Trombaka Noord and Trombaka Zuid. 8 Nov 1966. J.-P. Gosse & Léopold III (Sta. 156).
ISNB 756. 12, 14.6-90.8. Same data as holotype. Paratypes.
NRM A87/1966452.3943. 2, 78.1-85.0. Same data as holotype. Paratypes.

Krobia guianensis

SURINAME

Drainage unknown or uncertain

NMW 33650. 1. Paramaribo. Heller.
NMW 33818. 1. Surinam 1910. Heller.

Corantijn R. drainage

USNM 226017. 37. Distr. Nickerie, Corantijn R. at Km 180, side channel of main R. along Surinamese shore. 8 Sep 1980. R.P. Vari (RPV 80-19).

USNM 227341. 14. Distr. Nickerie, small kreek close to Mataway. 18 May 1980. H.M. Madaire (M 80-20).
USNM 225852. 2. Distr. Nickerie, kreek opposite Guyanese logging camp, 2 1/4 h S Matapi, ca 2 km downstream of Cow Falls. 11 Sep 1980. R.P. Vari (RPV 80-24).
USNM 225856. 2. Distr. Nickerie, Kamp Kreek 100 m N of turnoff to Camp Geology. 13 Sep 1980. R.P. Vari (RPV 80-27).
USNM 225857. 1. Distr. Nickerie, Steen Kreek where crossed by Amotopo-Paramaribo road. 13 Sep 1980. R.P. Vari (RPV 80-28).
USNM 226016. 39. Distr. Nickerie, small stream behind Kamp Infrastructure (opposite Avanavero airfield). 13 Sep 1980. R.P. Vari (RPV 80-29).
USNM 289525. 1. Distr. Nickerie, small stream behind Kamp Infrastructure (opposite Avanavero airfield). 13 Sep 1980. R.P. Vari (RPV 80-29).
USNM 225853. 3. Distr. Nickerie, small stream entering stream located about 200 m upriver of Dalbana Kreek mouth. 14 Sep 1980. R.P. Vari (RPV 80-30).
USNM 225854. 19. Distr. Nickerie, tributary to Dalbana Kreek ca 3 km upstream from where Dalbana Kreek is crossed by Amatopo-Camp Geologie road. 15 Sep 1980. R.P. Vari (RPV 80-32).
USNM 225855. 1. Distr. Nickerie, stream at Km 212 of Amatopo-Camp Geologie road, at Machine Park-Camp 212. 15 Sep 1980. R.P. Vari (RPV 80-33A).
RMNH 31067. 6, 54.4-81.7. Distr. Nickerie, right bank tributary to Kabalebo R., below Avanavero Vallen (Camp BWKW). 10 Apr 1971. M. Boeseman.
RMNH 31068 (3) + NRM A83/1971134.3021 (3). 76.1-105.7. Distr. Nickerie Winanna Kreek, left bank tributary of middle Kaboeri Kreek. 1 Apr 1971. M. Boeseman.
RMNH 31069 (2) + NRM A83/1971134.3021 (3). 38.6-74.9. Distr. Nickerie, Winanna Kreek, left bank tributary of middle Kaboeri Kreek. 1 Apr 1971. M. Boeseman.
RMNH 31070. 11, 24.7-84.9. Distr. Nickerie, Avanavero Vallen, Kabalebo R. 9 Apr 1971. M. Boeseman.
RMNH 31071. 7, 66.0-88.7. Distr. Nickerie, Sipaliwini. 23 Jan 1966. G.F. Mees.
RMNH 31072. 1, 63. Distr. Nickerie, right bank tributary kreek to Kaboeri Kreek below Fajastik Kreek. 2 Apr 1971. M. Boeseman.
RMNH 31073. 3, 38.7-66.8. Distr. Nickerie, right bank tributary to Kabalebo R., ca 8 km below Avanavero Vallen. 7 Apr 1971. M. Boeseman.
RMNH 31074. 2, 41.8-70.2. Distr. Nickerie, Kaboeri Kreek near mouth of Winanna Kreek. 1 Apr 1971. M. Boeseman.
RMNH 31075. 3, 35.2-42.5. Distr. Nickerie, Sipaliwini. 4 Feb 1966. G.F. Mees.
RMNH 31076. 10, 16.0-41.7. Distr. Nickerie, Sipaliwini. 23 Jan 1966. G.F. Mees.
RMNH 31077. 3, 54.6-66.4. Distr. Nickerie, Upper (SE) Kaboeri Kreek. 29 Mar 1971. M. Boeseman.
RMNH 31078. 14, 37.5-64.1. Distr. Nickerie, Avanavero Vallen. 24 Sep 1965. G.F. Mees.
RMNH 31079. 2, 53.2-75.2. Distr. Nickerie, Zandvallen. 23 Sep 1965. G.F. Mees.
RMNH 31080. 2, 42.7-45.8. Distr. Nickerie, right bank tributary kreek to Kabalebo R., just below Avanavero Vallen. 6 Apr 1971. M. Boeseman.
RMNH 31081. 7, 24.6-42.3. Distr. Nickerie, Sipaliwini. 23 Jan 1966. G.F. Mees.
RMNH 31082. 5, 61.1-92.6. Distr. Nickerie, Sipaliwini. 9 Feb 1966. G.F. Mees.
RMNH 31083. 41, 18.5-53.0. Distr. Nickerie, Sipaliwini. 9 Feb 1966. G.F. Mees.
RMNH 31084. 5, 55.2-70.4. Distr. Nickerie, Winanna Kreek, right bank tributary to middle Kaboeri Kreek. 1 Apr 1971. M. Boeseman.
RMNH 31085. 4, 78.7-95.2. Distr. Nickerie, right bank tributary of Kabalebo, below Avanavero Vallen (Camp BWKW). 10 Apr 1971. M. Boeseman.
RMNH 31086. 5, 45.8-74.6. Distr. Nickerie, Kabalebo R., Avanavero Vallen. 13 Sep 1965. G.F. Mees.
RMNH 31087. 15, 24.8-56.4. Distr. Nickerie, Sipaliwini. 6 Feb 1966. G.F. Mees.
RMNH 31088. 8, 51.1-85.2. Distr. Nickerie, Sipaliwini. 6 Feb 1966. G.F. Mees.

Nickerie R. drainage

ZMA 106.694. 2. Distr. Nickerie, Stondansie Val in Nickerie R. 5 Apr 1967. H. Nijssen (HN 121).
ZMA 106.693. 56. Distr. Nickerie, right bank tributary to the Nickerie R., 12.5 km WSW Stondansie Val. 5 Apr 1967. H. Nijssen (HN 120).
ZMA 106.699. 12. Distr. Nickerie, Fallawatra R., rapids 5 km SW Stondandie. 6 Apr 1967. H. Nijssen (HN 122).
RMNH 31089. 3, 27-52. Distr. Nickerie, in and around falls in right bank tributary to Nickerie R., behind meteo-camp, below Blanche-Marie Vallen. 15 Feb 1971. M. Boeseman.
RMNH 31090. 6, 72-111. Distr. Nickerie, right bank tributary of Maratakka R. just above Cupido (Bigi Birri Kreek). 8 Mar 1971. M. Boeseman.
NRM A83/1971101.3020. 2. Same data as RMNH 31090.
RMNH 31091. 16, 49-101. Distr. Nickerie, left bank tributary of upper Nickerie R., above Camp Stondansie. 30 Jan 1971. M. Boeseman.
RMNH 31092. 4, 39-99. Distr. Nickerie, forest kreek, tributary of Nickerie R. ca 1 km behind Camp Waterloopk, below Stondansie Vallen. 3 Feb 1971. M. Boeseman.
RMNH 31093. 11, 24-94. Distr. Nickerie, Mozeskreek at Reynoldskamp. 20 Jul 1975. W.N. Polder.
NRM A83/1975297.3023. 3. Same data as RMNH 31093.
RMNH 31094. 2, 91-105. Distr. Nickerie, kreek a few km above woodcutters camp, right bank tributary of Maratakka R. 28 Feb 1971. M. Boeseman.
RMNH 31095. 2, 43-68. Distr. Nickerie, right bank tributary of Maratakka R., just above Cupido (Bigi Birri Kreek). 8 Mar 1971. M. Boeseman.
RMNH 31096. 15, 20-56. Distr. Nickerie, left bank tributary of Nickerie R. below Blanche Marie Vallen. 16 Feb 1971. M. Boeseman.
RMNH 31097. 1, 48. Distr. Nickerie, right bank tributary of Nickerie R., ca 1/2 h below Blanche Marie Vallen. 11 Feb 1971. M. Boeseman.
RMNH 31098. 1, 31. Distr. Nickerie, in and around falls in right bank tributary of Nickerie R., behind meteo-camp, below Blanche Marie Vallen. 15 Feb 1971. M. Boeseman.
RMNH 31099. 1, 14. Same data as RMNH 31095.
RMNH 31100. 1, 58. Same data as RMNH 31095.
RMNH 31101. 13, 11-26. Distr. Nickerie, shore before bivouac on middle course of Maratakka R. 2 Mar 1971. M. Boeseman.

Coppename R. drainage

ZMA 106.703. 10. Distr. Saramacca, kreek on right bank of Linker Coppename R. in Wilhelminagebergte (03°51'N 56°45'W). 10 May 1967. H. Nijssen (HN 133).
ZMA 106.702. 37. Distr. Saramacca, kreek on left bank of Linker Coppename R. in Wilhelminagebergte (03°54'N 56°46'W). 9 May 1967. H. Nijssen (HN 132).
ZMA 106.704. 1. Distr. Saramacca, right bank tributary kreek of Coppename R. in Wilhelminagebergte (03°52'N 56°53'W). 19 May 1967. H. Nijssen (HN 136).
ZMA 106.701. 17. Distr. Saramacca, left bank tributary kreek of Coppename R. in Wilhelminagebergte (03°52'N 56°55'W). 18 May 1967. H. Nijssen (HN 135).
RMNH 31102. 8, 14.9-48.3. Distr. Saramacca, Raleigh Vallen, forest stream. 20 Mar 1972. G.F. Mees.

Saramacca R. drainage

ZMA 105.635. 9. Distr. Brokopondo, tributary kreek on left bank of Kleine Saramacca R., ca 13 km (geogr.) ESE mouth. 28 Feb 1967. H. Nijssen (HN 107).
ZMA 105.615. 80. Distr. Brokopondo, tributary kreek on right bank of Kleine Saramacca R., ca 11 km (geogr.) ESE mouth. 27 Feb 1967. H. Nijssen (HN 106).

ZMA 112.549. 1. Distr. Saramacca, Gordijn Kreek near bridge of Saramacca R., ca 30 km W Zanderij. 10 Nov 1972. P.J.H. van Bree.
RMNH 31103. 1. Distr. Brokopondo, pool below Feddiprati, soela. 9 Apr 1964. M. Boeseman (169).

Suriname R. drainage

RMNH 31104. 4, 66.5-94.2. Distr. Brokopondo, Tapoeripa Kreek. 8 Mar 1966. G.F. Mees.
RMNH 31105. 5, 58.8-87.9. Distr. Brokopondo, Soela in lower course of Gran Kreek,ca 4 km from mouth. 6 Mar 1964.
RMNH 31106. 9, 21.6-60.7. Distr. Brokopondo, Marchall Kreek. 28 Dec. G.F. Mees.
RMNH 31107. 5, 44.4-54.1. Distr. Brokopondo, Kajana. 16 Jul 1965.
RMNH 31108. 8, 20.7-40.0. Distr. Brokopondo, tributary kreek to Mama Kreek at Berg en Dal. 12 Mar 1966. G.F. Mees.
RMNH 31109. 1, 78.4. Distr. Brokopondo, stream at Brokopondo. 18 Dec 1963.
RMNH 31110. 1, 70.4. Distr. Brokopondo, Suriname R. at Brokopondo. 3-4 Feb 1964.
RMNH 31111. 25, 30.5-83.6. Distr. Brokopondo, Marchall Kreek. 28 Dec 1965. G.F. Mees.
RMNH 31112. 15, 42.5-79.2. Distr. Brokopondo, Afobaka. 31 Jul 1965.
RMNH 31113. 27, 12.6-32.8. Distr. Para, Saramacca Kreek. 21 May 1965. G.F. Mees.
RMNH 31114 (3) + NRM A83/1965143.3022 (3). 21.7-45.2. Distr. Para, Saramacca Kreek. 7 Apr 1965. G.F. Mees.
RMNH 31115. 4, 12.3-28.8. Distr. Para, Saramacca Kreek. 22 Apr 1965. G.F. Mees.
RMNH 31116. 1, 23.2. Distr. Para, Para (= Saramacca) Kreek. 29 Jul 1965. G.F. Mees.
RMNH 31117. 70, 16-38. Distr. Brokopondo, Makambie Kreek. 18 Feb 1966. G.F. Mees.
RMNH 31118. 2, 46.9-98.6. Distr. Para, Saramacca Kreek. 7 Apr 1965. G.F. Mees.
NRM A83/1965143.3019. 2, 61.9-66.6. Same data as RMNH 31118.
RMNH 31119. 12, 14.7-55.6. Distr. Para, Saramacca Kreek. 20 Apr 1965. G.F. Mees.
RMNH 31120. 4, 90.3-97.7. Distr. Brokopondo, Compagnie Kreek. 15 Mar 1966. G.F. Mees.
RMNH 31121. 2, 80.6-109.9. Distr. Para, Zanderij. 3 Mar 1966. G.F. Mees.
RMNH 31122. 2, 72.9-101.8. Distr. Brokopondo, Gansee, Brokopondomeer. 21 April 1965.
RMNH 31123. 1, 106.4. Distr. Brokopondo, tributary kreek of Gran Kreek ca 12 km from mouth of Gran Kreek. 21 Jul 1964. G.F. Mees.
RMNH 31124. 4, 79.9-91.3. Distr. Brokopondo, kreek 1/4 h upstream from Ligorio. 16 Jul 1965. G.F. Mees.
RMNH 31125. 7, 42.0-107.4. Distr. Brokopondo, Makambie Kreek. 28 Feb 1966. G.F. Mees.
RMNH 31126. 1, 97. Distr. Brokopondo, Brokopondomeer, shore ca 6 km SE of Afobaka, open space in drowned forest. 11 Apr 1965. G.F. Mees.
RMNH 31127. 1, 57.4. Distr. Brokopondo, Marchall Kreek. 28 Dec 1965. G.F. Mees.
RMNH 31128. 4, 26.6-53.3. Distr. Brokopondo, E tributary of Gran Kreek ca 12 km from mouth of Gran Kreek. 21 Apr 1964. M. Boeseman.
RMNH 31129. 2, 38.4-57.0. Distr. Brokopondo, Grandam. 18 Jul 1965. G.F. Mees.
RMNH 31130. 1, 28.2. Distr. Brokopondo, Compagnie Kreek. 5 Sep 1965. G.F. Mees.
RMNH 31131. 3, 34.2-46.0. Distr. ?, Desiong-Kondre. 10 Dec 1964. D. C. Geijskes (21).
ZMA 107.605. 5. Distr. Brokopondo, Brokopondomeer, 2 km SSW of former Komkomjee Island. 14 Dec 1978. C.J.J. Richter.
NMW 33349. 3. Suriname R. Reg. 1889 9.
USNM 289718. 2. Distr. Para, 43 km S Paramaribo. 3/4 mi W 's Lands Bosbeheer Hq, Zanderij edge of bay in a black water kreek in savanna rain forest. 13 Apr 1969. D.W. Dunham.

ZMA 105.282. 7. Distr. Brokopondo, Marchall Kreek at road Paranam-Afobaka, 1.5 km N Marchall, 7.5 km N Klaas Kreek. 12 May 1966. H. Nijssen (HN 12).
ZMA 106.684. 4. Distr. Brokopondo, Mamakreek at road Paranam-Afobaka, 1.5 km N divertion to Berg en Dal. 2 May 1966. H. Nijssen (HN 3).
ZMA 106.685. 3. Distr. Brokopondo, Compagnie Kreek, ca 7 km NW Brokopondo. 22 Aug 1966. H. Nijssen (HN 36).
ZMA 105.033. 5. Distr. Brokopondo, Mama Kreek on road Paranam-Afobaka, 1.5 km N divertion to Berg en Dal. 17 May 1966. H. Nijssen (HN 14).
ZMA 106.687. 51. Distr. Brokopondo, Tapoeripa Kreek at Hermansdorp, 1 km N Brokopondo. 25 Aug 1966. H. Nijssen (HN 43).
ZMA 106.686. 4. Distr. Brokopondo, Marchall Kreek, 1.5 km N Marchall village. 23 Aug 1966. H. Nijssen (HN 37).
ZMA 105.085. 5. Distr. Brokopondo, Brokopondomeer, shore between 1st and 2nd eastern auxiliary dams. 9 Sep 1966. H. Nijssen (HN 57).
ZMA 105.146. 3. Distr. Brokopondo, Mama kreek, 1.5 km N Berg en Dal on road Paranam-Afobaka. 2 Sep 1966. H. Nijssen (HN 50).
ZMA 106.688. 58. Distr. Brokopondo, Dabikwen (= Baling) Kreek at Brokobaka, ca 5 km N Afobaka. 19 Sep 1966. H. Nijssen (HN 61).
ZMA 105.119. 25. Distr. Brokopondo, Makami Kreek (= Makambi Kreek), ca 8 km S Brownsweg. 13 Sep 1966. H. Nijssen (HN 59).
ZMA 106.692. 11. Distr. Brokopondo, Sara Kreek, 27 km S Dam village. 14 Oct 1966. H. Nijssen (HN 76).
ZMA 106.691. 15. Distr. Brokopondo, Sara Kreek, ca 31 km S Dam village. 12 Oct 1966. H. Nijssen (HN 73). (1: 80 mm SL, depth 38.4 mm).
ZMA 106.690. 44. Distr. Brokopondo, Witte Kreek, 12 km S Brownsweg. 23 Sep 1966. H. Nijssen (HN 84).
ZMA 105.275. 20. Distr. Brokopondo, Marowijne (= Gran) Kreek, ca 63 km S Afobaka reservoir. 20 Oct 1966. H. Nijssen (HN 79).
ZMA 105.487. 18. Distr. Brokopondo, tributary kreek at right bank of Gran Rio, ca 600 m SW Ligorio. 27 Jan 1967. H. Nijssen (HN 95).
ZMA 106.689. 17. Distr. Brokopondo, Marchall Kreek, 1.5 km N Marchall village. 8 Dec 1966. H. Nijssen (HN 87).
ZMA 105.518. 3. Distr. Brokopondo, Gran Mau Kreek, right bank tributary to Gran Rio, ca 1 km NE Dombaai (= Bendi Watra). 30 Jan 1967. H. Nijssen (HN 99).
ZMA 105.443. 26. Distr. Brokopondo, tributary kreek at right bank of Gran Rio, ca 2.5 km NE of NE part of Awadam (= Awaradam). 28 Jan 1967. H. Nijssen (HN 96).
ZMA 105.434. 17. Distr. Brokopondo, tributary keek on left bank of Gran Rio, ca 3 km NE of NE part of Awadam (= Awaradam). 1 Feb 1967. H. Nijssen (HN 102).
ZMA 105.485. 10. Distr. Brokopondo, tributary keek on right bank of Gran Rio, ca 4 km NE of NE part of Awadam (= Awaradam). 31 Jan 1967. H. Nijssen (HN 101).
ZMA 105.746. 16. Distr. Brokopondo, Parwapa (= Paba) Kreek on left bank of Suriname R., 2.5 km N Botopasi at Foetoenakaba (= Voetoekaba). 20 Mar 1967. H. Nijssen (HN 113).
ZMA 105.786. 4. Distr. Brokopondo, Kwati Watra Kreek on left bank of Suriname R., 500 m S Botopasi. 19 Mar 1967. H. Nijssen (HN 112).
ZMA 105.688. 15. Distr. Brokopondo, Jenjee Kreek on right bank of Suriname R., 7.5 km N Botopasi. 21 Mar 1967. H. Nijssen (HN 114).
ZMA 106.977. 3. Distr. Brokopondo, Klaas Kreek at road Paranam-Afobaka. 2 May 1966. H. Nijssen (HN 4).
ZMA 106.974. 6. Distr. Brokopondo, Compagnie Kreek at road Paranam-Afobaka. 1 May 1966. H. Nijssen (HN 2).
ZMA 106.979. 3. Distr. Brokopondo, Compagnie Kreek at road Paranam-Afobaka. 9 May 1966. H. Nijssen (HN 8).
ZMA 106.978. 11. Distr. Brokopondo, Marowijne (= Gran) Kreek, 50.5 km S Afobaka along bank. 6 May 1966. H. Nijssen (HN 7).

ZMA 106.985. 2. Distr. Brokopondo, Marowijne (= Gran) Kreek, 55 km S Afobaka. 24 May 1966. H. Nijssen (HN 16).
ZMA 106.981. 1. Distr. Brokopondo, Brokopondomeer along bank near Afobaka. 10 May 1966. H. Nijssen (HN 9).
ZMA 106.987. 9. Distr. Brokopondo, Marowijne (= Gran) Kreek, 60 km S Afobaka. 8 Jun 1966. H. Nijssen (HN 24).
ZMA 106.986. 9. Distr. Brokopondo, Marowijne (= Gran) Kreek, 51 km S Afobaka. 27 May 1966. H. Nijssen (HN 19).
ZMA 106.989. 2. Distr. Brokopondo, kreek at right bank of Sara Kreek, 31 km S Dam village. 12 Oct 1966. H. Nijssen (HN 74).
ZMA 106.988. 2. Distr. Brokopondo, Marowijne (= Gran) Kreek, 58 km S Afobaka. 9 Jun 1966. H. Nijssen (HN 25).
ZMA 106.991. 29. Distr. Para, Carolina or Para Kreek, 10 km SE Zanderij airfield. 14 Dec 1966. H. Nijssen (HN 92).
ZMA 106.990. 2. Distr. Brokopondo, Tapoeripa Kreek at Brokopondo. 15 Sep 1966. H. Nijssen (HN 60).
ZMA 106.994. 6. Distr. Brokopondo, swamp at N side of 4th eastern auxiliary dam, 5 km E Afobaka. 7 Aug 1966. H. Nijssen (HN 86).
ZMA 106.993. 2. Distr. Brokopondo, kreek at left bank of Sara Kreek, 19 km S Dam village. 13 Oct 1966. H. Nijssen (HN 75).
ZMA 106.996. 1. Distr. Brokopondo, railway Km 114, kreek at Brownsweg. Oct 1960. Pijpers.
ZMA uncat. 1. Distr. Brokopondo, Brokopondo Meer along barrage at Afobaka and along eastern auxiliary dams. 18-19 May 1966. H. Nijssen.
RMNH 31132. 1, 52. Distr. Brokopondo, Hermansdorp. 5 Mar 1966. G.F. Mees.
ZMA 106.995. 6. Distr. Suriname, Tout-Lui-Faut Canal, 12 km S Paramaribo, at road Paramaribo-Lelydorp. 12 Dec 1966. H. Nijssen (HN 91).
ZMA 108.337. 3. Distr. Para, Coropina Kreek at Republiek. Oct 1956. J. v.d. Kamp.
ZMA 108.324. 4. Distr. Para, pool at Zanderij. 10 May 1956. J. v.d. Kamp.
ZMA 112.548. 9. Distr. Para, tributary of Coropina Kreek near Republiek. 3 Nov 1972. P.J.H. van Bree.

Cottica R. drainage

ZMA 106.180. 1. Distr. Marowijne, Ricanau Kreek, ca 5 km E Moengo at road Albina-Moengo. 4 Jun 1966. H. Nijssen (HN 23).
RMNH 31133. 7, 12-25. Distr. Marowijne, Ricanau Kreek, ca 5 km E Moengo. 1966. W. Vervoort.
RMNH 31134. 1, 38. Distr. Marowijne, 2nd Kamp Suralco, 10 km N Wanekreek. 4 Mar 1975. M.S. Hoogmoed.

Krobia itanyi

SURINAME

Marowijne R. drainage

ZMA 119.531. Neotype. Adult male, 89.6 mm SL. Distr. Marowijne, kreek on left bank of Marowijne R., W Manbari Val, 6 km N Stoelmanseiland. 20 Apr 1967. H. Nijssen (HN 127).
ISNB 19491. 2, 45-46. Distr. Marowijne, small brook tributary to right bank of Paloemeu R. on road between Trombaka Noord and Trombaka Zuid. 8 Nov 1966. J.-P. Gosse & Léopold III (Sta. 156).
ISNB 19492. 15. Distr. Marowijne, small left bank tributary of Oelemari R. upstream of Oelemari airfield. 10 Nov 1966. J.-P. Gosse & Léopold III (Sta. 159).

ISNB 19493. 37. Distr. Marowijne, soela of the Oelemari R., one day's boat journey upstream of Oelemari airfield. 15 Nov 1966. Léopold III & J.P. Gosse (Sta. 166).
ISNB 19494. 38. Distr. Marowijne, small river left bank tributary of Marowijne R. at Tabbetje Hede. 24 Nov 1966. J.-P. Gosse & Léopold III (Sta. 171).
ISNB 19509. 198. Distr. Marowijne, small stream tributary of upper Oelemari R. 14 Nov 1966. J.-P. Gosse & Léopold III (Sta. 163).
ZMA 106.695 (23) + NRM A85/1967164.3451 (5). 32-93.9. Same data as neotype.
ZMA 106.697. 21. Distr. Marowijne, Maka Kreek on left bank of Lawa R., 10 km S Stoelmanseiland. 21 Apr 1967. H. Nijssen (HN 128).
ZMA 106.698. 15, 34-125. Distr. Marowijne, Soea Kisi Kreek on left bank of Tapanahony R., 12 km SW Stoelmanseiland. 22 Apr 1967. H. Nijssen (HN 129).
ZMA 107.551. 4. Distr. Marowijne, Oelemari R., 2 km N Oelemari airstrip. 13 Oct 1977. H. Heyde (no. 12).
ZMA 107.750. 6. Distr. Marowijne, Tapanahony R. ca 2 km downstreams mouth of Paloemeu R. 17 Nov 1965. G.F. Mees.
RMNH 31135. 1, 46.0. Distr. Marowijne, Oelemari airstrip. 29 January 1975. M.S. Hoogmoed & W.N. Polder (1467).
RMNH 31136. 2, 23.5-60.5. Distr. Marowijne, R. Tapanahony, 2.5 km downstream of airfield. 30 Nov 1965. G.F. Mees.
RMNH 31137. 4, 21.5-100.8. Distr. Marowijne, Tapanahony R. behind Paloemeu airfield. 27 Nov 1965. G.F. Mees.
RMNH 31160. Several. Distr. Marowijne, Tapanahony. Nov 1965. G.F. Mees.

GUYANE FRANÇAISE

R. Maroni drainage

ISNB 19523. 21. Maripa soela, Crique Daouda near airfield. 12 Nov 1969. J.-P. Gosse & Léopold III (Sta. 1).
ISNB 19524. 7. Crique Bois-blanc, right bank tributary of R. Inini. 16 Nov 1969. J.-P. Gosse & Léopold III (Sta. 4).
ISNB 19525. 9. 44.5-61.4. R. Ouaqui, right bank tributary of R. Tampok, Saut Bali. 18 Nov 1969. J.-P. Gosse & Léopold III (Sta. 6).
ISNB 19526. 30. Small tributary on right bank of R. Ouaqui upstream of Saut Bali. 18 Nov 1960. J.-P. Gosse & Léopold III (Sta. 7).
ISNB 19527. 4. R. Marouini, downstream of village Epoia. 22 Nov 1969. J.-P. Gosse & Léopold III (Sta. 9).
ISNB 19530. 4. R. Inini, Saut Sonnelle. 16 Nov 1969. J.-P. Gosse & Léopold III (Sta. 3).
NRM A84/1978386.3837. 1. Canal St Laurent. 23 Sep 1978. P. Planquette.
NRM A84/1980041.3841. 1. Crique Cascade, St Jean. 21 Jan 1980. P. Planquette.
NRM A84/1979412.3839. 2. Crique Balaté. 9 Oct 1970. P. Planquette.
NRM A84/1979412.3832. 1. Crique Balaté. 9 Oct 1970. P. Planquette.
NRM A84/1979248.3836. 2. Canal St Laurent. Jun 1979. P. Planquette.
ZMA 106.696. 18, 23-99. Crique on right bank of R. Maroni E Mooisantitabbetje (36 km N Stoelmanseiland). 19 Apr 1967. H. Nijssen (HN 125).
ZMA 106.700. 46. Kamaloea or Saloea Kreek on right bank of R. Maroni, 9 km SE mouth of Gran Kreek. 24 Apr 1967. H. Nijssen (HN 131).

Cleithracara maronii

SURINAME

Drainage unknown or uncertain

NMW 33682. 2. Oberer Surinam. Registered 1901. Michaelis.
BMNH 1870.3.10:41-42. 2. Surinam. No date. Kappler.

Corantijn R. drainage

RMNH 31138. 1, 49.3. Distr. Nickerie, Winanna Kreek, left bank tributary to middle Kaboeri Kreek. 1 Apr 1971. M. Boeseman.
USNM 225858. 2. Distr. Nickerie, Dalbane Kreek at Camp McClemmen. 6 Sep 1980. R.P. Vari (RPV-80-13).
USNM 225859. 1. Distr. Nickerie, Lana Kreek, ca 4 km upstream from mouth. 7 Sep 1980. R.P. Vari (RPV-80-15).
USNM 225871. 2. Distr. Nickerie, Kapoeri Kreek, ca 4 km from mouth. 8 Sep 1980. R.P. Vari (RPV-80-18).

Nickerie R. drainage

RMNH 31139. 5, 47.1-63.2. Distr. Nickerie, 3 km behind Camp Waterloopk., below Stondansi Val. 1 Feb 1971. M. Boeseman.
RMNH 31140. 1, 48.6. Distr. Nickerie, shore before bivouac at middle course of Maratakka R. 1 Mar 1971. M. Boeseman.

Suriname R. drainage

RMNH 31141. 2, 52.8-60.2. Distr. Para, Saramacca Kreek. 7 Apr 1965. G.F. Mees.
NRM A83/1965144.3011. 1, 66.3. Same data as RMNH 31141.
RMNH 31142. 1, 36.8. Distr. Para, Saramacca Kreek. 7 Apr 1965. G.F. Mees.
RMNH 18396. 2. Distr. Para, Republiek, Coropina kreek. Oct 1946. Leg. D.C. Geijskes.
ZMA 107.022. 3, 43.9-70.9. Distr. Para, Carolina (= Malasie) Kreek, 10 km SE Zanderij airfield. 14 Dec 1966. H. Nijssen.
ISNB 19503. 3. Distr. Brokopondo, small brook on new road Zanderij-Matta. 29 Nov 1966. J.-P. Gosse & Léopold III (Sta. 174).

Cottica R. drainage

RMNH 19541. 5, 32.4-38.8. Distr. Marowijne, Moengotapoe, Laikreek. 21 Sep 1948. P.H. Creutzberg.
NRM A83/1948392.3010. 2, 36.9-48.3. Same data as RMNH 19541. (1 specimen cleared and stained.)

Commewijne R. drainage

NMW 33548-33551pt. 1. Tributary to the upper Commewijne R. Registered 1909.

Marowijne R. drainage

NMW 33683. Holotype. 58.4 mm SL. Maroni-Fluss, Surinam. Registered 1882.
RMNH 31143. 2, 63.0-69.8. Distr. Marowijne, second camp of Suralco, 10 km N Wanekreek. 4 Mar 1975. M.S. Hoogmoed (686).
ZMA 114.340. 1, 67.3. Distr. Marowijne, kreek N Paloemeu airstrip. 17 Oct 1976. H. Heyde.
ZMA 108.339. 1, 36.5. Distr. Marowijne, Marowijne R. at Albina. 7 Aug 1960. H.P. Pijpers.
ISNB 19500. 32. Distr. Marowijne, small brook tributary to the upper Oelemari R. 14 Nov 1966. J.-P. Gosse & Léopold III (Sta. 163).
ISNB 20850. 1. Distr. Marowijne, small tributary to left bank of R. Oelemari upstream of Oelemari airfield. 10 Nov 1966. J.-P. Gosse & Léopold III (Sta. 159).
ISNB 20851. 4. Distr. Marowijne, small brook tributary to the upper Oelemari R. 14 Nov 1966. J.-P. Gosse & Léopold III. (Sta. 164).
ISNB 20852. 4. Distr. Marowijne, small kreek tributary to the upper Oelemari R. 13 Nov 1966. J.-P. Gosse & Léopold III (Sta. 162).
ISNB 20853. 5. Distr. Marowijne, small brook at Albina. 26 Nov 1966. J.-P. Gosse & Léopold III (Sta. 173).

GUYANE FRANÇAISE

R. Maroni draiage

MNHN 95-131-132. 2. Maroni. No date. E. Ducaux.
NRM A84/1979412.3059. 1, 31.6. Crique Balaté. 9 Oct 1979. P. Planquette.
NRM A84/1979411.3060. 6, 36.4-54.8. Crique Vampire. 8 Oct 1979. P. Planquette.
NRM A84/1979248.3546. 2, ca 50.1-57.5. Canal St. Laurent. June 1979. P. Planquette.

Nannacara anomala

SURINAME

Corantijn R. drainage

RMNH 31144. Male, 30.4. Distr. Nickerie, Winanna Kreek, left bank tributary to middle Kapoeri Kreek. 1 Apr 1971. M. Boeseman.
USNM 225845. 1. Distr. Nickerie, kreek tributary to Corantijn R., 5°32'N 57°10'W. 15 May 1980. H.M. Madaire (80-7).
USNM 225846. 7. Distr. Nickerie, Koekwie Kreek. 15 May 1980. H.M. Madaire (80-8).
USNM 225847. 2. Distr. Nickerie, Makilikabroe Kreek. 15 May 1980. Leg. H.M. Madaire (80-9).
USNM 225841. 105. Distr. Nickerie, kreek at Camp McClemmen, Dalbane Kreek. 6 Sep 1980. R.P. Vari (RPV 80-13).
USNM 225842. 15. Distr. Nickerie, Lana Kreek, ca 4 km upstream from intersection with Corantijn R. 7 Sep 1980. R.P. Vari (RPV 80-15).
USNM 225843. 4. Distr. Nickerie, Kapoeri Kreek, ca 4 km N from intersection with Corantijn R. 8 Sep 1980. R.P. Vari (RPV 80-18).
USNM 225844. 1. Distr. Nickerie, small kreek entering Lucie R. about 3 km upstream from junction of Lucie and Corantijn rivers. 18 Sep 1980. R.P. Vari (RPV 80-37).
RMNH 31161. 8. Distr. Nickerie, right tributary kreek, Kapoeri kreek below Fajaslik Kreek. 2 Apr 1971. M. Boeseman.
RMNH 31162. 1. Distr. Nickerie, lower Nanni Kreek, just above sluice of van Wouw canal. 18 Mar 1971. D.C. Geijskes.

Nickerie R. drainage

RMNH 31145. 3 males, 24.2-33.6, 3 females, 21.3-25.1, 2 not sexable, 20.7-22.4. Distr. Nickerie, left bank tributary of Nickerie R., 3 km behind Camp Waterloopk., below Stondansie Val. 1 February 1971. M. Boeseman.
NRM A83/1971051.3018. Male, 35.4, female, 25.9. Same data as RMNH 31145.
RMNH 31163. Not counted. Distr. Nickerie, kreek at second savannah, right tributary of middle Maratakka R. (at woodcutterscamp). 6 Mar 1971. M. Boeseman.

Saramacca R. drainage

ZMA 105.659. 3. Distr. Brokopondo, tributary kreek on right bank of Kleine Saramacca R., ca 11 km ESE mouth. 27 Feb 1967. H. Nijssen (HN 106).
RMNH 31146. 14. Distr. Brokopondo, brook 3 km E Saramacca bridge. 28 Nov 1971. V.J. Mees.

Suriname R. drainage

ISNB 20854. 3 males, 37.9-44.9, female, 29.0. Distr. Brokopondo, small brook on road Zanderij-Matta. 29 Nov 1966. J.-P. Gosse & Léopold III (Sta 174).
RMNH 31147. 6 males, 13.5-29.3, 9 females, 13.2-26.8. Distr. Para, road near Overtoom. 27 Dec 1974. M.S. Hoogmoed (276).
RMNH 31148. Male, 14.0. Distr. Para, road near Overtoom. 20 Dec 1974. M.S. Hoogmoed (267).

RMNH 31149. 12 males, 16.5-46.9, 17 females, 15.7-37.2, one juv., ca 13.0. Distr. Para, Para (= Saramacca) Kreek. 29 July 1965. G. F. Mees.
RMNH 31150. 1. Distr. Para, Para (= Saramacca) Kreek. 29 July 1965. G. F. Mees.
RMNH 31151. 8. Distr. Para, Republiek. 2 July 1965. G. F. Mees.
RMNH 31152. 5. Distr. Para, Saramacca Kreek. 20 Apr 1965. Leg. G. F. Mees.
RMNH 31153. 6. Distr. Para, Saramacca Kreek. 21 May 1965. Leg. G. F. Mees.
RMNH 31154. Female, 19.0. Distr. Brokopondo, Gansee. 22 Jun 1965. Leg. G. F. Mees.
RMNH 18731. 1. Distr. Brokopondo, Kabelstation, Suriname R. 21-28 Sep 1938. D. C. Geijskes.
ZMA 105.328. 2. Distr. Suriname, 'Tout Lui Faut' canal, ca 12 km S Paramaribo at road to Lelydorp. 12 Dec 1966. H. Nijssen (HN 91).
ZMA 105.327. 3. Distr. Brokopondo, Marchall Kreek, E road Paranam-Afobaka, ca 1.5 km N Marchall village. 8 Dec 1966. H. Nijssen (HN 87).
ZMA 105.362. 3. Distr. Brokopondo, Marowijne (= Gran) Kreek, ca 51 km S Afobaka reservoir. 27 May 1966. H. Nijssen (HN 19).
ZMA 105.344 + NRM A82/1966503.3454. 35 males, 19.6-44.1, 31 females, 17.7-39.1. Distr. Para, Carolina (= Malasie) Kreek, ca 10 km SE airfield Zanderij on Road Zanderij-Kraka. 14 Dec 1966. H. Nijssen (HN 92). 2 specimens, 27.1-30.5 mm, cleared and stained.
ZMA 105.547. 6. Distr. Brokopondo, Tapoeripa Kreek, ca 3 km N Brokopondo. 13 Nov 1966. H. Nijssen.
ZMA 105.535. 13. Distr. Brokopondo, tributary kreek on right bank of Gran Rio, ca 4 km NE Awara dam. 31 Jan 1967. H. Nijssen (HN 101).
ZMA 105.702. 24. Distr. Brokopondo, Jenjee Kreek on left bank of Suriname R., 7.5 km N Botopasi. 21 Mar 1967. H. Nijssen (HN 114).
ZMA 107.298. 1. Distr. Brokopondo, Marchall Kreek on road Paranam-Afobaka, 1.5 km N Marchall village. 12 May 1966. H. Nijssen (HN 12).
ZMA 107.297. 3. Distr. Brokopondo, Marowijne (Gran) Kreek, 50.5 km S Afobaka reservoir. 6 May 1966. H. Nijssen (HN 7).
ZMA 107.300. 1. Distr. Brokopondo, Marowijne Kreek 50.5 km S Afobaka reservoir. 26 May 1966. H. Nijssen (HN 19).
ZMA 107.299. 4. Distr. Brokopondo, Marowijne Kreek ca 55 km S Afobaka reservoir. 24 May 1966. H. Nijssen (HN 16).
ZMA 107.302. 4. Distr. Brokopondo, Marowijne Kreek ca 60 km S Afobaka reservoir. 8 Jun 1966. H. Nijssen (HN 24).
ZMA 107.031. 3. Distr. Brokopondo, Marowijne Kreek ca 58 km S Afobaka reservoir. 9 Jun 1966. H. Nijssen (HN 25).
ZMA 107.304. 32. Distr. Brokopondo, Tapoeripa Kreek at Hermansdorp, 1 km N Brokopondo. 25 Aug 1966. H. Nijssen (HN 43).
ZMA 107.303. 8, incl. male 56.4. Distr. Brokopondo, Tapoeripa Kreek at Hermansdorp, 1 km N Brokopondo. 23 Aug 1966. H. Nijssen (HN 40).
ZMA 107.306. 2. Distr. Brokopondo, tributary kreek (Gran Mau) on right bank of Gran Rio, ca 1 km NE Dombaai (Bendi Watra). 30 Jan 1967. H. Nijssen (HN 99).
ZMA 107.305. 16. Distr. Brokopondo, Tapoeripa Kreek at Brokopondo. 15 Sep 1966. H. Nijssen (HN 60).
ZMA 107.308. 9. Distr. Brokopondo, Kwati Watra, kreek on left bank of Suriname R., 500 m S Botopasi. 19 Mar 1966. H. Nijssen (HN 112).
ZMA 107.307. 3. Distr. Brokopondo, tributary kreek on left bank of Gran Rio, 3 km NE Awaradam. 1 February 1967. H. Nijssen (HN 102).
ZMA 107.310. 1. Distr. Para, rain pools in sand, 2 km N Zanderij airport. 22 May 1966. H. Nijssen.
ZMA 107.309. 4. Distr. Brokopondo, Tapoeripa Kreek at Brokopondo. 24 Mar 1967. H. Nijssen (HN 118).
ZMA 108.340. 2. Distr. Para, Coropina Kreek. 20 May 1956. J. v. d. Kamp.

ZMA 108.323. 10. Distr. Para, pool at Zanderij. 10 May 1956. J. v.d. Kamp.
ZMA 108.341. 6. Distr. Para, pool in Coropina Kreek. 20 May 1956. J. v.d. Kamp.
ZMA 113.607. 5. Distr. Brokopondo, kreek at Brownsweg at railroad km 114. Oct 1960. Pijpers.
RMNH 31164. Not counted. Distr. Para, Saramacca Kreek. 7 Apr 1965. G.F. Mees.
RMNH 18639. 3. Distr. Para, Zanderij I, savanna streamlet. 2 Sep 1942. D.C. Geijskes.
RMNH 18370. 1. Distr. Brokopondo, Makambi Kreek. 27 Sep 1938. D.C. Geijskes.
RMNH 18372. 1. Distr. Suriname, Paramaribo, in ditch. Jun 1938. D.C. Geijskes.
RMNH 18449. 3. Distr. Suriname, Plantation Kwatta, Paramaribo. Aug 1911. W.C. van Heurn.
RMNH 31165. Several. Distr. Brokopondo, Tapoeripa Kreek. 8 Mar 1966. G.F. Mees.

Marowijne R. drainage

RMNH 31155. Male, 25.5. Distr. Marowijne, Albina. 6 Apr 1966. G.F. Mees.
ISNB 20855. 44. Distr. Marowijne, small brook at Albina. 26 Nov 1966. J.-P. Gosse & Léopold III (Sta. 173).
ISNB 20856. 13 males, 15.8-36.3, 13 females, 16.4-30.7. Distr. Marowijne, Nengre Kreek on road Paramaribo-Albina, near Albina. 26 Nov 1966. J.-P. Gosse & Léopold III (Sta. 172).
ISNB 20857. Male, 30.0. Distr. Marowijne, small river on right bank of R. Tapanahony at Paloemeu airstrip. 9 Nov 1966. J.-P. Gosse & Léopold III. (Sta. 158).

GUYANE FRANÇAISE

R. Maroni drainage

NRM A84/1980041.3551. 2, 53.5-54.1. Crique Cascade. 21 Jan 1980. P. Planquette.
NRM A84/1979411.3552. 2, 39.5-39.8. Crique Vampire. 8 Oct 1979. P. Planquette.

Crenicichla saxatilis

See Ploeg 1986a, 1986b, 1987.

Crenicichla lugubris

See Ploeg 1987.

Crenicichla multispinosa

See Ploeg 1986b, 1987.

Crenicichla albopunctata

See Ploeg 1986b, 1987.

Crenicichla coppenamensis

See Ploeg 1987.

Crenicichla nickeriensis

See Ploeg 1987.

Crenicichla sipaliwini

See Ploeg 1987.

LITERATURE CITED

AHL, E. 1936. Ueber eine kleine Sammlung von Süsswasserfischen aus dem Gebiet des Amazonas. Mitt. zool. Mus. Berl. 21: 264-269.

ALLGAYER, R. 1983. *Nannacara aureocephalus*, espèce nouvelle de Guyane française. Revue fr. Cichlidophiles 1983 (Nov.): 13-16, 21-24.

ARNOLD, J. P. 1912. *Nannacara taenia* Regan. Ein neuer Zwergcichlide aus dem Amazonenstrom. Wschr. Aquar. Terrark. 9: 521-524.

AXELROD, H. R. 1958. In passing... Trop. Fish Hobby. 7 (2): 25-39.

BLOCH, M. E. 1791. Naturgeschichte der ausländischen Fische. Fünfter Theil. Berlin, [6 pp.] + 152 pp.

BLOCH, M. E. 1792. Naturgeschichte der ausländischen Fische. Sechster Theil. Berlin, [4 pp.] + 126 pp.

BOESEMAN, M. 1952. A preliminary list of Surinam fishes not included in Eigenmann's enumeration of 1912. Zool. Meded. Leiden 31: 179-200.

BOESEMAN, M. 1954. On a small collection of Surinam fishes. Zool. Meded. Leiden 33: 17-24.

BOESEMAN, M. 1956. On recent accessions of Surinam fishes. Zool. Meded. Leiden 34: 183-199.

BOESEMAN, M. 1968. The genus *Hypostomus* Lacépède, 1803, and its Surinam representatives. Zool. Verh. Leiden (99): 1-89.

BOESEMAN, M. 1971. The "comb-toothed" Loricariinae of Surinam, with reflections on the phylogenetic tendencies within the family Loricariidae. Zool. Verh. Leiden (116): 1-56.

CASTELNAU, F. de. 1855. Animaux nouveaux ou rares recueillis pendant l'expédition dans les parties centrales de l'Amérique du sud, de Rio de Janeiro à Lima, et de Lima au Para; exécutée par ordre du gouvernement français pendant les années 1843 à 1847, sous la direction du Comte Francis de Castelnau. Poissons. Paris, XII + 112 pp.

COLLIN, G. D. 1822. Fauna Surinamensis. Upsala, 8 pp.

COPE, E. D. 1872. On the fishes of the Ambyiacu river. Proc. Acad. nat. Sci. Philad. 1871: 250-294.

EIGENMANN, C. H. 1910. Catalogue of the fresh-water fishes of tropical and south temperate America. Repts Princeton Univ. Exped. Patagonia 1896-1899. Zool. 3: 375-512.

EIGENMANN, C. H. 1912. The freshwater fishes of British Guiana, including a study of the ecological grouping of species and the relation of the fauna of the plateau to that of the lowlands. Mem. Carneg. Mus. 5: xxii + 578 pp.

EIGENMANN, C. H. & W. L. BRAY. 1894. A revision of the American Cichlidae. Ann. N.Y. Acad. Sci. 7: 607-624.

EIGENMANN, C. H. & R. S. EIGENMANN. 1891. A catalogue of the fresh-water fishes of South America. Proc. U.S. natl Mus. 15 (842): 1-81.

EIGENMANN, C. H. & C. H. KENNEDY. 1903. On a collection of fishes from Paraguay, with a synopsis of the American genera of Cichlids. Proc. Acad. nat. Sci. Philad. 55: 497-537.

FERNANDEZ YEPEZ, A. 1951. Presencia de *Chaetobranchus* en Venezuela. Evencias (San Fernando) (11): 4 pp., unpag.

FERNHOLM, B. & A. WHEELER. 1983. Linnaean fish specimens in the Swedish Museum of Natural History, Stockholm. Zool. J. Linn. Soc. 78: 199-286.

FOWLER, H. W. 1914. Fishes from the Rupununi River, British Guiana. Proc. Acad. nat. Sci. Philad. 66: 229-284.

Fowler, H.W. 1954. Os peixes de agua doce do Brasil. Volume II. Archos Zool. S. Paulo 9: 1-400.

Gaemers, P.A.M. 1984. Taxonomic position of the Cichlidae as demonstrated by the morphology of their otoliths. Netherl. J. Zool. 34: 566-595.

Geijskes, D.C. 1973. Reisverslag van de Expeditie West Suriname 1971. Zool. Bijdr. (15): 1-41.

Géry, J. 1972. Poissons characoïdes des Guyanes. I. Généralités. II. Famille des Serrasalmidae. Zool. Verh. Leiden (122): 1-250.

Gosse, J.P. 1971. Révision du genre *Retroculus* (Castelnau, 1855), designation d'un neotype de *Retroculus lapidifer* (Castelnau, 1855) et description de deux espèces nouvelles. Bull. Inst. r. Sci. nat. Belg. 47 (43): 1-13.

Gosse, J.-P. 1976. Révision du genre *Geophagus*. Mém. Acad. r. Sci. Outre-mer. Cl. Sci. nat. méd. (N.S.) 19 (3): 1-172.

Gray, J.E. (ed.) 1854. Catalogue of fish collected and described by Laurence Theodore Gronow, now in the British Museum. London, vii + 196 pp.

Greenwood, P.H. 1976. A review of the family Centropomidae. Bull. Br. Mus. (nat. Hist.) (Zool.) 29:1-81.

Greenwood, P.H. 1985a. Notes on the anatomy and phyletic relationships of *Hemichromis* Peters, 1858. Bull. Br. Mus. nat. Hist. (Zool.) 48: 131-171.

Greenwood, P.H. 1985b. The generic status and affinities of *Paratilapia thomasi* Blgr 1915. Bull. Br. Mus. nat. Hist. (Zool.) 49: 257-272.

Greenwood P.H. 1986. The *pars jugularis* and the intrarelationships of cichlid fishes. J. nat. Hist. 20: 949-974.

Gronovius, L.T. 1754. Museum Ichthyologicum sistens Piscium indigenorum & quorumdam exoticarum qui in Museo Laurentii Theodorii Gronovii, J. U. D. adservantur, descriptiones ordine systematico. Lugduni Batavorum, [8 pp. +] 70 pp.

Gronovius, L.T. 1756. Musei Ichthyologici tomus secundus sistens Piscium indigenorum & nonnullarum exoticorum, quorum maxima pars in Museo Laurentii Theodori Gronovii, J. U. D. adservantur, nec non quorumdam in aliis Museis observatorum descriptiones. Lugduni Batavorum, [6pp.] + 88pp. [Fish section 46 pp. + pp. 86-88].

Günther, A. 1862. Catalogue of the fishes in the British Museum. Volume fourth. London, 534 pp.

Heckel, J. 1840. Johann Natterer's neue Flussfische Brasilien's nach den Beobachtungen und Mittheilungen des Entdeckers beschrieben. (Erste Abtheilung, die Labroiden). Annln wien. Mus. Natges. 2: 327-470.

Hoedeman, J.J. 1951. Notes on the fishes of the cichlid family I. *Apistogramma cacatuoides* sp.n. Beaufortia (4): 1-4.

Holthuis, L.B. 1959. The Crustacea Decapoda of Suriname (Dutch Guiana). Zool. Verh. Leiden (44): 1-296.

Hoogmoed, M.S. 1973. Notes on the herpetofauna of Surinam IV. The lizards and amphisbaenians of Surinam. Den Haag, 419 pp.

Jardine, W. (ed.). 1843. The naturalist's library [vol. 38]. Ichthyology Vol. V. Fishes of Guiana. Part II. Edinburgh, 214 pp. Also as Vol. XXXI. Ichthyology. Fishes of British Guiana. Part 2nd. Two pl. 18 listed in text and Contents (one present).

Jordan, D.S. & B.W. Evermann. 1898. The fishes of North and Middle America: A descriptive catalogue of the species of fish-like vertebrates found in the waters of North America, north of the Isthmus of Panama. Part II. Bull. U.S. natl Mus. (47): 1241-2183.

Kanazawa, R.H. 1966. The fishes of the genus *Osteoglossum* with a description of a new species from the Rio Negro. Ichthyologica Aquar. J. 37: 161-172.

Kappler, A. 1887. Surinam, sein Land, seine Natur, Bevölkerung und seine Kultur-Verhältnisse mit Bezug auf Kolonisation. Stuttgart, 383 pp.

Kaufman, L.S. & K.F. Liem. 1982. Fishes of the suborder Labroidei: phylogeny, ecology, and evolutionary significance. Breviora (472): 1-19.

KEENLEYSIDE, M. H. A. & B. F. BIETZ. 1981. The reproductive behaviour of *Aequidens vittatus* in Surinam, South America. Env. Biol. Fish 6: 87-94.

KENNY, J. S. 1979. Some recent animal colonizations. Living World, J. Trinidad Tobago Field Nat. Club 1978-1979: 27.

KLINCKOWSTRÖM, A. 1892. Zoologischen Untersuchungen während eines Aufenthaltes in Suriname. Biol. Fören. Förhandl. 4: 101-104.

KNÖPPEL, H.-A. 1970. Food of central Amazonian fishes. Contribution to the nutrient-ecology of Amazonian rain-forest-streams. Amazoniana 2: 257-352.

KOSLOWSKI, I. 1985a. Die Buntbarsche der Neuen Welt. Zwergcichliden. Essen, 192 pp.

KOSLOWSKI, I. 1985b. Descriptions of new species of *Apistogramma* from the Rio Mamoré system in Bolivia. Bonn. zool. Beitr. 36: 145-162.

KULLANDER, S. O. 1978. A redescription of *Crenicara filamentosa* Ladiges, 1958. Mitt. hamb. zool. Mus. Inst. 75: 267-278.

KULLANDER, S. O. 1979a. Species of *Apistogramma* from the Orinoco drainage basin, South America, with descriptions of four new species. Zool. Scr. 8: 69-79.

KULLANDER, S. O. 1979b. Description of a new species of the genus *Apistogramma* from Peru. Revue suisse Zool. 86: 937-945.

KULLANDER, S. O. 1980a. A taxonomical study of the genus *Apistogramma* Regan, with a revision of Brazilian and Peruvian species. Bonn. zool. Monogr. 14: 1-152.

KULLANDER, S. O. 1980b. A redescription of the South American cichlid fish *Papiliochromis ramirezi* (Myers & Harry, 1948). Stud. neotrop. Fauna Envir. 15: 91-108.

KULLANDER, S. O. 1980c. Description of a new species of *Apistogramma* from the Rio Madeira system in Brazil. Bull. zool. Mus. Univ. Amsterd. 7: 157-164.

KULLANDER, S. O. 1981a. Description of a new species of *Apistogramma* from the upper Amazonas basin. Bonn. zool. Beitr. 32: 183-194.

KULLANDER, S. O. 1981b. Cichlid fishes from the La Plata basin. Part I. Collections from Paraguay in the Muséum d'Histoire naturelle de Genève. Revue suisse Zool. 88: 675-692.

KULLANDER, S. O. 1982b. Cichlid fishes from the La Plata basin. Part II. *Apistogramma commbrae* (Regan, 1906). Revue suisse Zool. 89: 33-48.

KULLANDER, S. O. 1982a. Beschreibung einer neuen *Apistogramma*-Art aus Zentral-Amazonien. DCG-Informn 13: 181-193.

KULLANDER, S. O. 1982c. Description of a new species of *Apistogramma* Regan, from the Oyapock and Approuague river systems. Cybium (3) 6: 65-72.

KULLANDER, S. O. 1983a. A revision of the South American cichlid genus *Cichlasoma*. Stockholm, 296 pp.

KULLANDER, S. O. 1983b. Cichlid fishes from the La Plata basin. Part IV. Review of the *Apistogramma* species, with description of a new species. Zool. Scr. 11: 307-313.

KULLANDER, S. O. 1984. Cichlid fishes from the La Plata basin. Part V. Description of *Aequidens plagiozonatus* sp. n. from the Paraguay River system. Zool. Scr. 13: 155-159.

KULLANDER, S. O. 1986. Cichlid fishes of the Amazon River drainage of Peru. Stockholm, 431 pp.

KULLANDER, S. O. 1987. A new *Apistogramma* species from the Rio Negro in Brazil and Venezuela. Zool. Scr. 16: 259-270.

KULLANDER, S. O. 1988. *Teleocichla*, a new genus of South American rheophilic cichlid fishes with six new species. Copeia 1988: 196-230.

LE BAIL, P. Y., P. PLANQUETTE & J. GERY. 1984. Clé de determination de Poissons continentaux et côtiers de Guyane. Espèces dulceaquicoles non Siluriformes. Bull. Liaison Centre Rech. agron. Antilles Guyane (9): 1-97.

LE BAIL, P. Y., P. PLANQUETTE & J. GERY. 1984. Clé de determination de Poissons continentaux et côtiers de Guyane. Fascicule No. I. Bull. Liaison Centre Rech. agron. Antilles Guyane (6): 1-63.

Le Bail, P. Y., P. Planquette & J. Gery. 1984. Clé de determination de Poissons continentaux et côtiers de Guyane. Fascicule No. III. Bull. Liaison Centre Rech. agron. Antilles Guyane (8): 1-67.

Leentvaar, P. 1973. Further developments in Lake Brokopondo, Surinam. Amazoniana 4: 1-8.

Lijding, H. W. 1959a. Voorlopige lijst van garnalen, krabben en vissen in Suriname. Surinam. Landbouw 7: 70-97.

Lijding, H. W. 1959b. Proeven met *Tilapia* in Suriname. Surinam. Landbouw 6: 183-194.

Linnaeus, C. 1754. Hans Maj:ts Adolf Frideriks vår allernådigste konungs naturalie samling innehållande sällsynte och främmande djur, som bevaras på kongl. lustslottet Ulriksdahl; beskrefne och afritade samt på nådig befallning utgifne af Carl Linnaeus. Stockholm, I-XXX, 1-96 pp. + 7 pp.

Linnaeus, C. 1758. Systema naturae per regna tria naturae, secundum classes, ordines, genera, species, cum characteribus, differentiis, synonymis, locis. Tomus I. Editio decima, reformata. Holmiae, 823 pp.

Lowe-McConnell, R. H. 1969. The cichlid fishes of Guyana, South America, with notes on their ecology and breeding behaviour. Zool. J. Linn. Soc. 48: 255-302.

Machado-Allison, A. 1971. Contribución al conocimiento de la taxonomia del género *Cichla* en Venezuela. Parte I. Acta biol. Venez. 7: 459-497.

Mago Leccia, F. 1970. Lista de los peces de Venezuela, incluyendo un estudio preliminar sobre la ictiogeografia del pais. Caracas, 283 pp.

Mees, G. F. 1974. The Auchenipteridae and Pimelodidae of Suriname. Zool. Verh. Leiden (132). 1-256.

Meinken, H. 1960. Eine neue *Apistogramma*-Art. Intl. Revue ges. Hydrobiol. 45: 655-661.

Meinken, H. 1963. *Chela caeruleostigmata, Aequidens itani* und andere Neuheiten. Aquar. Terrar. Z. 16: 261-264.

Nijssen, H. 1966. Ichthyological observations: May and June 1966. In: Progress Reports, Biological Brokopondo Research Project, Surinam 3: 172-180.

Nijssen, H. 1967a. Ichthyological observations: January-March 1967. In: Progress Reports, Biological Brokopondo Research Project, Surinam 4: 234-237.

Nijssen, H. 1967b. Final remarks and tentative list of fish species. In: Progress Reports, Biological Brokopondo Research Project, Surinam 4:240-246.

Nijssen, H. 1970. Revision of the Surinam catfishes of the genus *Corydoras* Lacépède, 1803. Beaufortia 18 (230): 1-75.

Pellegrin, J. 1904. Contribution à l'étude anatomique, biologique et taxinomique des Poissons de la famille des Cichlidés. Mém. Soc. zool. France 16: 41-399.

Pellegrin, J. 1908. Les Poissons d'eau douce de la Guyane française. Revue Coloniale (N. sér.) (67): 557-591.

Pellegrin, J. 1909. Poissons d'eau douce de la Guyane française. Bull. Soc. natl. Acclim. Fr. 56: 179-185, 219-227, 271-274, 303-312.

Peters, W. 1866. Mittheilung über Fische (*Protopterus*, *Auliscops,*, *Labrax*, *Labracoglossa*, *Nematocentris*, *Serranus*, *Scorpis*, *Opisthognathus*, *Scombresox*, *Acharnes*, *Anguilla*, *Gymnomuraena*, *Chilorhinus*, *Ophichthys*, *Helmichthys*). Mber. k. preuss. Akad. Wiss. Berl. 1866: 509-526.

Ploeg, A. 1986a. Occurrence and variability of *Crenicichla saxatilis* (Linnaeus, 1758) in Surinam, and restriction of its type-locality. Bijdr. Dierk. 56 (1): 47-59.

Ploeg, A. 1986b. The fishes of the cichlid genus *Crenicichla* in French Guiana. Bijdr. Dierk. 56 (2): 221-231.

Ploeg, A. 1987. Review of the cichlid genus *Crenicichla* Heckel, 1840 from Surinam, with descriptions of three new species. Beaufortia 37 (5): 73-98.

Popta, C. M. L. 1901. Les appendices des arcs branchiaux des Poissons. Arch Sci. nat. (8, Zool.) 12: 139-216.

POPTA, C. 1914-1917. Articles on fishes. In: H.D. Benjamins & J.F. Snelleman (eds.) Encyclopaedie van Nederlandsch West-Indië. 's-Gravenhage & Leiden, 782 pp.

PRANCE, G.T. (ed.) 1982. Biological diversification in the tropics. New York, 714 pp.

PRICE, J.L. 1955. A survey of the freshwater fishes of the Island of Trinidad. J. agricult. Soc. Trinidad Tobago 55: 390-416.

PUYO, J. 1943. Nouveaux poissons d'eau douce de la Guyane française. Bull. Soc. Hist. nat. Toulouse 78: 141-149.

PUYO, J. 1948. Deux poissons de la Guyane. Bull. Soc. Hist. nat. Toulouse 83: 78-82.

PUYO, J. 1949. Poissons de la Guyane française. Faune de l'Empire français. XII. Paris, 280 pp.

REGAN, C.T. 1905a. A revision of the fishes of the South-American cichlid genera *Crenacara*, *Batrachops*, and *Crenicichla*. Proc. zool. Soc. Lond. 1905: 152-168.

REGAN, C.T. 1905b. A revision of the fishes of the South-American cichlid genera *Acara*, *Nannacara*, *Acaropsis*, and *Astronotus*. Ann. Mag. nat. Hist. (7) 15: 329-347.

REGAN, C.T. 1905c. A revision of the fishes of the American cichlid genus *Cichlosoma* and of the allied genera. Ann. Mag. nat. Hist. (7) 16: 60-77, 225-243, 316-340, 433-445.

REGAN, C.T. 1906a. A revision of the South-American cichlid genera *Retroculus*, *Geophagus*, *Heterogramma*, and *Biotoecus*. Ann. Mag. nat. Hist. (7) 17: 49-66.

REGAN, C.T. 1906b. A revision of the fishes of the South-American cichlid genera *Cichla*, *Chaetobranchus*, and *Chaetobranchopsis*, with notes on the genera of American Cichlidae. Ann. Mag. nat. Hist. (7) 17: 230-239.

REGAN, C.T. 1908. Description of a new cichlid fish of the genus *Heterogramma* from Demerara. Ann. Mag. nat. Hist. (8) 1: 370-371.

REGAN, C.T. 1912. Descriptions of new cichlid fishes from South America in the British Museum. Ann. Mag. nat. Hist. (8) 9: 505-507.

REGAN, C.T. 1913a. A synopsis of the cichlid fishes of the genus *Crenicichla*. Ann. Mag. nat. Hist. (8) 11: 498-540.

REGAN, C.T. 1913b. Fishes from the River Ucayali, Peru, collected by Mr. Mounsey. Ann. Mag. nat. Hist. (8) 12: 281-283.

REID, M.J. & J.W. ATZ. 1958. Oral incubation in the cichlid fish *Geophagus jurupari* Heckel. Zoologica N.Y. 43: 77-88.

RIBEIRO, A. DE MIRANDA. 1918. Cichlidae. Commissão Linhas telegr. estrat. Matto Grosso Amazonas Publ. (46): 1-18.

RICHTER, C.J.J. & H. NIJSSEN. 1980. Notes on the fishery potential and fish fauna of the Brokopondo reservoir (Surinam). Fish. Mgmt 11: 119-130.

SAUVAGE, H.E. 1882. Description de quelques poissons de la collection du Muséum d'histoire naturelle. Bull. Soc. philom. Paris (7) 5: 168-176.

SCHMETTKAMP, W. 1982. Die Zwergcichliden Südamerikas. Hannover, 176 pp.

SCHNEIDER, J.G. (ed.) 1801. M.E. Blochii systema ichthyologiae. Berolini, LX + 584 pp.

SCHOMBURGK, R. 1849. Reisen in Britisch-Guiana in den Jahren 1840-1844. Im Auftrag Sr. Mäjestat des Königs von Preussen ausgeführt von Richard Schomburgk. Dritter Theil. Versuch einer Fauna und Flora von Britisch-Guiana. Nach Vorlagen von Johannes Müller, Ehrenberg, Erichson, Klotzsch, Troschel, Cabanis und Andern. Leipzig, I-VIII, 533-1260.

SCHULTZ, L.P. 1949. A further contribution to the ichthyology of Venezuela. Proc. U.S. natl Mus. 99 (3235): 1-211.

SMITT, F.A. 1892. Skandinaviens fiskar. Andra uppl. Förra delen. Stockholm, 566 pp.

STEINDACHNER, F. 1875. Beiträge zur Kenntniss der Chromiden des Amazonenstromes. Sber. k. Akad. Wiss. Wien Math.-natw. Cl. 71: 61-137.

STEINDACHNER, F. 1882. Beiträge zur Kenntniss der Flussfische Südamerika's. II. Denkschr. k. Akad. Wiss. Wien Math.-natw. Cl. 43: 103-146.

STEINDACHNER, F. 1915a. Beiträge zur Kenntnis der Flussfische Südamerikas V. Denkschr. k. Akad. Wiss. Wien Math.-natw. Kl. 93: 15-106.

STEINDACHNER, F. 1915b. Ichthyologische Beiträge (XVIII). Sber. k. Akad. Wiss. Wien Math.-natw. Kl. 124: 567-591.

Stiassny, M. L. J. 1981. The phyletic status of the family Cichlidae: a comparative anatomical investigation. Netherl. J. Zool. 31: 275-314.
Stiassny, M. L. J. 1982. The relationships of the neotropical genus *Cichla*: a phyletic analysis including some functional considerations. J. Zool. Lond. 197: 427-453.
Stiassny, M. L. J. 1987. Cichlid familial intrarelationships and the placement of the neotropical genus *Cichla*. J. nat. Hist. 21: 1311-1331.
Stiassny, M. L. J. & J. S. Jensen. 1987. Labroid intrarelationships revisited: morphological complexity, key innovations, and the study of comparative diversity. Bull. Mus. comp. Zool. Harv. 151: 269-319.
Swain, J. 1883. A review of Swainson's genera of fishes. Proc. Acad. nat. Sci. Philad. 34: 272-284.
Swainson, W. 1839. The natural history of fishes, amphibians & reptiles or monocardian animals, vol. II. Cabinet Encyclopedia, London, 452 pp.
Thys van den Audenaerde, D. F. E. 1968. An annotated bibliography of *Tilapia*. Doc. zool., Mus. r. Afr. centr. (14): 1-406.
Trewavas, E. 1947. An example of 'mimicry' in fishes. Nature, Lond. 160: 120.
Trewavas. E. 1983. Tilapiine fishes of the genera *Sarotherodon*, *Oreochromis*, and *Danakilia*. London, 583 pp.
Trewavas, E. 1987. *Sarotherodon melanotheron* Rüppell, 1852: proposed conservation of the specific name. Bull. zool. Nom. 44:190-191.
Zaret, T. M. 1977. Inhibition of cannibalism in *Cichla ocellaris* and hypothesis of predator mimicry among South American fishes. Evolution 31: 421-437.
Zihler, F. 1982. Gross morphology and configuration of digestive tracts of Cichlidae: phylogenetic and functional significance. Netherl. J. Zool. 32: 544-571.

www.ingramcontent.com/pod-product-compliance
Lightning Source LLC
LaVergne TN
LVHW102226300926
849160LV00005B/42